T0261342

Isuzu Pick-ups & Trooper Automotive Repair Manual

by Larry Warren, Robert Maddox and John H Haynes

Member of the Guild of Motoring Writers

Models covered:
Isuzu Pick-ups - 1981 through 1993
Isuzu Trooper - 1984 through 1991
Does not include Amigo or diesel engine information

(47020-4K22)

Haynes Group Limited
Haynes North America, Inc.
www.haynes.com

Acknowledgements

We are grateful for the help and cooperation of the Isuzu Motor Company for their assistance with technical information.

© **Haynes North America, Inc. 1990, 1991, 1993**

With permission from Haynes Group Limited

A book in the Haynes Automotive Repair Manual Series

ISBN-10: 1-56392-033-6

ISBN-13: 978-1-56392-033-2

Library of Congress Catalog Card Number 92-70512

While every attempt is made to ensure that the information in this manual is correct, no liability can be accepted by the authors or publishers for loss, damage or injury caused by any errors in, or omissions from, the information given.

Contents

Haynes author, mechanic and photographer with Isuzu Trooper II

About this manual

Its purpose

The purpose of this manual is to help you get the best value from your vehicle. It can do so in several ways. It can help you decide what work must be done, even if you choose to have it done by a dealer service department or a repair shop; it provides information and procedures for routine maintenance and servicing; and it offers diagnostic and repair procedures to follow when trouble occurs.

We hope you use the manual to tackle the work yourself. For many simpler jobs, doing it yourself may be quicker than arranging an appointment to get the vehicle into a shop and making the trips to leave it and pick it up. More importantly, a lot of money can be saved by avoiding the expense the shop must pass on to you to cover its labor and overhead costs. An added benefit is the sense of satisfaction and accomplishment that you feel after doing the job yourself.

Using the manual

The manual is divided into Chapters. Each Chapter is divided into numbered Sections, which are headed in bold type between horizontal lines. Each Section consists of consecutively numbered paragraphs.

At the beginning of each numbered Section you will be referred to any illustrations which apply to the procedures in that Section. The reference numbers used in illustration captions pinpoint the pertinent Section and the Step within that Section. That is, illustration 3.2 means the illustration refers to Section 3 and Step (or paragraph) 2 within that Section.

Procedures, once described in the text, are not normally repeated. When it's necessary to refer to another Chapter, the reference will be given as Chapter and Section number. Cross references given without use of the word "Chapter" apply to Sections and/or paragraphs in the same Chapter. For example, "see Section 8" means in the same Chapter.

References to the left or right side of the vehicle assume you are sitting in the driver's seat, facing forward.

Even though we have prepared this manual with extreme care, neither the publisher nor the author can accept responsibility for any errors in, or omissions from, the information given.

NOTE

A **Note** provides information necessary to properly complete a procedure or information which will make the procedure easier to understand.

CAUTION

A **Caution** provides a special procedure or special steps which must be taken while completing the procedure where the Caution is found. Not heeding a Caution can result in damage to the assembly being worked on.

WARNING

A **Warning** provides a special procedure or special steps which must be taken while completing the procedure where the Warning is found. Not heeding a Warning can result in personal injury.

Introduction to the Isuzu Pick-ups and Trooper

Isuzu Trooper models are available in two and four-door station wagon body styles. Pick-up models are available in short and long bed styles.

Early four-cylinder engines are equipped with a carburetor. Some later four-cylinder engines are equipped with port fuel injection. All V6 engines are equipped with throttle body fuel injection.

The engine drives the rear wheels through either a four or five-speed manual or four-speed automatic transmission via a driveshaft and solid rear axle. A transfer case and driveshaft are used to drive the front driveaxles on 4WD models.

The suspension features independent control arms, torsion bars and shock absorbers at the front. A solid axle at the rear is suspended by leaf springs and shock absorbers.

The steering box is mounted to the left of the engine and is connected to the steering arms through a series of rods. Power assist is optional on some models.

The brakes are disc at the front and drums at the rear, with power assist standard. Some later models are equipped with disc brakes in the rear, instead of drums.

Vehicle identification numbers

Modifications are a continuing and unpublicized process in vehicle manufacturing. Since spare parts manuals and lists are compiled on a numerical basis, the individual vehicle numbers are essential to correctly identify the component required.

Vehicle Identification Number (VIN)

This very important identification number is stamped on a plate attached to the left side of the dashboard just inside the windshield on the driver's side of the vehicle (see illustration). The VIN also appears on the Vehicle Certificate of Title and Registration. It contains information such as where and when the vehicle was manufactured, the model year and the body style.

Engine identification number

The engine ID number on four-cylinder engines is located in one of two places: 1) on a machined surface on the left rear corner of the block (see illustration) or 2) in the center of the right side of the block, near the top. On V6 engines, the ID number is located on a pad at the front of the block (see illustration).

Paint code identification plate

This plate is located in the engine compartment, usually on the radiator support. It tells what color and type of paint was originally applied to the vehicle.

Safety Certification label

The Safety Certification label is affixed to the left front door pillar. The plate contains the name of the manufacturer, the month and year of production, the Gross Vehicle Weight Rating (GVWR) and the certification statement.

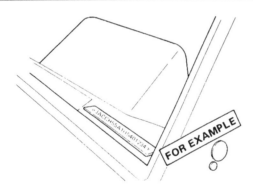

The Vehicle Identification Number (VIN) is visible from outside the vehicle through the driver's side of the windshield

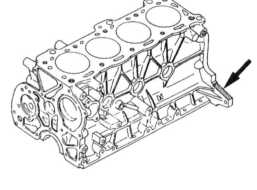

On some four-cylinder engines, like the one shown here, the identification number is located on a machined surface on the left rear corner of the block - on others, it's in the center of the right side of the block, towards the top

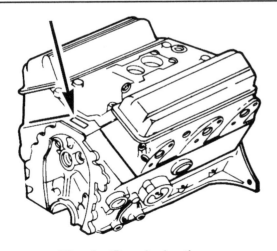

V6 engine ID number location

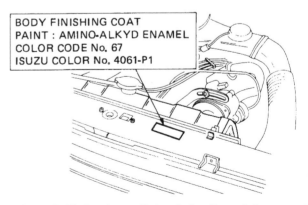

The paint code ID plate is usually located on the radiator support

Buying parts

Replacement parts are available from many sources, which generally fall into one of two categories - authorized dealer parts departments and independent retail auto parts stores. Our advice concerning these parts is as follows:

Retail auto parts stores: Good auto parts stores will stock frequently needed components which wear out relatively fast, such as clutch components, exhaust systems, brake parts, tune-up parts, etc. These stores often supply new or reconditioned parts on an exchange basis, which can save a considerable amount of money. Discount auto parts stores are often very good places to buy materials and parts needed for general vehicle maintenance such as oil, grease, filters, spark plugs, belts, touch-up paint, bulbs, etc. They also usually sell tools and general accessories, have convenient hours, charge lower prices and can often be found not far from home.

Authorized dealer parts department: This is the best source for parts which are unique to the vehicle and not generally available elsewhere (such as major engine parts, transmission parts, trim pieces, etc.).

Warranty information: If the vehicle is still covered under warranty, be sure that any replacement parts purchased - regardless of the source - do not invalidate the warranty!

To be sure of obtaining the correct parts, have engine and chassis numbers available and, if possible, take the old parts along for positive identification.

Maintenance techniques, tools and working facilities

Maintenance techniques

There are a number of techniques involved in maintenance and repair that will be referred to throughout this manual. Application of these techniques will enable the home mechanic to be more efficient, better organized and capable of performing the various tasks properly, which will ensure that the repair job is thorough and complete.

Fasteners

Fasteners are nuts, bolts, studs and screws used to hold two or more parts together. There are a few things to keep in mind when working with fasteners. Almost all of them use a locking device of some type, either a lockwasher, locknut, locking tab or thread adhesive. All threaded fasteners should be clean and straight, with undamaged threads and undamaged corners on the hex head where the wrench fits. Develop the habit of replacing all damaged nuts and bolts with new ones. Special locknuts with nylon or fiber inserts can only be used once. If they are removed, they lose their locking ability and must be replaced with new ones.

Rusted nuts and bolts should be treated with a penetrating fluid to ease removal and prevent breakage. Some mechanics use turpentine in a spout-type oil can, which works quite well. After applying the rust penetrant, let it work for a few minutes before trying to loosen the nut or bolt. Badly rusted fasteners may have to be chiseled or sawed off or removed with a special nut breaker, available at tool stores.

If a bolt or stud breaks off in an assembly, it can be drilled and removed with a special tool commonly available for this purpose. Most automotive machine shops can perform this task, as well as other repair procedures, such as the repair of threaded holes that have been stripped out.

Flat washers and lockwashers, when removed from an assembly, should always be replaced exactly as removed. Replace any damaged washers with new ones. Never use a lockwasher on any soft metal surface (such as aluminum), thin sheet metal or plastic.

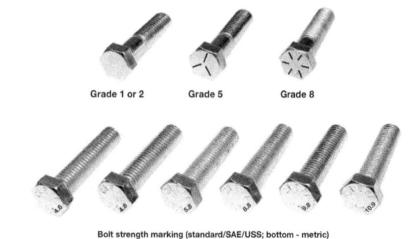

Grade 1 or 2   Grade 5   Grade 8

Bolt strength marking (standard/SAE/USS; bottom - metric)

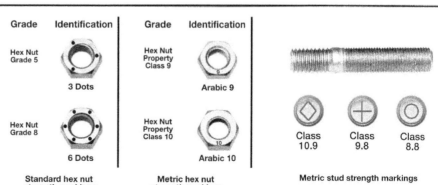

Grade	Identification
Hex Nut Grade 5	3 Dots
Hex Nut Grade 8	6 Dots

Standard hex nut strength markings

Grade	Identification
Hex Nut Property Class 9	Arabic 9
Hex Nut Property Class 10	Arabic 10

Metric hex nut strength markings

Class 10.9   Class 9.8   Class 8.8

Metric stud strength markings

Fastener sizes

For a number of reasons, automobile manufacturers are making wider and wider use of metric fasteners. Therefore, it is important to be able to tell the difference between standard (sometimes called U.S. or SAE) and metric hardware, since they cannot be interchanged.

All bolts, whether standard or metric, are sized according to diameter, thread pitch and length. For example, a standard 1/2 - 13 x 1 bolt is 1/2 inch in diameter, has 13 threads per inch and is 1 inch long. An M12 - 1.75 x 25 metric bolt is 12 mm in diameter, has a thread pitch of 1.75 mm (the distance between threads) and is 25 mm long. The two bolts are nearly identical, and easily confused, but they are not interchangeable.

In addition to the differences in diameter, thread pitch and length, metric and standard bolts can also be distinguished by examining the bolt heads. To begin with, the distance across the flats on a standard bolt head is measured in inches, while the same dimension on a metric bolt is sized in millimeters (the same is true for nuts). As a result, a standard wrench should not be used on a metric bolt and a metric wrench should not be used on a standard bolt. Also, most standard bolts have slashes radiating out from the center of the head to denote the grade or strength of the bolt, which is an indication of the amount of torque that can be applied to it. The greater the number of slashes, the greater the strength of the bolt. Grades 0 through 5 are commonly used on automobiles. Metric bolts have a property class (grade) number, rather than a slash, molded into their heads to indicate bolt strength. In this case, the higher the number, the stronger the bolt. Property class numbers 8.8, 9.8 and 10.9 are commonly used on automobiles.

Strength markings can also be used to distinguish standard hex nuts from metric hex nuts. Many standard nuts have dots stamped into one side, while metric nuts are marked with a number. The greater the number of dots, or the higher the number, the greater the strength of the nut.

Metric studs are also marked on their ends according to property class (grade). Larger studs are numbered (the same as metric bolts), while smaller studs carry a geometric code to denote grade.

It should be noted that many fasteners, especially Grades 0 through 2, have no distinguishing marks on them. When such is the case, the only way to determine whether it is standard or metric is to measure the thread pitch or compare it to a known fastener of the same size.

Standard fasteners are often referred to as SAE, as opposed to metric. However, it should be noted that SAE technically refers to a non-metric fine thread fastener only. Coarse thread non-metric fasteners are referred to as USS sizes.

Since fasteners of the same size (both standard and metric) may have different

Metric thread sizes	Ft-lbs	Nm
M-6	6 to 9	9 to 12
M-8	14 to 21	19 to 28
M-10	28 to 40	38 to 54
M-12	50 to 71	68 to 96
M-14	80 to 140	109 to 154

Pipe thread sizes	Ft-lbs	Nm
1/8	5 to 8	7 to 10
1/4	12 to 18	17 to 24
3/8	22 to 33	30 to 44
1/2	25 to 35	34 to 47

U.S. thread sizes	Ft-lbs	Nm
1/4 - 20	6 to 9	9 to 12
5/16 - 18	12 to 18	17 to 24
5/16 - 24	14 to 20	19 to 27
3/8 - 16	22 to 32	30 to 43
3/8 - 24	27 to 38	37 to 51
7/16 - 14	40 to 55	55 to 74
7/16 - 20	40 to 60	55 to 81
1/2 - 13	55 to 80	75 to 108

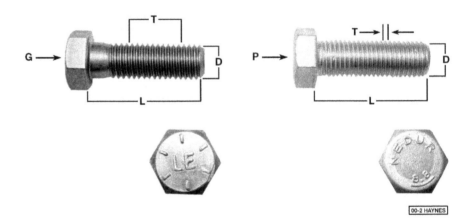

Standard (SAE and USS) bolt dimensions/grade marks

G   Grade marks (bolt strength)
L   Length (in inches)
T   Thread pitch (number of threads per inch)
D   Nominal diameter (in inches)

Metric bolt dimensions/grade marks

P   Property class (bolt strength)
L   Length (in millimeters)
T   Thread pitch (distance between threads in millimeters)
D   Diameter

strength ratings, be sure to reinstall any bolts, studs or nuts removed from your vehicle in their original locations. Also, when replacing a fastener with a new one, make sure that the new one has a strength rating equal to or greater than the original.

Tightening sequences and procedures

Most threaded fasteners should be tightened to a specific torque value (torque is the twisting force applied to a threaded component such as a nut or bolt). Overtightening the fastener can weaken it and cause it to break, while undertightening can cause it to eventually come loose. Bolts, screws and studs, depending on the material they are made of and their thread diameters, have

specific torque values, many of which are noted in the Specifications at the beginning of each Chapter. Be sure to follow the torque recommendations closely. For fasteners not assigned a specific torque, a general torque value chart is presented here as a guide. These torque values are for dry (unlubricated) fasteners threaded into steel or cast iron (not aluminum). As was previously mentioned, the size and grade of a fastener determine the amount of torque that can safely be applied to it. The figures listed here are approximate for Grade 2 and Grade 3 fasteners. Higher grades can tolerate higher torque values.

Fasteners laid out in a pattern, such as cylinder head bolts, oil pan bolts, differential cover bolts, etc., must be loosened or tightened in sequence to avoid warping the com-

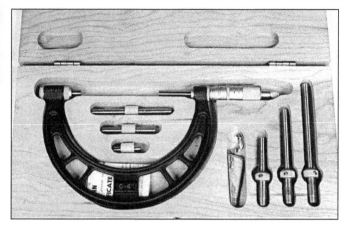

Micrometer set

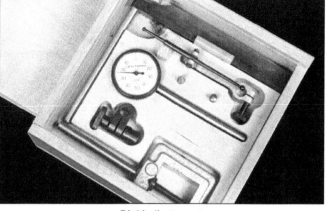

Dial indicator set

ponent. This sequence will normally be shown in the appropriate Chapter. If a specific pattern is not given, the following procedures can be used to prevent warping.

Initially, the bolts or nuts should be assembled finger-tight only. Next, they should be tightened one full turn each, in a criss-cross or diagonal pattern. After each one has been tightened one full turn, return to the first one and tighten them all one-half turn, following the same pattern. Finally, tighten each of them one-quarter turn at a time until each fastener has been tightened to the proper torque. To loosen and remove the fasteners, the procedure would be reversed.

Component disassembly

Component disassembly should be done with care and purpose to help ensure that the parts go back together properly. Always keep track of the sequence in which parts are removed. Make note of special characteristics or marks on parts that can be installed more than one way, such as a grooved thrust washer on a shaft. It is a good idea to lay the disassembled parts out on a clean surface in the order that they were removed. It may also be helpful to make sketches or take instant photos of components before removal.

When removing fasteners from a component, keep track of their locations. Sometimes threading a bolt back in a part, or putting the washers and nut back on a stud, can prevent mix-ups later. If nuts and bolts cannot be returned to their original locations, they should be kept in a compartmented box or a series of small boxes. A cupcake or muffin tin is ideal for this purpose, since each cavity can hold the bolts and nuts from a particular area (i.e. oil pan bolts, valve cover bolts, engine mount bolts, etc.). A pan of this type is especially helpful when working on assemblies with very small parts, such as the carburetor, alternator, valve train or interior dash and trim pieces. The cavities can be marked with paint or tape to identify the contents.

Whenever wiring looms, harnesses or connectors are separated, it is a good idea to

identify the two halves with numbered pieces of masking tape so they can be easily reconnected.

Gasket sealing surfaces

Throughout any vehicle, gaskets are used to seal the mating surfaces between two parts and keep lubricants, fluids, vacuum or pressure contained in an assembly.

Many times these gaskets are coated with a liquid or paste-type gasket sealing compound before assembly. Age, heat and pressure can sometimes cause the two parts to stick together so tightly that they are very difficult to separate. Often, the assembly can be loosened by striking it with a soft-face hammer near the mating surfaces. A regular hammer can be used if a block of wood is placed between the hammer and the part. Do not hammer on cast parts or parts that could be easily damaged. With any particularly stubborn part, always recheck to make sure that every fastener has been removed.

Avoid using a screwdriver or bar to pry apart an assembly, as they can easily mar the gasket sealing surfaces of the parts, which must remain smooth. If prying is absolutely necessary, use an old broom handle, but keep in mind that extra clean up will be necessary if the wood splinters.

After the parts are separated, the old gasket must be carefully scraped off and the gasket surfaces cleaned. Stubborn gasket material can be soaked with rust penetrant or treated with a special chemical to soften it so it can be easily scraped off. A scraper can be fashioned from a piece of copper tubing by flattening and sharpening one end. Copper is recommended because it is usually softer than the surfaces to be scraped, which reduces the chance of gouging the part. Some gaskets can be removed with a wire brush, but regardless of the method used, the mating surfaces must be left clean and smooth. If for some reason the gasket surface is gouged, then a gasket sealer thick enough to fill scratches will have to be used during reassembly of the components. For most applications, a non-drying (or semi-drying) gasket sealer should be used.

Hose removal tips

Warning: *If the vehicle is equipped with air conditioning, do not disconnect any of the A/C hoses without first having the system depressurized by a dealer service department or a service station.*

Hose removal precautions closely parallel gasket removal precautions. Avoid scratching or gouging the surface that the hose mates against or the connection may leak. This is especially true for radiator hoses. Because of various chemical reactions, the rubber in hoses can bond itself to the metal spigot that the hose fits over. To remove a hose, first loosen the hose clamps that secure it to the spigot. Then, with slip-joint pliers, grab the hose at the clamp and rotate it around the spigot. Work it back and forth until it is completely free, then pull it off. Silicone or other lubricants will ease removal if they can be applied between the hose and the outside of the spigot. Apply the same lubricant to the inside of the hose and the outside of the spigot to simplify installation.

As a last resort (and if the hose is to be replaced with a new one anyway), the rubber can be slit with a knife and the hose peeled from the spigot. If this must be done, be careful that the metal connection is not damaged.

If a hose clamp is broken or damaged, do not reuse it. Wire-type clamps usually weaken with age, so it is a good idea to replace them with screw-type clamps whenever a hose is removed.

Tools

A selection of good tools is a basic requirement for anyone who plans to maintain and repair his or her own vehicle. For the owner who has few tools, the initial investment might seem high, but when compared to the spiraling costs of professional auto maintenance and repair, it is a wise one.

To help the owner decide which tools are needed to perform the tasks detailed in this manual, the following tool lists are offered: *Maintenance and minor repair, Repair/overhaul* and *Special*.

The newcomer to practical mechanics

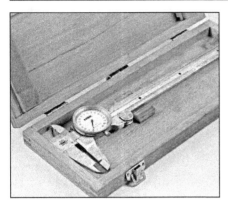

Dial caliper

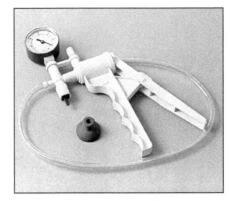

Hand-operated vacuum pump

Timing light

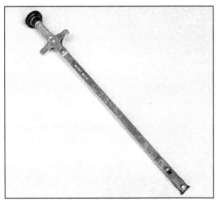

Compression gauge with spark plug
hole adapter

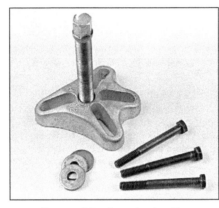

Damper/steering wheel puller

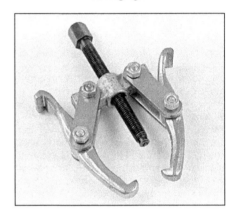

General purpose puller

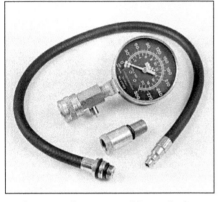

Hydraulic lifter removal tool

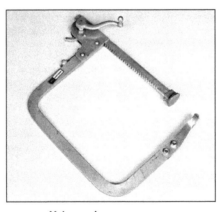

Valve spring compressor

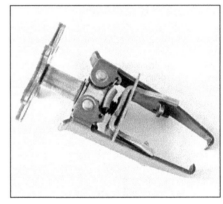

Valve spring compressor

Ridge reamer

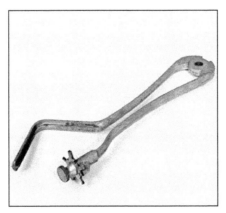

Piston ring groove cleaning tool

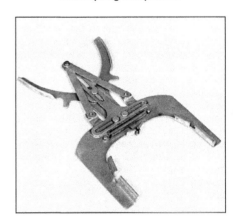

Ring removal/installation tool

Ring compressor

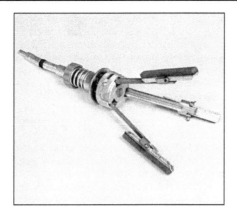

Cylinder hone

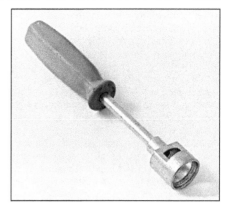

Brake hold-down spring tool

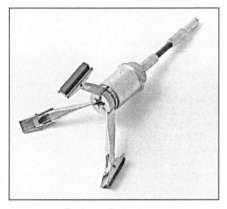

Brake cylinder hone

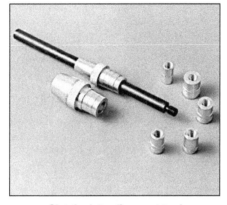

Clutch plate alignment tool

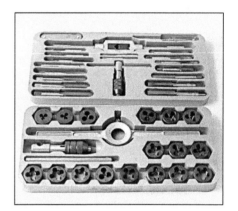

Tap and die set

should start off with the *maintenance and minor repair* tool kit, which is adequate for the simpler jobs performed on a vehicle. Then, as confidence and experience grow, the owner can tackle more difficult tasks, buying additional tools as they are needed. Eventually the basic kit will be expanded into the *repair and overhaul* tool set. Over a period of time, the experienced do-it-yourselfer will assemble a tool set complete enough for most repair and overhaul procedures and will add tools from the special category when it is felt that the expense is justified by the frequency of use.

Maintenance and minor repair tool kit

The tools in this list should be considered the minimum required for performance of routine maintenance, servicing and minor repair work. We recommend the purchase of combination wrenches (box-end and open-end combined in one wrench). While more expensive than open end wrenches, they offer the advantages of both types of wrench.

> *Combination wrench set (1/4-inch to 1 inch or 6 mm to 19 mm)*
> *Adjustable wrench, 8 inch*
> *Spark plug wrench with rubber insert*
> *Spark plug gap adjusting tool*
> *Feeler gauge set*
> *Brake bleeder wrench*
> *Standard screwdriver (5/16-inch x 6 inch)*

> *Phillips screwdriver (No. 2 x 6 inch)*
> *Combination pliers - 6 inch*
> *Hacksaw and assortment of blades*
> *Tire pressure gauge*
> *Grease gun*
> *Oil can*
> *Fine emery cloth*
> *Wire brush*
> *Battery post and cable cleaning tool*
> *Oil filter wrench*
> *Funnel (medium size)*
> *Safety goggles*
> *Jackstands (2)*
> *Drain pan*

Note: *If basic tune-ups are going to be part of routine maintenance, it will be necessary to purchase a good quality stroboscopic timing light and combination tachometer/dwell meter. Although they are included in the list of special tools, it is mentioned here because they are absolutely necessary for tuning most vehicles properly.*

Repair and overhaul tool set

These tools are essential for anyone who plans to perform major repairs and are in addition to those in the maintenance and minor repair tool kit. Included is a comprehensive set of sockets which, though expensive, are invaluable because of their versatility, especially when various extensions and drives are available. We recommend the 1/2-inch drive over the 3/8-inch drive. Although the larger drive is bulky and more expensive,

it has the capacity of accepting a very wide range of large sockets. Ideally, however, the mechanic should have a 3/8-inch drive set and a 1/2-inch drive set.

> *Socket set(s)*
> *Reversible ratchet*
> *Extension - 10 inch*
> *Universal joint*
> *Torque wrench (same size drive as sockets)*
> *Ball peen hammer - 8 ounce*
> *Soft-face hammer (plastic/rubber)*
> *Standard screwdriver (1/4-inch x 6 inch)*
> *Standard screwdriver (stubby - 5/16-inch)*
> *Phillips screwdriver (No. 3 x 8 inch)*
> *Phillips screwdriver (stubby - No. 2)*
> *Pliers - vise grip*
> *Pliers - lineman's*
> *Pliers - needle nose*
> *Pliers - snap-ring (internal and external)*
> *Cold chisel - 1/2-inch*
> *Scribe*
> *Scraper (made from flattened copper tubing)*
> *Centerpunch*
> *Pin punches (1/16, 1/8, 3/16-inch)*
> *Steel rule/straightedge - 12 inch*
> *Allen wrench set (1/8 to 3/8-inch or 4 mm to 10 mm)*
> *A selection of files*
> *Wire brush (large)*
> *Jackstands (second set)*
> *Jack (scissor or hydraulic type)*

Note: *Another tool which is often useful is an electric drill with a chuck capacity of 3/8-inch and a set of good quality drill bits.*

Special tools

The tools in this list include those which are not used regularly, are expensive to buy, or which need to be used in accordance with their manufacturer's instructions. Unless these tools will be used frequently, it is not very economical to purchase many of them. A consideration would be to split the cost and use between yourself and a friend or friends. In addition, most of these tools can be obtained from a tool rental shop on a temporary basis.

This list primarily contains only those tools and instruments widely available to the public, and not those special tools produced by the vehicle manufacturer for distribution to dealer service departments. Occasionally, references to the manufacturer's special tools are included in the text of this manual. Generally, an alternative method of doing the job without the special tool is offered. However, sometimes there is no alternative to their use. Where this is the case, and the tool cannot be purchased or borrowed, the work should be turned over to the dealer service department or an automotive repair shop.

> *Valve spring compressor*
> *Piston ring groove cleaning tool*
> *Piston ring compressor*
> *Piston ring installation tool*
> *Cylinder compression gauge*
> *Cylinder ridge reamer*
> *Cylinder surfacing hone*
> *Cylinder bore gauge*
> *Micrometers and/or dial calipers*
> *Hydraulic lifter removal tool*
> *Balljoint separator*
> *Universal-type puller*
> *Impact screwdriver*
> *Dial indicator set*
> *Stroboscopic timing light (inductive pick-up)*
> *Hand operated vacuum/pressure pump*
> *Tachometer/dwell meter*
> *Universal electrical multimeter*
> *Cable hoist*
> *Brake spring removal and installation tools*
> *Floor jack*

Buying tools

For the do-it-yourselfer who is just starting to get involved in vehicle maintenance and repair, there are a number of options available when purchasing tools. If maintenance and minor repair is the extent of the work to be done, the purchase of individual tools is satisfactory. If, on the other hand, extensive work is planned, it would be a good idea to purchase a modest tool set from one of the large retail chain stores. A set can usually be bought at a substantial savings over the individual tool prices, and they often come with a tool box. As additional tools are

needed, add-on sets, individual tools and a larger tool box can be purchased to expand the tool selection. Building a tool set gradually allows the cost of the tools to be spread over a longer period of time and gives the mechanic the freedom to choose only those tools that will actually be used.

Tool stores will often be the only source of some of the special tools that are needed, but regardless of where tools are bought, try to avoid cheap ones, especially when buying screwdrivers and sockets, because they won't last very long. The expense involved in replacing cheap tools will eventually be greater than the initial cost of quality tools.

Care and maintenance of tools

Good tools are expensive, so it makes sense to treat them with respect. Keep them clean and in usable condition and store them properly when not in use. Always wipe off any dirt, grease or metal chips before putting them away. Never leave tools lying around in the work area. Upon completion of a job, always check closely under the hood for tools that may have been left there so they won't get lost during a test drive.

Some tools, such as screwdrivers, pliers, wrenches and sockets, can be hung on a panel mounted on the garage or workshop wall, while others should be kept in a tool box or tray. Measuring instruments, gauges, meters, etc. must be carefully stored where they cannot be damaged by weather or impact from other tools.

When tools are used with care and stored properly, they will last a very long time. Even with the best of care, though, tools will wear out if used frequently. When a tool is damaged or worn out, replace it. Subsequent jobs will be safer and more enjoyable if you do.

How to repair damaged threads

Sometimes, the internal threads of a nut or bolt hole can become stripped, usually from overtightening. Stripping threads is an all-too-common occurrence, especially when working with aluminum parts, because aluminum is so soft that it easily strips out.

Usually, external or internal threads are only partially stripped. After they've been cleaned up with a tap or die, they'll still work. Sometimes, however, threads are badly damaged. When this happens, you've got three choices:

1) *Drill and tap the hole to the next suitable oversize and install a larger diameter bolt, screw or stud.*
2) *Drill and tap the hole to accept a threaded plug, then drill and tap the plug to the original screw size. You can also buy a plug already threaded to the original size. Then you simply drill a hole to the specified size, then run the threaded plug into the hole with a bolt and jam*

nut. Once the plug is fully seated, remove the jam nut and bolt.
3) *The third method uses a patented thread repair kit like Heli-Coil or Slimsert. These easy-to-use kits are designed to repair damaged threads in straight-through holes and blind holes. Both are available as kits which can handle a variety of sizes and thread patterns. Drill the hole, then tap it with the special included tap. Install the Heli-Coil and the hole is back to its original diameter and thread pitch.*

Regardless of which method you use, be sure to proceed calmly and carefully. A little impatience or carelessness during one of these relatively simple procedures can ruin your whole day's work and cost you a bundle if you wreck an expensive part.

Working facilities

Not to be overlooked when discussing tools is the workshop. If anything more than routine maintenance is to be carried out, some sort of suitable work area is essential.

It is understood, and appreciated, that many home mechanics do not have a good workshop or garage available, and end up removing an engine or doing major repairs outside. It is recommended, however, that the overhaul or repair be completed under the cover of a roof.

A clean, flat workbench or table of comfortable working height is an absolute necessity. The workbench should be equipped with a vise that has a jaw opening of at least four inches.

As mentioned previously, some clean, dry storage space is also required for tools, as well as the lubricants, fluids, cleaning solvents, etc. which soon become necessary.

Sometimes waste oil and fluids, drained from the engine or cooling system during normal maintenance or repairs, present a disposal problem. To avoid pouring them on the ground or into a sewage system, pour the used fluids into large containers, seal them with caps and take them to an authorized disposal site or recycling center. Plastic jugs, such as old antifreeze containers, are ideal for this purpose.

Always keep a supply of old newspapers and clean rags available. Old towels are excellent for mopping up spills. Many mechanics use rolls of paper towels for most work because they are readily available and disposable. To help keep the area under the vehicle clean, a large cardboard box can be cut open and flattened to protect the garage or shop floor.

Whenever working over a painted surface, such as when leaning over a fender to service something under the hood, always cover it with an old blanket or bedspread to protect the finish. Vinyl covered pads, made especially for this purpose, are available at auto parts stores.

Booster battery (jump) starting

Observe these precautions when using a booster battery to start a vehicle:

a) *Before connecting the booster battery, make sure the ignition switch is in the Off position.*
b) *Turn off the lights, heater and other electrical loads.*
c) *Your eyes should be shielded. Safety goggles are a good idea.*
d) *Make sure the booster battery is the same voltage as the dead one in the vehicle.*
e) *The two vehicles MUST NOT TOUCH each other!*
f) *Make sure the transaxle is in Neutral (manual) or Park (automatic).*
g) *If the booster battery is not a maintenance-free type, remove the vent caps and lay a cloth over the vent holes.*

Connect the red jumper cable to the positive (+) terminals of each battery **(see illustration)**.

Connect one end of the black jumper cable to the negative (-) terminal of the booster battery. The other end of this cable should be connected to a good ground on the vehicle to be started, such as a bolt or bracket on the body.

Start the engine using the booster battery, then, with the engine running at idle speed, disconnect the jumper cables in the reverse order of connection.

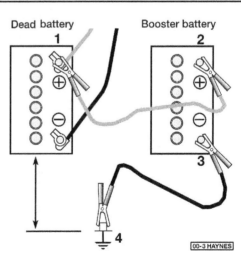

Make the booster battery cable connections in the numerical order shown (note that the negative cable of the booster battery is NOT attached to the negative terminal of the dead battery)

Jacking and towing

Jacking

The jack supplied with the vehicle should only be used for raising the vehicle when changing a tire or placing jackstands under the frame. **Warning:** *Never work under the vehicle or start the engine while this jack is being used as the only means of support.*

The vehicle should be on level ground with the hazard flashers on, the wheels blocked, the parking brake applied and the transmission in Park (automatic) or Reverse (manual). If a tire is being changed, loosen the lug nuts one-half turn and leave them in place until the wheel is raised off the ground.

Place the jack under the vehicle suspension in the indicated position **(see illustrations)**. Operate the jack with a slow, smooth motion until the wheel is raised off the ground. Remove the lug nuts, pull off the wheel, install the spare and thread the lug nuts back on with the beveled sides facing in. Tighten them snugly, but wait until the vehicle is lowered to tighten them completely.

Lower the vehicle, remove the jack and tighten the nuts (if loosened or removed) in a criss-cross pattern.

Towing

As a general rule, vehicles should be towed with all four wheels on the ground. If necessary, the front or rear wheels may be raised for towing. Do not exceed 35 MPH or tow the vehicle farther than 50 miles.

Equipment specifically designed for towing should be used and should be attached to the main structural members of the vehicle, not the bumper or brackets. Tow hooks are attached to the frame at both ends of the vehicle. However, they are for emergency use only and should not be used for highway towing. Stand clear of vehicles when using the tow hooks - tow straps and chains may break, causing serious injury.

Safety is a major consideration when towing and all applicable state and local laws must be obeyed. A safety chain must be used for all towing (in addition to the tow bar).

While towing, the parking brake must be released, the transmission must be in Neutral and the transfer case (if equipped) must be in 2H. The steering must be unlocked (ignition switch in the Off position). If you're towing a 4WD model with the front wheels on the ground, the front hubs must be unlocked. Remember that power steering and power brakes will not work with the engine off.

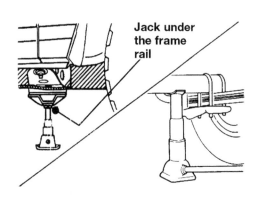

Front and rear jacking points for Trooper models - the front jacking point is shown at the top

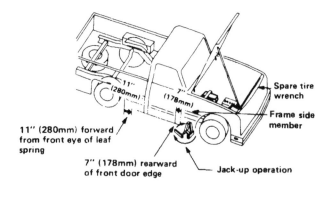

Jacking points for pick-up models

Automotive chemicals and lubricants

A number of automotive chemicals and lubricants are available for use during vehicle maintenance and repair. They include a wide variety of products ranging from cleaning solvents and degreasers to lubricants and protective sprays for rubber, plastic and vinyl.

Cleaners

Carburetor cleaner and choke cleaner is a strong solvent for gum, varnish and carbon. Most carburetor cleaners leave a dry-type lubricant film which will not harden or gum up. Because of this film it is not recommended for use on electrical components.

Brake system cleaner is used to remove grease and brake fluid from the brake system, where clean surfaces are absolutely necessary. It leaves no residue and often eliminates brake squeal caused by contaminants.

Electrical cleaner removes oxidation, corrosion and carbon deposits from electrical contacts, restoring full current flow. It can also be used to clean spark plugs, carburetor jets, voltage regulators and other parts where an oil-free surface is desired.

Demoisturants remove water and moisture from electrical components such as alternators, voltage regulators, electrical connectors and fuse blocks. They are non-conductive, non-corrosive and non-flammable.

Degreasers are heavy-duty solvents used to remove grease from the outside of the engine and from chassis components. They can be sprayed or brushed on and, depending on the type, are rinsed off either with water or solvent.

Lubricants

Motor oil is the lubricant formulated for use in engines. It normally contains a wide variety of additives to prevent corrosion and reduce foaming and wear. Motor oil comes in various weights (viscosity ratings) from 0 to 50. The recommended weight of the oil depends on the season, temperature and the demands on the engine. Light oil is used in cold climates and under light load conditions. Heavy oil is used in hot climates and where high loads are encountered. Multi-viscosity oils are designed to have characteristics of both light and heavy oils and are available in a number of weights from 5W-20 to 20W-50.

Gear oil is designed to be used in differentials, manual transmissions and other areas where high-temperature lubrication is required.

Chassis and wheel bearing grease is a heavy grease used where increased loads and friction are encountered, such as for wheel bearings, balljoints, tie-rod ends and universal joints.

High-temperature wheel bearing grease is designed to withstand the extreme temperatures encountered by wheel bearings

in disc brake equipped vehicles. It usually contains molybdenum disulfide (moly), which is a dry-type lubricant.

White grease is a heavy grease for metal-to-metal applications where water is a problem. White grease stays soft under both low and high temperatures (usually from -100 to +190-degrees F), and will not wash off or dilute in the presence of water.

Assembly lube is a special extreme pressure lubricant, usually containing moly, used to lubricate high-load parts (such as main and rod bearings and cam lobes) for initial start-up of a new engine. The assembly lube lubricates the parts without being squeezed out or washed away until the engine oiling system begins to function.

Silicone lubricants are used to protect rubber, plastic, vinyl and nylon parts.

Graphite lubricants are used where oils cannot be used due to contamination problems, such as in locks. The dry graphite will lubricate metal parts while remaining uncontaminated by dirt, water, oil or acids. It is electrically conductive and will not foul electrical contacts in locks such as the ignition switch.

Moly penetrants loosen and lubricate frozen, rusted and corroded fasteners and prevent future rusting or freezing.

Heat-sink grease is a special electrically non-conductive grease that is used for mounting electronic ignition modules where it is essential that heat is transferred away from the module.

Sealants

RTV sealant is one of the most widely used gasket compounds. Made from silicone, RTV is air curing, it seals, bonds, waterproofs, fills surface irregularities, remains flexible, doesn't shrink, is relatively easy to remove, and is used as a supplementary sealer with almost all low and medium temperature gaskets.

Anaerobic sealant is much like RTV in that it can be used either to seal gaskets or to form gaskets by itself. It remains flexible, is solvent resistant and fills surface imperfections. The difference between an anaerobic sealant and an RTV-type sealant is in the curing. RTV cures when exposed to air, while an anaerobic sealant cures only in the absence of air. This means that an anaerobic sealant cures only after the assembly of parts, sealing them together.

Thread and pipe sealant is used for sealing hydraulic and pneumatic fittings and vacuum lines. It is usually made from a Teflon compound, and comes in a spray, a paint-on liquid and as a wrap-around tape.

Chemicals

Anti-seize compound prevents seizing, galling, cold welding, rust and corrosion in

fasteners. High-temperature anti-seize, usually made with copper and graphite lubricants, is used for exhaust system and exhaust manifold bolts.

Anaerobic locking compounds are used to keep fasteners from vibrating or working loose and cure only after installation, in the absence of air. Medium strength locking compound is used for small nuts, bolts and screws that may be removed later. High-strength locking compound is for large nuts, bolts and studs which aren't removed on a regular basis.

Oil additives range from viscosity index improvers to chemical treatments that claim to reduce internal engine friction. It should be noted that most oil manufacturers caution against using additives with their oils.

Gas additives perform several functions, depending on their chemical makeup. They usually contain solvents that help dissolve gum and varnish that build up on carburetor, fuel injection and intake parts. They also serve to break down carbon deposits that form on the inside surfaces of the combustion chambers. Some additives contain upper cylinder lubricants for valves and piston rings, and others contain chemicals to remove condensation from the gas tank.

Miscellaneous

Brake fluid is specially formulated hydraulic fluid that can withstand the heat and pressure encountered in brake systems. Care must be taken so this fluid does not come in contact with painted surfaces or plastics. An opened container should always be resealed to prevent contamination by water or dirt.

Weatherstrip adhesive is used to bond weatherstripping around doors, windows and trunk lids. It is sometimes used to attach trim pieces.

Undercoating is a petroleum-based, tar-like substance that is designed to protect metal surfaces on the underside of the vehicle from corrosion. It also acts as a sound-deadening agent by insulating the bottom of the vehicle.

Waxes and polishes are used to help protect painted and plated surfaces from the weather. Different types of paint may require the use of different types of wax and polish. Some polishes utilize a chemical or abrasive cleaner to help remove the top layer of oxidized (dull) paint on older vehicles. In recent years many non-wax polishes that contain a wide variety of chemicals such as polymers and silicones have been introduced. These non-wax polishes are usually easier to apply and last longer than conventional waxes and polishes.

Conversion factors

Length (distance)

Inches (in)	X	25.4	= Millimetres (mm)	X	0.0394	= Inches (in)
Feet (ft)	X	0.305	= Metres (m)	X	3.281	= Feet (ft)
Miles	X	1.609	= Kilometres (km)	X	0.621	= Miles

Volume (capacity)

Cubic inches (cu in; in^3)	X	16.387	= Cubic centimetres (cc; cm^3)	X	0.061	= Cubic inches (cu in; in^3)
Imperial pints (Imp pt)	X	0.568	= Litres (l)	X	1.76	= Imperial pints (Imp pt)
Imperial quarts (Imp qt)	X	1.137	= Litres (l)	X	0.88	= Imperial quarts (Imp qt)
Imperial quarts (Imp qt)	X	1.201	= US quarts (US qt)	X	0.833	= Imperial quarts (Imp qt)
US quarts (US qt)	X	0.946	= Litres (l)	X	1.057	= US quarts (US qt)
Imperial gallons (Imp gal)	X	4.546	= Litres (l)	X	0.22	= Imperial gallons (Imp gal)
Imperial gallons (Imp gal)	X	1.201	= US gallons (US gal)	X	0.833	= Imperial gallons (Imp gal)
US gallons (US gal)	X	3.785	= Litres (l)	X	0.264	= US gallons (US gal)

Mass (weight)

Ounces (oz)	X	28.35	= Grams (g)	X	0.035	= Ounces (oz)
Pounds (lb)	X	0.454	= Kilograms (kg)	X	2.205	= Pounds (lb)

Force

Ounces-force (ozf; oz)	X	0.278	= Newtons (N)	X	3.6	= Ounces-force (ozf; oz)
Pounds-force (lbf; lb)	X	4.448	= Newtons (N)	X	0.225	= Pounds-force (lbf; lb)
Newtons (N)	X	0.1	= Kilograms-force (kgf; kg)	X	9.81	= Newtons (N)

Pressure

Pounds-force per square inch (psi; lbf/in^2; lb/in^2)	X	0.070	= Kilograms-force per square centimetre (kgf/cm^2; kg/cm^2)	X	14.223	= Pounds-force per square inch (psi; lbf/in^2; lb/in^2)
Pounds-force per square inch (psi; lbf/in^2; lb/in^2)	X	0.068	= Atmospheres (atm)	X	14.696	= Pounds-force per square inch (psi; lbf/in^2; lb/in^2)
Pounds-force per square inch (psi; lbf/in^2; lb/in^2)	X	0.069	= Bars	X	14.5	= Pounds-force per square inch (psi; lbf/in^2; lb/in^2)
Pounds-force per square inch (psi; lbf/in^2; lb/in^2)	X	6.895	= Kilopascals (kPa)	X	0.145	= Pounds-force per square inch (psi; lbf/in^2; lb/in^2)
Kilopascals (kPa)	X	0.01	= Kilograms-force per square centimetre (kgf/cm^2; kg/cm^2)	X	98.1	= Kilopascals (kPa)

Torque (moment of force)

Pounds-force inches (lbf in; lb in)	X	1.152	= Kilograms-force centimetre (kgf cm; kg cm)	X	0.868	= Pounds-force inches (lbf in; lb in)
Pounds-force inches (lbf in; lb in)	X	0.113	= Newton metres (Nm)	X	8.85	= Pounds-force inches (lbf in; lb in)
Pounds-force inches (lbf in; lb in)	X	0.083	= Pounds-force feet (lbf ft; lb ft)	X	12	= Pounds-force inches (lbf in; lb in)
Pounds-force feet (lbf ft; lb ft)	X	0.138	= Kilograms-force metres (kgf m; kg m)	X	7.233	= Pounds-force feet (lbf ft; lb ft)
Pounds-force feet (lbf ft; lb ft)	X	1.356	= Newton metres (Nm)	X	0.738	= Pounds-force feet (lbf ft; lb ft)
Newton metres (Nm)	X	0.102	= Kilograms-force metres (kgf m; kg m)	X	9.804	= Newton metres (Nm)

Vacuum

Inches mercury (in. Hg)	X	3.377	= Kilopascals (kPa)	X	0.2961	= Inches mercury
Inches mercury (in. Hg)	X	25.4	= Millimeters mercury (mm Hg)	X	0.0394	= Inches mercury

Power

Horsepower (hp)	X	745.7	= Watts (W)	X	0.0013	= Horsepower (hp)

Velocity (speed)

Miles per hour (miles/hr; mph)	X	1.609	= Kilometres per hour (km/hr; kph)	X	0.621	= Miles per hour (miles/hr; mph)

Fuel consumption*

Miles per gallon, Imperial (mpg)	X	0.354	= Kilometres per litre (km/l)	X	2.825	= Miles per gallon, Imperial (mpg)
Miles per gallon, US (mpg)	X	0.425	= Kilometres per litre (km/l)	X	2.352	= Miles per gallon, US (mpg)

Temperature

Degrees Fahrenheit = (°C x 1.8) + 32

Degrees Celsius (Degrees Centigrade; °C) = (°F - 32) x 0.56

*It is common practice to convert from miles per gallon (mpg) to litres/100 kilometres (l/100km), where mpg (Imperial) x l/100 km = 282 and mpg (US) x l/100 km = 235

Safety first!

Regardless of how enthusiastic you may be about getting on with the job at hand, take the time to ensure that your safety is not jeopardized. A moment's lack of attention can result in an accident, as can failure to observe certain simple safety precautions. The possibility of an accident will always exist, and the following points should not be considered a comprehensive list of all dangers. Rather, they are intended to make you aware of the risks and to encourage a safety conscious approach to all work you carry out on your vehicle.

Essential DOs and DON'Ts

DON'T rely on a jack when working under the vehicle. Always use approved jackstands to support the weight of the vehicle and place them under the recommended lift or support points.

DON'T attempt to loosen extremely tight fasteners (i.e. wheel lug nuts) while the vehicle is on a jack - it may fall.

DON'T start the engine without first making sure that the transmission is in Neutral (or Park where applicable) and the parking brake is set.

DON'T remove the radiator cap from a hot cooling system - let it cool or cover it with a cloth and release the pressure gradually.

DON'T attempt to drain the engine oil until you are sure it has cooled to the point that it will not burn you.

DON'T touch any part of the engine or exhaust system until it has cooled sufficiently to avoid burns.

DON'T siphon toxic liquids such as gasoline, antifreeze and brake fluid by mouth, or allow them to remain on your skin.

DON'T inhale brake lining dust - it is potentially hazardous (see *Asbestos* below).

DON'T allow spilled oil or grease to remain on the floor - wipe it up before someone slips on it.

DON'T use loose fitting wrenches or other tools which may slip and cause injury.

DON'T push on wrenches when loosening or tightening nuts or bolts. Always try to pull the wrench toward you. If the situation calls for pushing the wrench away, push with an open hand to avoid scraped knuckles if the wrench should slip.

DON'T attempt to lift a heavy component alone - get someone to help you.

DON'T rush or take unsafe shortcuts to finish a job.

DON'T allow children or animals in or around the vehicle while you are working on it.

DO wear eye protection when using power tools such as a drill, sander, bench grinder, etc. and when working under a vehicle.

DO keep loose clothing and long hair well out of the way of moving parts.

DO make sure that any hoist used has a safe working load rating adequate for the job.

DO get someone to check on you periodically when working alone on a vehicle.

DO carry out work in a logical sequence and make sure that everything is correctly assembled and tightened.

DO keep chemicals and fluids tightly capped and out of the reach of children and pets.

DO remember that your vehicle's safety affects that of yourself and others. If in doubt on any point, get professional advice.

Asbestos

Certain friction, insulating, sealing, and other products - such as brake linings, brake bands, clutch linings, torque converters, gaskets, etc. - may contain asbestos. Extreme care must be taken to avoid inhalation of dust from such products, since it is hazardous to health. If in doubt, assume that they do contain asbestos.

Fire

Remember at all times that gasoline is highly flammable. Never smoke or have any kind of open flame around when working on a vehicle. But the risk does not end there. A spark caused by an electrical short circuit, by two metal surfaces contacting each other, or even by static electricity built up in your body under certain conditions, can ignite gasoline vapors, which in a confined space are highly explosive. Do not, under any circumstances, use gasoline for cleaning parts. Use an approved safety solvent.

Always disconnect the battery ground (-) cable at the battery before working on any part of the fuel system or electrical system. Never risk spilling fuel on a hot engine or exhaust component. It is strongly recommended that a fire extinguisher suitable for use on fuel and electrical fires be kept handy in the garage or workshop at all times. Never try to extinguish a fuel or electrical fire with water.

Fumes

Certain fumes are highly toxic and can quickly cause unconsciousness and even death if inhaled to any extent. Gasoline vapor falls into this category, as do the vapors from some cleaning solvents. Any draining or pouring of such volatile fluids should be done in a well ventilated area.

When using cleaning fluids and solvents, read the instructions on the container carefully. Never use materials from unmarked containers.

Never run the engine in an enclosed space, such as a garage. Exhaust fumes contain carbon monoxide, which is extremely poisonous. If you need to run the engine, always do so in the open air, or at least have the rear of the vehicle outside the work area.

If you are fortunate enough to have the use of an inspection pit, never drain or pour gasoline and never run the engine while the vehicle is over the pit. The fumes, being heavier than air, will concentrate in the pit with possibly lethal results.

The battery

Never create a spark or allow a bare light bulb near a battery. They normally give off a certain amount of hydrogen gas, which is highly explosive.

Always disconnect the battery ground (-) cable at the battery before working on the fuel or electrical systems.

If possible, loosen the filler caps or cover when charging the battery from an external source (this does not apply to sealed or maintenance-free batteries). Do not charge at an excessive rate or the battery may burst.

Take care when adding water to a non maintenance-free battery and when carrying a battery. The electrolyte, even when diluted, is very corrosive and should not be allowed to contact clothing or skin.

Always wear eye protection when cleaning the battery to prevent the caustic deposits from entering your eyes.

Household current

When using an electric power tool, inspection light, etc., which operates on household current, always make sure that the tool is correctly connected to its plug and that, where necessary, it is properly grounded. Do not use such items in damp conditions and, again, do not create a spark or apply excessive heat in the vicinity of fuel or fuel vapor.

Secondary ignition system voltage

A severe electric shock can result from touching certain parts of the ignition system (such as the spark plug wires) when the engine is running or being cranked, particularly if components are damp or the insulation is defective. In the case of an electronic ignition system, the secondary system voltage is much higher and could prove fatal.

Troubleshooting

Contents

This Section provides an easy reference guide to the more common problems that may occur during the operation of your vehicle. Various symptoms and their probable causes are grouped under headings denoting components or systems, such as Engine, Cooling system, etc. They also refer to the Chapter and/or Section that deals with the problem.

Remember that successful troubleshooting isn't a mysterious black art practiced only by professional mechanics, it's simply the result of knowledge combined with an intelligent, systematic approach to a problem. Always use a process of elimination starting with the simplest solution and working through to the most complex - and never overlook the obvious. Anyone can run the gas tank dry or leave the lights on overnight, so don't assume that you're exempt from such oversights.

Finally, always establish a clear idea why a problem has occurred and take steps to ensure that it doesn't happen again. If the electrical system fails because of a poor connection, check all other connections in the system to make sure they don't fail as well. If a particular fuse continues to blow, find out why - don't just go on replacing fuses. Remember, failure of a small component can often be indicative of potential failure or incorrect functioning of a more important component or system.

Engine and performance

1 Engine will not rotate when attempting to start

1   Battery terminal connections loose or corroded. Check the cable terminals at the battery; tighten cable clamp and/or clean off corrosion as necessary (see Chapter 1).
2   Battery discharged or faulty. If the cable ends are clean and tight on the battery posts, turn the key to the On position and switch on the headlights or windshield wipers. If they won't run, the battery is discharged.
3   Automatic transmission not engaged in park (P) or Neutral (N).
4   Broken, loose or disconnected wires in the starting circuit. Inspect all wires and connectors at the battery, starter solenoid and ignition switch (on steering column).
5   Starter motor pinion jammed in flywheel ring gear. If manual transmission, place transmission in gear and rock the vehicle to manually turn the engine. Remove starter (Chapter 5) and inspect pinion and flywheel (Chapter 2) at earliest convenience.
6   Starter solenoid faulty (Chapter 5).
7   Starter motor faulty (Chapter 5).
8   Ignition switch faulty (Chapter 12).
9   Engine seized. Try to turn the crankshaft with a large socket and breaker bar on the pulley bolt.

2 Engine rotates but will not start

1   Fuel tank empty.
2   Battery discharged (engine rotates slowly). Check the operation of electrical components as described in previous Section.
3   Battery terminal connections loose or corroded. See previous Section.
4   Fuel not reaching carburetor or fuel injector. Check for clogged fuel filter or lines and defective fuel pump. Also make sure the tank vent lines aren't clogged (Chapter 4).
5   Choke not operating properly (Chapter 1).
6   Faulty distributor components. Check the cap and rotor (Chapter 1).
7   Low cylinder compression. Check as described in Chapter 2.
8   Valve clearances not properly adjusted - Chapter 1 (four-cylinder engine) or Chapter 2B (V6 engine).
9   Water in fuel. Drain tank and fill with new fuel.
10   Defective ignition coil (Chapter 5).
11   Dirty or clogged carburetor jets or fuel injector. Carburetor out of adjustment. Check the float level (Chapter 4).
12   Wet or damaged ignition components (Chapters 1 and 5).
13   Worn, faulty or incorrectly gapped spark plugs (Chapter 1).
14   Broken, loose or disconnected wires in the starting circuit (see previous Section).
15   Loose distributor (changing ignition timing). Turn the distributor body as necessary to start the engine, then adjust the ignition timing as soon as possible (Chapter 1).
16   Broken, loose or disconnected wires at the ignition coil or faulty coil (Chapter 5).
17   Timing chain or belt failure or wear affecting valve timing (Chapter 2).

3 Starter motor operates without turning engine

1   Starter pinion sticking. Remove the starter (Chapter 5) and inspect.
2   Starter pinion or flywheel/driveplate teeth worn or broken. Remove the inspection cover and inspect.

4 Engine hard to start when cold

1   Battery discharged or low. Check as described in Chapter 1.
2   Fuel not reaching the carburetor or fuel injectors. Check the fuel filter, lines and fuel pump (Chapters 1 and 4).
3   Choke inoperative (Chapters 1 and 4).
4   Defective spark plugs (Chapter 1).

5 Engine hard to start when hot

1   Air filter dirty (Chapter 1).

2   Fuel not reaching carburetor or fuel injectors (see Section 4). Check for a vapor lock situation, brought about by clogged fuel tank vent lines.
3   Bad engine ground connection.
4   Choke sticking (Chapter 1).
5   Defective pick-up coil in distributor (Chapter 5).
6   Float level too high (Chapter 4).

6 Starter motor noisy or engages roughly

1   Pinion or flywheel/driveplate teeth worn or broken. Remove the inspection cover on the left side of the engine and inspect.
2   Starter motor mounting bolts loose or missing.

7 Engine starts but stops immediately

1   Loose or damaged wire harness connections at distributor, coil or alternator.
2   Intake manifold vacuum leaks. Make sure all mounting bolts/nuts are tight and all vacuum hoses connected to the manifold are attached properly and in good condition.
3   Insufficient fuel flow (see Chapter 4).

8 Engine 'lopes' while idling or idles erratically

1   Vacuum leaks. Check mounting bolts at the intake manifold for tightness. Make sure that all vacuum hoses are connected and in good condition. Use a stethoscope or a length of fuel hose held against your ear to listen for vacuum leaks while the engine is running. A hissing sound will be heard. A soapy water solution will also detect leaks. Check the intake manifold gasket surfaces.
2   Leaking EGR valve or plugged PCV valve (see Chapters 1 and 6).
3   Air filter clogged (Chapter 1).
4   Fuel pump not delivering sufficient fuel (Chapter 4).
5   Leaking head gasket. Perform a cylinder compression check (Chapter 2).
6   Timing chain worn (Chapter 2).
7   Camshaft lobes worn (Chapter 2).
8   Valve clearance out of adjustment - Chapter 1 (four-cylinder engine) or Chapter 2B (V6 engine).
9   Valves burned or otherwise leaking (Chapter 2).
10   Ignition timing out of adjustment (Chapter 1).
11   Ignition system not operating properly (Chapters 1 and 5).
12   Thermostatic air cleaner not operating properly (Chapter 1).
13   Choke not operating properly (Chapters 1 and 4).
14   Dirty or clogged injector(s). Carburetor

dirty, clogged or out of adjustment. Check the float level (Chapter 4).
15 Idle speed out of adjustment (Chapter 1).

9 Engine misses at idle speed

1 Spark plugs faulty or not gapped properly (Chapter 1).
2 Faulty spark plug wires (Chapter 1).
3 Wet or damaged distributor components (Chapter 1).
4 Short circuits in ignition, coil or spark plug wires.
5 Sticking or faulty emissions systems (see Chapter 6).
6 Clogged fuel filter and/or foreign matter in fuel. Remove the fuel filter (Chapter 1) and inspect.
7 Vacuum leaks at intake manifold or hose connections. Check as described in Section 8.
8 Incorrect idle speed (Chapter 1) or idle mixture (Chapter 4).
9 Incorrect ignition timing (Chapter 1).
10 Low or uneven cylinder compression. Check as described in Chapter 2.
11 Choke not operating properly (Chapter 1).
12 Clogged or dirty fuel injectors (Chapter 4).

10 Excessively high idle speed

1 Sticking throttle linkage (Chapter 4).
2 Choke opened excessively at idle (Chapter 4).
3 Idle speed incorrectly adjusted (Chapter 1).
4 Valve clearances incorrectly adjusted - Chapter 1 (four-cylinder engine) or Chapter 2B (V6 engine).

11 Battery will not hold a charge

1 Alternator drivebelt defective or not adjusted properly (Chapter 1).
2 Battery cables loose or corroded (Chapter 1).
3 Alternator not charging properly (Chapter 5).
4 Loose, broken or faulty wires in the charging circuit (Chapter 5).
5 Short circuit causing a continuous drain on the battery.
6 Battery defective internally.

12 Alternator light stays on

1 Fault in alternator or charging circuit (Chapter 5).
2 Alternator drivebelt defective or not properly adjusted (Chapter 1).

13 Alternator light fails to come on when key is turned on

1 Faulty bulb (Chapter 12).
2 Defective alternator (Chapter 5).
3 Fault in the printed circuit, dash wiring or bulb holder (Chapter 12).

14 Engine misses throughout driving speed range

1 Fuel filter clogged and/or impurities in the fuel system. Check fuel filter (Chapter 1) or clean system (Chapter 4).
2 Faulty or incorrectly gapped spark plugs (Chapter 1).
3 Incorrect ignition timing (Chapter 1).
4 Cracked distributor cap, disconnected distributor wires or damaged distributor components (Chapter 1).
5 Defective spark plug wires (Chapter 1).
6 Emissions system components faulty (Chapter 6).
7 Low or uneven cylinder compression pressures. Check as described in Chapter 2.
8 Weak or faulty ignition coil (Chapter 5).
9 Weak or faulty ignition system (Chapter 5).
10 Vacuum leaks at intake manifold or vacuum hoses (see Section 8).
11 Dirty or clogged carburetor or fuel injector (Chapter 4).
12 Leaky EGR valve (Chapter 6).
13 Carburetor out of adjustment (Chapter 4).
14 Idle speed out of adjustment (Chapter 1).

15 Hesitation or stumble during acceleration

1 Ignition timing incorrect (Chapter 1).
2 Ignition system not operating properly (Chapter 5).
3 Dirty or clogged carburetor or fuel injector (Chapter 4).
4 Low fuel pressure. Check for proper operation of the fuel pump and for restrictions in the fuel filter and lines (Chapter 4).
5 Carburetor out of adjustment (Chapter 4).

16 Engine stalls

1 Idle speed incorrect (Chapter 1).
2 Fuel filter clogged and/or water and impurities in the fuel system (Chapter 1).
3 Choke not operating properly (Chapter 1).
4 Damaged or wet distributor cap and wires.
5 Emissions system components faulty

(Chapter 6).
6 Faulty or incorrectly gapped spark plugs (Chapter 1). Also check the spark plug wires (Chapter 1).
7 Vacuum leak at the carburetor, intake manifold or vacuum hoses. Check as described in Section 8.
8 Valve clearances incorrect - Chapter 1 (four-cylinder engine) or Chapter 2B (V6 engine).

17 Engine lacks power

1 Incorrect ignition timing (Chapter 1).
2 Excessive play in distributor shaft. At the same time check for faulty distributor cap, wires, etc. (Chapter 1).
3 Faulty or incorrectly gapped spark plugs (Chapter 1).
4 Air filter dirty (Chapter 1).
5 Faulty ignition coil (Chapter 5).
6 Brakes binding (Chapters 1 and 10).
7 Automatic transmission fluid level incorrect, causing slippage (Chapter 1).
8 Clutch slipping (Chapter 8).
9 Fuel filter clogged and/or impurities in the fuel system (Chapters 1 and 4).
10 EGR system not functioning properly (Chapter 6).
11 Use of sub-standard fuel. Fill tank with proper octane fuel.
12 Low or uneven cylinder compression pressures. Check as described in Chapter 2.
13 Air leak at carburetor or intake manifold (check as described in Section 8).
14 Dirty or clogged carburetor jets or malfunctioning choke (Chapters 1 and 4).
15 Restricted exhaust system (Chapter 4).

18 Engine backfires

1 EGR system not functioning properly (Chapter 6).
2 Ignition timing incorrect (Chapter 1).
3 Thermostatic air cleaner system not operating properly (Chapter 6).
4 Vacuum leak (refer to Section 8).
5 Valve clearances incorrect - Chapter 1 (four-cylinder engine) or Chapter 2B (V6 engine).
6 Damaged valve springs or sticking valves (Chapter 2).
7 Intake air leak (see Section 8).
8 Carburetor float level out of adjustment (Chapter 4).

19 Engine surges while holding accelerator steady

1 Intake air leak (see Section 8).
2 Fuel pump not working properly (Chapter 4).

20 Pinging or knocking engine sounds when engine is under load

1   Incorrect grade of fuel. Fill tank with fuel of the proper octane rating.
2   Ignition timing incorrect (Chapter 1).
3   Carbon build-up in combustion chambers. Remove cylinder head(s) and clean combustion chambers (Chapter 2).
4   Incorrect spark plugs (Chapter 1).

21 Engine diesels (continues to run) after being turned off

1   Idle speed too high (Chapter 1).
2   Ignition timing incorrect (Chapter 1).
3   Incorrect spark plug heat range (Chapter 1).
4   Intake air leak (see Section 8).
5   Carbon build-up in combustion chambers. Remove the cylinder head(s) and clean the combustion chambers (Chapter 2).
6   Valves sticking (Chapter 2).
7   Valve clearances incorrect - Chapter 1 (four-cylinder engine) or Chapter 2B (V6 engine).
8   EGR system not operating properly (Chapter 6).
9   Fuel shut-off system not operating properly (Chapter 6).
10   Check for causes of overheating (Section 27).

22 Low oil pressure

1   Improper grade of oil.
2   Oil pump worn or damaged (Chapter 2).
3   Engine overheating (refer to Section 27).
4   Clogged oil filter (Chapter 1).
5   Clogged oil strainer (Chapter 2).
6   Oil pressure gauge not working properly (Chapter 2).

23 Excessive oil consumption

1   Loose oil drain plug.
2   Loose bolts or damaged oil pan gasket (Chapter 2).
3   Loose bolts or damaged front cover gasket (Chapter 2).
4   Front or rear crankshaft oil seal leaking (Chapter 2).
5   Loose bolts or damaged rocker arm cover gasket (Chapter 2).
6   Loose oil filter (Chapter 1).
7   Loose or damaged oil pressure switch (Chapter 2).
8   Pistons and cylinders excessively worn (Chapter 2).
9   Piston rings not installed correctly on pistons (Chapter 2).
10   Worn or damaged piston rings (Chapter 2).
11   Intake and/or exhaust valve oil seals

worn or damaged (Chapter 2).
12   Worn valve stems.
13   Worn or damaged valves/guides (Chapter 2).

24 Excessive fuel consumption

1   Dirty or clogged air filter element (Chapter 1).
2   Incorrect ignition timing (Chapter 1).
3   Incorrect idle speed (Chapter 1).
4   Low tire pressure or incorrect tire size (Chapter 11).
5   Fuel leakage. Check all connections, lines and components in the fuel system (Chapter 4).
6   Choke not operating properly (Chapter 1).
7   Dirty or clogged carburetor jets or fuel injectors (Chapter 4).

25 Fuel odor

1   Fuel leakage. Check all connections, lines and components in the fuel system (Chapter 4).
2   Fuel tank overfilled. Fill only to automatic shut-off.
3   Charcoal canister filter in Evaporative Emissions Control system clogged (Chapter 1).
4   Vapor leaks from Evaporative Emissions Control system lines (Chapter 6).

26 Miscellaneous engine noises

1   A strong dull noise that becomes more rapid as the engine accelerates indicates worn or damaged crankshaft bearings or an unevenly worn crankshaft. To pinpoint the trouble spot, remove the spark plug wire from one plug at a time and crank the engine over. If the noise stops, the cylinder with the removed plug wire indicates the problem area. Replace the bearing and/or service or replace the crankshaft (Chapter 2).
2   A similar (yet slightly higher pitched) noise to the crankshaft knocking described in the previous paragraph, that becomes more rapid as the engine accelerates, indicates worn or damaged connecting rod bearings (Chapter 2). The procedure for locating the problem cylinder is the same as described in Paragraph 1.
3   An overlapping metallic noise that increases in intensity as the engine speed increases, yet diminishes as the engine warms up indicates abnormal piston and cylinder wear (Chapter 2). To locate the problem cylinder, use the procedure described in Paragraph 1.
4   A rapid clicking noise that becomes faster as the engine accelerates indicates a worn piston pin or piston pin hole. This sound will happen each time the piston hits the highest and lowest points in the stroke

(Chapter 2). The procedure for locating the problem piston is described in Paragraph 1.
5   A metallic clicking noise coming from the water pump indicates worn or damaged water pump bearings or pump. Replace the water pump with a new one (Chapter 3).
6   A rapid tapping sound or clicking sound that becomes faster as the engine speed increases indicates "valve tapping" or improperly adjusted valve clearances. This can be identified by holding one end of a section of hose to your ear and placing the other end at different spots along the rocker arm cover. The point where the sound is loudest indicates the problem valve. Adjust the valve clearance (Chapter 1 or 2B). If the problem persists, you likely have a collapsed valve lifter or other damaged valve train component. Changing the engine oil and adding a high viscosity oil treatment will sometimes cure a stuck lifter problem. If the problem still persists, the lifters, pushrods and rocker arms must be removed for inspection (see Chapter 2).
7   A steady metallic rattling or rapping sound coming from the area of the timing chain cover indicates a worn, damaged or out-of-adjustment timing chain. Service or replace the chain and related components (Chapter 2).

Cooling system

27 Overheating

1   Insufficient coolant in system (Chapter 1).
2   Drivebelt defective or not adjusted properly (Chapter 1).
3   Radiator core blocked or radiator grille dirty or restricted (Chapter 3).
4   Thermostat faulty (Chapter 3).
5   Fan not functioning properly (Chapter 3).
6   Radiator cap not maintaining proper pressure. Have cap pressure tested by gas station or repair shop.
7   Ignition timing incorrect (Chapter 1).
8   Defective water pump (Chapter 3).
9   Improper grade of engine oil.
10   Inaccurate temperature gauge (Chapter 12).

28 Overcooling

1   Thermostat faulty (Chapter 3).
2   Inaccurate temperature gauge (Chapter 12).

29 External coolant leakage

1   Deteriorated or damaged hoses. Loose clamps at hose connections (Chapter 1).
2   Water pump seals defective. If this is the case, water will drip from the weep hole in the water pump body (Chapter 3).

3   Leakage from radiator core or header tank. This will require the radiator to be professionally repaired (see Chapter 3 for removal procedures).
4   Engine drain plugs or water jacket freeze plugs leaking (see Chapters 1 and 2).
5   Leak from coolant temperature switch (Chapter 3).
6   Leak from damaged gaskets or small cracks (Chapter 2).
7   Damaged head gasket. This can be verified by checking the condition of the engine oil as noted in Section 30.

30   Internal coolant leakage

Note: *Internal coolant leaks can usually be detected by examining the oil. Check the dipstick and inside the rocker arm cover for water deposits and an oil consistency like that of a milkshake.*

1   Leaking cylinder head gasket. Have the system pressure tested or remove the cylinder head (Chapter 2) and inspect.
2   Cracked cylinder bore or cylinder head. Dismantle engine and inspect (Chapter 2).
3   Loose cylinder head bolts (tighten as described in Chapter 2).

31   Abnormal coolant loss

1   Overfilling system (Chapter 1).
2   Coolant boiling away due to overheating (see causes in Section 27).
3   Internal or external leakage (see Sections 29 and 30).
4   Faulty radiator cap. Have the cap pressure tested.
5   Cooling system being pressurized by engine compression. This could be due to a cracked head or block or leaking head gasket(s).

32   Poor coolant circulation

1   Inoperative water pump. A quick test is to pinch the top radiator hose closed with your hand while the engine is idling, then release it. You should feel a surge of coolant if the pump is working properly (Chapter 3).
2   Restriction in cooling system. Drain, flush and refill the system (Chapter 1). If necessary, remove the radiator (Chapter 3) and have it reverse flushed or professionally cleaned.
3   Loose water pump drivebelt (Chapter 1).
4   Thermostat sticking (Chapter 3).
5   Insufficient coolant (Chapter 1).

33   Corrosion

1   Excessive impurities in the water. Soft, clean water is recommended. Distilled or

rainwater is satisfactory.
2   Insufficient antifreeze solution (refer to Chapter 1 for the proper ratio of water to antifreeze).
3   Infrequent flushing and draining of system. Regular flushing of the cooling system should be carried out at the specified intervals as described in (Chapter 1).

Clutch

Note: *All clutch related service information is located in Chapter 8, unless otherwise noted.*

34   Fails to release (pedal pressed to the floor - shift lever does not move freely in and out of Reverse)

1   Freeplay incorrectly adjusted (see Chapter 1)
2   Clutch contaminated with oil. Remove clutch plate and inspect.
3   Clutch plate warped, distorted or otherwise damaged.
4   Diaphragm spring fatigued. Remove clutch cover/pressure plate assembly and inspect.
5   Broken, binding or damaged release cable or linkage (models with a cable-operated release system).
6   Leakage of fluid from clutch hydraulic system. Inspect master cylinder, operating cylinder and connecting lines.
7   Air in clutch hydraulic system. Bleed the system.
8   Insufficient pedal stroke. Check and adjust as necessary.
9   Piston seal in operating cylinder deformed or damaged.
10   Lack of grease on pilot bearing.

35   Clutch slips (engine speed increases with no increase in vehicle speed)

1   Worn or oil soaked clutch plate.
2   Clutch plate not broken in. It may take 30 or 40 normal starts for a new clutch to seat.
3   Diaphragm spring weak or damaged. Remove clutch cover/pressure plate assembly and inspect.
4   Flywheel warped (Chapter 2).
5   Debris in master cylinder preventing the piston from returning to its normal position.
6   Clutch hydraulic line damaged.
7   Binding in the release mechanism.

36   Grabbing (chattering) as clutch is engaged

1   Oil on clutch plate. Remove and inspect.

Repair any leaks.
2   Worn or loose engine or transmission mounts. They may move slightly when clutch is released. Inspect mounts and bolts.
3   Worn splines on transmission input shaft. Remove clutch components and inspect.
4   Warped pressure plate or flywheel. Remove clutch components and inspect.
5   Diaphragm spring fatigued. Remove clutch cover/pressure plate assembly and inspect.
6   Clutch linings hardened or warped.
7   Clutch lining rivets loose.

37   Squeal or rumble with clutch engaged (pedal released)

1   Improper pedal adjustment. Adjust pedal freeplay (Chapter 1).
2   Release bearing binding on transmission shaft. Remove clutch components and check bearing. Remove any burrs or nicks, clean and relubricate before reinstallation.
3   Pilot bushing worn or damaged.
4   Clutch rivets loose.
5   Clutch plate cracked.
6   Fatigued clutch plate torsion springs. Replace clutch plate.

38   Squeal or rumble with clutch disengaged (pedal depressed)

1   Worn or damaged release bearing.
2   Worn or broken pressure plate diaphragm fingers.

39   Clutch pedal stays on floor when disengaged

Binding linkage or release bearing. Inspect linkage or remove clutch components as necessary.

Manual transmission

Note: *All manual transmission service information is located in Chapter 7, unless otherwise noted.*

40   Noisy in Neutral with engine running

1   Input shaft bearing worn.
2   Damaged main drive gear bearing.
3   Insufficient transmission oil (Chapter 1).
4   Transmission oil in poor condition. Drain and fill with proper grade oil. Check old oil for water and debris (Chapter 1).
5   Noise can be caused by variations in engine torque. Change the idle speed and see if noise disappears.

41 Noisy in all gears

1   Any of the above causes, and/or:
2   Worn or damaged output gear bearings or shaft.

42 Noisy in one particular gear

1   Worn, damaged or chipped gear teeth.
2   Worn or damaged synchronizer.

43 Slips out of gear

1   Transmission loose on clutch housing.
2   Stiff shift lever seal.
3   Shift linkage binding.
4   Broken or loose input gear bearing retainer.
5   Dirt between clutch lever and engine housing.
6   Worn linkage.
7   Damaged or worn check balls, fork rod ball grooves or check springs.
8   Worn mainshaft or countershaft bearings.
9   Loose engine mounts (Chapter 2).
10   Excessive gear end play.
11   Worn synchronizers.

44 Oil leaks

1   Excessive amount of lubricant in transmission (see Chapter 1 for correct checking procedures). Drain lubricant as required.
2   Rear oil seal or speedometer oil seal damaged.
3   To pinpoint a leak, first remove all built-up dirt and grime from the transmission. Degreasing agents and/or steam cleaning will achieve this. With the underside clean, drive the vehicle at low speeds so the air flow will not blow the leak far from its source. Raise the vehicle and determine where the leak is located.

45 Difficulty engaging gears

1   Clutch not releasing completely.
2   Loose or damaged shift linkage. Make a thorough inspection, replacing parts as necessary.
3   Insufficient transmission oil (Chapter 1).
4   Transmission oil in poor condition. Drain and fill with proper grade oil. Check oil for water and debris (Chapter 1).
5   Worn or damaged striking rod.
6   Sticking or jamming gears.

46 Noise occurs while shifting gears

1   Check for proper operation of the clutch

(Chapter 8).
2   Faulty synchronizer assemblies.

Automatic transmission

Note: *Due to the complexity of the automatic transmission, it's difficult for the home mechanic to properly diagnose and service. For problems other than those following, the vehicle should be taken to a reputable mechanic.*

47 Fluid leakage

1   Automatic transmission fluid is a deep red color, and fluid leaks should not be confused with engine oil which can easily be blown by air flow to the transmission.
2   To pinpoint a leak, first remove all built-up dirt and grime from the transmission. Degreasing agents and/or steam cleaning will achieve this. With the underside clean, drive the vehicle at low speeds so the air flow will not blow the leak far from its source. Raise the vehicle and determine where the leak is located. Common areas of leakage are:

a) **Fluid pan:** *tighten mounting bolts and/or replace pan gasket as necessary (Chapter 1).*
b) **Rear extension:** *tighten bolts and/or replace oil seal as necessary.*
c) **Filler pipe:** *replace the rubber oil seal where pipe enters transmission case.*
d) **Transmission oil lines:** *tighten fittings where lines enter transmission case and/or replace lines.*
e) **Vent pipe:** *transmission overfilled and/or water in fluid (see checking procedures, Chapter 1).*
f) **Speedometer connector:** *replace the O-ring where speedometer cable enters transmission case.*

48 General shift mechanism problems

Chapter 7 deals with checking and adjusting the shift linkage on automatic transmissions. Common problems which may be caused by out of adjustment linkage are:

a) *Engine starting in gears other than P (park) or N (Neutral).*
b) *Indicator pointing to a gear other than the one actually engaged.*
c) *Vehicle moves with transmission in P (Park) position.*

49 Transmission will not downshift with the accelerator pedal pressed to the floor

Chapter 7 deals with adjusting the TV linkage to enable the transmission to downshift properly.

50 Engine will start in gears other than Park or Neutral

Chapter 7 deals with adjusting the Neutral start switch installed on automatic transmissions.

51 Transmission slips, shifts rough, is noisy or has no drive in forward or Reverse gears

1   There are many probable causes for the above problems, but the home mechanic should concern himself only with one possibility; fluid level.
2   Before taking the vehicle to a shop, check the fluid level and condition as described in Chapter 1. Add fluid, if necessary, or change the fluid and filter if needed. If problems persist, have a professional diagnose the transmission.

Driveshaft

Note: *Refer to Chapter 8, unless otherwise specified, for service information.*

52 Leaks at front of driveshaft

Defective transmission rear seal. See Chapter 7 for replacement procedure. As this is done, check the splined yoke for burrs or roughness that could damage the new seal. Remove burrs with a fine file or whetstone.

53 Knock or clunk when transmission is under initial load (just after transmission is put into gear)

1   Loose or disconnected rear suspension components. Check all mounting bolts and bushings (Chapters 7 and 10).
2   Loose driveshaft bolts. Inspect all bolts and nuts and tighten them securely.
3   Worn or damaged universal joint bearings. Inspect the universal joints (Chapter 8).
4   Worn sleeve yoke and mainshaft spline.

54 Metallic grating sound consistent with vehicle speed

Pronounced wear in the universal joint bearings. Replace U-joints or driveshafts, as necessary.

55 Vibration

Note: *Before blaming the driveshaft, make sure the tires are perfectly balanced and perform the following test.*

1   Install a tachometer inside the vehicle to monitor engine speed as the vehicle is driven. Drive the vehicle and note the engine speed at which the vibration (roughness) is most pronounced. Now shift the transmission to a different gear and bring the engine speed to the same point.

2   If the vibration occurs at the same engine speed (rpm) regardless of which gear the transmission is in, the driveshaft is NOT at fault since the driveshaft speed varies.

3   If the vibration decreases or is eliminated when the transmission is in a different gear at the same engine speed, refer to the following probable causes.

4   Bent or dented driveshaft. Inspect and replace as necessary.

5   Undercoating or built-up dirt, etc. on the driveshaft. Clean the shaft thoroughly.

6   Worn universal joint bearings. Replace the U-joints or driveshaft as necessary.

7   Driveshaft and/or companion flange out of balance. Check for missing weights on the shaft. Remove driveshaft and reinstall 180-degrees from original position, then recheck. Have the driveshaft balanced if problem persists.

8   Loose driveshaft mounting bolts/nuts.

9   Defective center bearing, if so equipped.

10   Worn transmission rear bushing (Chapter 7).

56 Scraping noise

Make sure the dust cover on the sleeve yoke isn't rubbing on the transmission extension housing.

57 Whining or whistling noise

Defective center bearing, if so equipped.

Rear axle and differential

Note: *For differential servicing information, refer to Chapter 8, unless otherwise specified.*

58 Noise - same when in drive as when vehicle is coasting

1   Road noise. No corrective action available.

2   Tire noise. Inspect tires and check tire pressures (Chapter 1).

3   Front wheel bearings loose, worn or damaged (Chapter 1).

4   Insufficient differential oil (Chapter 1).

5   Defective differential.

59 Knocking sound when starting or shifting gears

Defective or incorrectly adjusted differential.

60 Noise when turning

Defective differential.

61 Vibration

See probable causes under Driveshaft. Proceed under the guidelines listed for the driveshaft. If the problem persists, check the rear wheel bearings by raising the rear of the vehicle and spinning the wheels by hand. Listen for evidence of rough (noisy) bearings. Remove and inspect (Chapter 8).

62 Oil leaks

1   Pinion oil seal damaged (Chapter 8).

2   Axleshaft oil seals damaged (Chapter 8).

3   Differential cover leaking. Tighten mounting bolts or replace the gasket as required.

4   Loose filler or drain plug on differential (Chapter 1).

5   Clogged or damaged breather on differential.

Transfer case (4WD models)

Note: *Unless otherwise specified, refer to Chapter 7C for service and repair information.*

63 Gear jumping out of mesh

1   Incorrect control lever freeplay.

2   Interference between the control lever and the console.

3   Play or fatigue in the transfer case mounts.

4   Internal wear or incorrect adjustments.

64 Difficult shifting

1   Lack of oil.

2   Internal wear, damage or incorrect adjustment.

65 Noise

1   Lack of oil in transfer case.

2   Noise in 4H and 4L, but not in 2H indicates cause is in the front differential or front axle.

3   Noise in 2H, 4H and 4L indicates cause is in rear differential or rear axle.

4   Noise in 2H and 4H but not in 4L, or in 4L only, indicates internal wear or damage in transfer case.

Brakes

Note: *Before assuming a brake problem exists, make sure the tires are in good condition and inflated properly, the front end alignment is correct and the vehicle is not loaded with weight in an unequal manner. All service procedures for the brakes are included in Chapter 9, unless otherwise noted.*

66 Vehicle pulls to one side during braking

1   Defective, damaged or oil contaminated brake pad on one side. Inspect as described in Chapter 1. Refer to Chapter 10 if replacement is required.

2   Excessive wear of brake pad material or disc on one side. Inspect and repair as necessary.

3   Loose or disconnected front suspension components. Inspect and tighten all bolts securely (Chapters 1 and 11).

4   Defective caliper assembly. Remove caliper and inspect for stuck piston or damage.

5   Brake pad-to-rotor adjustment needed. Inspect automatic adjusting mechanism for proper operation.

6   Scored or out of round rotor.

7   Loose caliper mounting bolts.

8   Incorrect wheel bearing adjustment.

67 Noise (high-pitched squeal)

1   Front brake pads worn out. This noise comes from the wear sensor rubbing against the disc. Replace pads with new ones immediately!

2   Glazed or contaminated pads.

3   Dirty or scored rotor.

4   Bent support plate.

68 Excessive brake pedal travel

1   Partial brake system failure. Inspect entire system (Chapter 1) and correct as required.

2   Insufficient fluid in master cylinder. Check (Chapter 1) and add fluid - bleed system if necessary.

3   Air in system. Bleed system.

4   Excessive lateral rotor play.

5   Brakes out of adjustment. Check the operation of the automatic adjusters.

6   Defective proportioning valve. Replace valve and bleed system.

69 Brake pedal feels spongy when depressed

1   Air in brake lines. Bleed the brake system.

2   Deteriorated rubber brake hoses. Inspect all system hoses and lines. Replace parts as necessary.

3   Master cylinder mounting nuts loose. Inspect master cylinder bolts (nuts) and tighten them securely.
4   Master cylinder faulty.
5   Incorrect shoe or pad clearance.
6   Defective check valve. Replace valve and bleed system.
7   Clogged reservoir cap vent hole.
8   Deformed rubber brake lines.
9   Soft or swollen caliper seals.
10   Poor quality brake fluid. Bleed entire system and fill with new approved fluid.

70   Excessive effort required to stop vehicle

1   Power brake booster not operating properly.
2   Excessively worn linings or pads. Check and replace if necessary.
3   One or more caliper pistons seized or sticking. Inspect and rebuild as required.
4   Brake pads or linings contaminated with oil or grease. Inspect and replace as required.
5   New pads or linings installed and not yet seated. It'll take a while for the new material to seat against the rotor or drum.
6   Worn or damaged master cylinder or caliper assemblies. Check particularly for frozen pistons.
7   Also see causes listed under Section 69.

71   Pedal travels to the floor with little resistance

Little or no fluid in the master cylinder reservoir caused by leaking caliper piston(s) or loose, damaged or disconnected brake lines. Inspect entire system and repair as necessary.

72   Brake pedal pulsates during brake application

1   Wheel bearings damaged, worn or out of adjustment (Chapter 1).
2   Caliper not sliding properly due to improper installation or obstructions. Remove and inspect.
3   Rotor not within specifications. Remove the rotor and check for excessive lateral runout and parallelism. Have the rotors resurfaced or replace them with new ones. Also make sure that all rotors are the same thickness.
4   Out of round rear brake drums. Remove the drums and have them turned or replace them with new ones.

73   Brakes drag (indicated by sluggish engine performance or wheels being very hot after driving)

1   Output rod adjustment incorrect at the

brake pedal.
2   Obstructed master cylinder compensator. Disassemble master cylinder and clean.
3   Master cylinder piston seized in bore. Overhaul master cylinder.
4   Caliper assembly in need of overhaul.
5   Brake pads or shoes worn out.
6   Piston cups in master cylinder or caliper assembly deformed. Overhaul master cylinder.
7   Rotor not within specifications (Section 72).
8   Parking brake assembly will not release.
9   Clogged brake lines.
10   Wheel bearings out of adjustment (Chapter 1).
11   Brake pedal height improperly adjusted.
12   Wheel cylinder needs overhaul.
13   Improper shoe to drum clearance. Adjust as necessary.

74   Rear brakes lock up under light brake application

1   Tire pressures too high.
2   Tires excessively worn (Chapter 1).

75   Rear brakes lock up under heavy brake application

1   Tire pressures too high.
2   Tires excessively worn (Chapter 1).
3   Front brake pads contaminated with oil, mud or water. Clean or replace the pads.
4   Front brake pads excessively worn.
5   Defective master cylinder or caliper assembly.

Suspension and steering

Note: *All service procedures for the suspension and steering systems are included in Chapter 10, unless otherwise noted.*

76   Vehicle pulls to one side

1   Tire pressures uneven (Chapter 1).
2   Defective tire (Chapter 1).
3   Excessive wear in suspension or steering components (Chapter 1).
4   Wheel alignment incorrect.
5   Front brakes dragging. Inspect as described in Section 73.
6   Wheel bearings improperly adjusted (Chapter 1 or 8).
7   Wheel lug nuts loose.

77   Shimmy, shake or vibration

1   Tire or wheel out of balance or out of round. Have them balanced on the vehicle.
2   Loose, worn or out of adjustment wheel

bearings (Chapter 1 or 8).
3   Shock absorbers and/or suspension components worn or damaged. Check for worn bushings in the upper and lower links.
4   Wheel lug nuts loose.
5   Incorrect tire pressures.
6   Excessively worn or damaged tire.
7   Loosely mounted steering gear housing.
8   Steering gear improperly adjusted.
9   Loose, worn or damaged steering components.
10   Damaged idler arm.
11   Worn balljoint.

78   Excessive pitching and/or rolling around corners or during braking

1   Defective shock absorbers. Replace as a set.
2   Broken or weak leaf springs and/or suspension components.
3   Worn or damaged stabilizer bar or bushings.

79   Wandering or general instability

1   Improper tire pressures.
2   Worn or damaged upper and lower link or tension rod bushings.
3   Incorrect front end alignment.
4   Worn or damaged steering linkage or suspension components.
5   Improperly adjusted steering gear.
6   Out of balance wheels.
7   Loose wheel lug nuts.
8   Worn rear shock absorbers.
9   Fatigued or damaged rear leaf springs.

80   Excessively stiff steering

1   Lack of lubricant in power steering fluid reservoir, where appropriate (Chapter 1).
2   Incorrect tire pressures (Chapter 1).
3   Lack of lubrication at balljoints (Chapter 1).
4   Front end out of alignment.
5   Steering gear out of adjustment or lacking lubrication.
6   Improperly adjusted wheel bearings.
7   Worn or damaged steering gear.
8   Interference of steering column with turn signal switch.
9   Low tire pressures.
10   Worn or damaged balljoints.
11   Worn or damaged steering linkage.
12   See also Section 79.

81   Excessive play in steering

1   Loose wheel bearings (Chapter 1 or 8).
2   Excessive wear in suspension bushings (Chapter 1).
3   Steering gear improperly adjusted.
4   Incorrect wheel alignment.

5   Steering gear mounting bolts loose.
6   Worn steering linkage.

82 Lack of power assistance

1   Steering pump drivebelt faulty or not adjusted properly (Chapter 1).
2   Fluid level low (Chapter 1).
3   Hoses or pipes restricting the flow. Inspect and replace parts as necessary.
4   Air in power steering system. Bleed system.
5   Defective power steering pump.

83 Steering wheel fails to return to straight-ahead position

1   Incorrect front end alignment.
2   Tire pressures low.
3   Steering gears improperly engaged.
4   Steering column out of alignment.
5   Worn or damaged balljoint.
6   Worn or damaged steering linkage.
7   Improperly lubricated idler arm.
8   Insufficient oil in steering gear.
9   Lack of fluid in power steering pump.

84 Steering effort not the same in both directions (power system)

1   Leaks in steering gear.
2   Clogged fluid passage in steering gear.

85 Noisy power steering pump

1   Insufficient oil in pump.
2   Clogged hoses or oil filter in pump.
3   Loose pulley.
4   Improperly adjusted drivebelt (Chapter 1).
5   Defective pump.

86 Miscellaneous noises

1   Improper tire pressures.
2   Insufficiently lubricated balljoint or steering linkage.
3   Loose or worn steering gear, steering linkage or suspension components.
4   Defective shock absorber.
5   Defective wheel bearing.
6   Worn or damaged suspension bushings.
7   Damaged leaf spring.
8   Loose wheel lug nuts.
9   Worn or damaged rear axleshaft spline.
10   Worn or damaged rear shock absorber mounting bushing.
11   Incorrect rear axle end play.
12   See also causes of noises at the rear axle and driveshaft.

87 Excessive tire wear (not specific to one area)

1   Incorrect tire pressures.

2   Tires out of balance. Have them balanced on the vehicle.
3   Wheels damaged. Inspect and replace as necessary.
4   Suspension or steering components worn (Chapter 1).

88 Excessive tire wear on outside edge

1   Incorrect tire pressure.
2   Excessive speed in turns.
3   Front end alignment incorrect (excessive toe-in).

89 Excessive tire wear on inside edge

1   Incorrect tire pressure.
2   Front end alignment incorrect (toe-out).
3   Loose or damaged steering components (Chapter 1).

90 Tire tread worn in one place

1   Tires out of balance. Have them balanced on the vehicle.
2   Damaged or buckled wheel. Inspect and replace if necessary.
3   Defective tire.

Chapter 1
Tune-up and routine maintenance

Contents

Specifications

Recommended lubricants and fluids

Note: *Listed here are manufacturer recommendations at the time this manual was written. Manufacturers occasionally upgrade their fluid and lubricant specifications, so check with your local auto parts store for current recommendations.*

Engine oil

Type	API SE or SF
Viscosity	See accompanying chart

ENGINE OIL VISCOSITY CHART

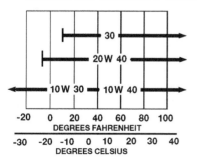

MANUAL TRANSMISSION LUBRICANT VISCOSITY CHART

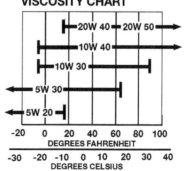

For best performance, select the lowest SAE viscosity grade of oil and lubricant for the expected temperature range

1-A1 HAYNES

Recommended lubricants and fluids (continued)

Engine oil (continued)
    Capacity (with new oil filter)
        Four-cylinder engine
            1981 through 1985 .... 3.7 qts
            1986-on .... 4.4 qts
        V6 engine .... 4.5 qts
Automatic transmission fluid
    Type.. Dexron II ATF
    Capacity (approximate) .... 6 qts
Manual transmission lubricant
    Type.. API grade SF engine oil
    Viscosity .... SAE 5W-30
Differential lubricant.. SAE 80/90W GL-4 gear lubricant
Transfer case lubricant
    1981 through 1987 .... Same as manual transmission lubricant*
    1988-on
        With V6 engine .... SAE 5W-30 engine oil, API grade SF
        With four-cylinder engine .... Dexron II ATF
Power steering fluid type .... Dexron II ATF
Brake fluid type... DOT 3 brake fluid
Clutch fluid type.. DOT 3 brake fluid
Coolant
    Type.. 50/50 mixture of ethylene glycol-based antifreeze and water
    Capacity (approximate)
        1981 and 1982 .... 7.5 qts
        1983 .... 8.5 qts
        1984 through 1987.. 6.5 qts
        1988-on
            Pick-up
                Four-cylinder engine.............................. 9.5 qts
                V6 engine... 11.4 qts
            Trooper
                Four-cylinder engine
                    1988 .... 10.3 qts
                    1989-on .... 8.5 qts
                V6 engine... 10 qts
Chassis grease .... NLGI no. 2 chassis grease
Hood and door hinges... Engine oil

The transfer case and transmission housings are integral and the lubricant is shared.

Tune-up information

Spark plug type
    All 4 cylinder engines .... NGK BPR 6ES11 or equivalent
    2.8L V6 engine... AC R45TS or equivalent
    3.1L V6 engine... AC R43TSK or equivalent
Spark plug gap
    All 4 cylinder engines .... .040 in.
    2.8L V6 engine... .044 in.
    3.1L V6 engine... .045 in.
Firing order
    Four-cylinder engine 1-3-4-2
    V6 engine.. 1-2-3-4-5-6

**4ZD1 and 4ZE1
four-cylinder engines**

**G180Z and G200Z
four-cylinder engines**

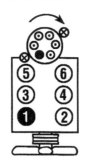

V6 engine

**Cylinder locations and
distributor rotation**

*The blackened terminal shown on the
distributor cap indicates the Number
One spark plug wire position*

Cylinder numbers/locations... See Chapter 2
Ignition timing .... Refer to the Vehicle Emission Control Information label
 in the engine compartment
Engine idle speed .... Refer to the Vehicle Emission Control Information label
 in the engine compartment
Valve clearances (engine cold)
    4ZE1 engine (1987-on) only
        Intake .... 0.008 in (0.20 mm)
        Exhaust .... 0.008 in (0.20 mm)
    All others
        Intake .... 0.006 in (0.15 mm)
        Exhaust .... 0.010 in (0.25 mm)

Clutch

Pedal freeplay .... 3/4 in (20 mm)
Cable freeplay (measured at firewall).. 15/64 in (5 mm)
Pedal height
    Hydraulic clutch .... 7 in (174 mm)
    Cable-operated clutch
        Trooper .... 9 to 9-1/2 in (231 to 241 mm)
        Pick-up
            1981 through 1983 .... 6-7/8 to 7-1/4 in (174 to 184 mm)
            1984 through 1987 .... 8-3/64 to 8-1/2 in (206 to 216 mm)
            1988-on .... 7-1/3 to 7-11/16 in (185 to 195 mm)

Brakes

Pedal height
    Pick-up
        1981 and 1982 .... 6.1 to 6.5 in (154 to 164 mm)
        1983 through 1987.. 6.5 to 6.9 in (164 to 174 mm)
        1988-on.. 6.9 to 7.24 in (174 to 184 mm)
    Trooper
        1985 through 1987.. 7.8 to 8.2 in (198 to 208 mm)
        1988-on.. 6.9 to 7.24 in (184 to 184 mm)
Pedal freeplay
    1981 through 1983 .... 0.02 to 0.04 in (0.5 to 1.0 mm)
    1984-on .... 0.28 to 0.40 in (7 to 10 mm)
Brake lining wear limit (disc and drum brakes)................................... 3/64 in (1 mm)
Front hub (2WD only) wheel bearing preload (measured with spring scale)
    1981 through 1984 .... 1.8 to 2.5 lbs
    1985-on .... 2.2 lbs

Torque specifications

                                                                                      Ft-lbs
Automatic transmission pan bolts 4.3
Spark plugs.. 14 to 22
Engine oil drain plug 18
Rocker arm bracket nuts .... 12 to 16
Manual transmission check/fill and drain plugs .... 18 to 25
Differential check/fill and drain plugs 60
Transfer case check/fill and drain plugs.. 14 to 29
Oxygen sensor
    Four-cylinder engine .... 13 to 17
    V6 engine.. 30 to 37
Wheel lug nuts
    Steel wheels .... 60 to 85
    Aluminum wheels .... 80 to 94

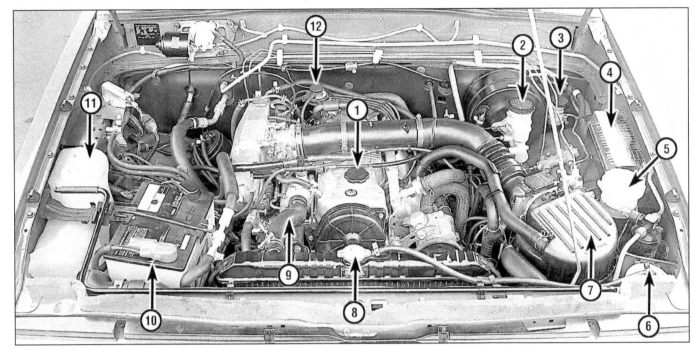

Typical engine compartment component layout (Trooper shown)

1	Engine oil filter cap	5	Power steering fluid reservoir	9	Radiator hose
2	Brake fluid reservoir	6	Coolant reservoir	10	Battery
3	Clutch fluid reservoir	7	Air cleaner assembly	11	Windshield washer fluid reservoir
4	Fuse box	8	Radiator cap	12	EGR valve transducer

Typical under side view of engine/transmission (Trooper shown)

1	Engine oil filter	4	Front disc brake	7	Engine oil drain plug
2	Front differential check/fill plug	5	Transmission	8	Front axle CV joint
3	Shock absorber	6	Exhaust pipe		

Typical rear under side component layout

1	Driveplate U-joint	4	Exhaust pipe	7	Rear disc brake caliper
2	Muffler	5	Rear differential check/fill plug	8	Rear leaf spring
3	Shock absorber	6	Rear differential drain plug	9	Fuel filter

1 Introduction

This Chapter is designed to help the home mechanic maintain the Isuzu pick-up and Trooper with the goals of maximum performance, economy, safety and reliability in mind.

Included is a master maintenance schedule (page 35), followed by procedures dealing specifically with every item on the schedule. Visual checks, adjustments, component replacement and other helpful items are included. Refer to the accompanying illustrations of the engine compartment and the under side of the vehicle for the locations of various components.

Servicing your vehicle in accordance with the planned mileage/time maintenance schedule and the step-by-step procedures should result in maximum reliability and extend the life of your vehicle. Keep in mind that it's a comprehensive plan - maintaining some items but not others at the specified intervals will not produce the same results.

As you perform routine maintenance procedures, you'll find that many can, and should, be grouped together because of the nature of the procedures or because of the proximity of two otherwise unrelated components or systems.

For example, if the vehicle is raised for chassis lubrication, you should inspect the exhaust, suspension, steering and fuel systems while you're under the vehicle. When you're rotating the tires, it makes good sense to check the brakes since the wheels are already removed. Finally, let's suppose you have to borrow or rent a torque wrench. Even if you only need it to tighten the spark plugs, you might as well check the torque of as many critical fasteners as time allows.

The first step in this maintenance program is to prepare yourself before the actual work begins. Read through all the procedures you're planning to do, then gather up all the parts and tools needed. If it looks like you might run into problems during a particular job, seek advice from a mechanic or experienced do-it-yourselfer.

2 Isuzu Trooper and Pick-up Maintenance schedule

The following maintenance intervals are based on the assumption that the vehicle owner will be doing the maintenance or service work, as opposed to having a dealer service department do the work. Although the time/mileage intervals are loosely based on factory recommendations, most have been shortened to ensure, for example, that such items as lubricants and fluids are checked/changed at intervals that promote maximum engine/driveline service life. Also, subject to the preference of the individual owner interested in keeping his or her vehicle in peak condition at all times, and with the vehicle's ultimate resale in mind, many of the maintenance procedures may be performed more often than recommended in the following schedule. We encourage such owner initiative.

When the vehicle is new it should be serviced initially by a factory authorized dealer service department to protect the factory warranty. In many cases the initial maintenance check is done at no cost to the owner (check with your dealer service department for more information).

Every 250 miles or weekly, whichever comes first

Check the engine oil level (Section 4)
Check the engine coolant level (Section 4)
Check the windshield washer fluid level (Section 4)
Check the brake and clutch fluid levels (Section 4)
Check the automatic transmission fluid level (Section 5)
Check the power steering fluid level (Section 6)
Check the tires and tire pressures (Section 7)

Every 5000 miles or 4 months, whichever comes first

All items listed above plus:
Check and service the battery (Section 8)
Check the cooling system (Section 9)
Inspect and replace if necessary the windshield wiper blades (Section 10)
Inspect and replace if necessary all underhood hoses (Section 11)
Check and lubricate the accelerator linkage (Section 12)
Change the engine oil and filter (Section 13)*
Rotate the tires (Section 14)

Every 15,000 miles or 12 months, whichever comes first

All items listed above plus:
Adjust the valves (Section 15)
Lubricate the chassis components (Section 16)
Inspect the suspension and steering components(Section 17)*
Inspect the exhaust system (Section 18)*
Check and adjust if necessary, the clutch pedal freeplay (Section 19)
Check the manual transmission lubricant (Section 20)*
Check the transfer case lubricant level (Section 21)*
Check the differential lubricant level (Section 22)*
Check the brakes (Section 23)*
Inspect the fuel system (Section 24)
Check the operation of the thermostatic air cleaner system (Section 25)

Check the engine drivebelts (Section 26)
Check the seat belts (Section 27)

Every 22,500 miles or 24 months, whichever comes first

Change the power steering fluid (1981 through 1985 models) (Section 28)

Every 30,000 miles or 24 months, whichever comes first

Change the power steering fluid (1986 and later models) (Section 28)
Replace the air filter (Section 29)
Replace the fuel filter (Section 30)
Check and adjust, if necessary, the brake pedal height (Section 31)
Check and lubricate the distributor advance mechanism (four-cylinder engines) (Section 32)
Inspect the evaporative emissions control (Section 33)
Check the operation of the carburetor choke (Section 34)
Check and adjust if necessary, the idle speed (four-cylinder models only) (Section 35)
Change the manual transmission lubricant (Section 37)
Change the transfer case lubricant (4WD models) (Section 36)
Change the differential lubricant (Section 38)
Change the automatic transmission fluid and filter (Section 39)**
Check, repack and adjust the front wheel bearings (40)
Service the cooling system (drain, flush and refill) (Section 41)
Inspect the PCV system (Section 42)
Replace the spark plugs (Section 43)
Inspect the spark plug wires, distributor cap and rotor(Section 44)
Check and adjust, if necessary, the ignition timing (Section 45)
Check the Exhaust Gas Recirculation (EGR) system (Section 46)

Every 60,000 miles or 24 months, whichever comes first

Replace the timing belt (four-cylinder engines) (Chapter 2)

Every 90,000 miles or 24 months

Replace the oxygen sensor (1988 and later models) (Section 47)
* *This item is affected by "severe" operating conditions as described below. If your vehicle is operated under severe conditions, perform all maintenance indicated with an asterisk (*) at 3000 mile/3 month intervals.*
Severe conditions are indicated if you mainly operate your vehicle under one or more of the following:
In dusty areas
Towing a trailer
Idling for extended periods and/or low speed operation
When outside temperatures remain below freezing and most trips are less than 4 miles
** *If operated under one or more of the following conditions, change the automatic transmission fluid every 15,000 miles:*
In heavy city traffic where the outside temperature regularly reaches 90-degrees F (32-degrees C) or higher
In hilly or mountainous terrain
Frequent trailer pulling

3   Tune-up general information

The term tune-up is used in this manual to represent a combination of individual operations rather than one specific procedure.

If, from the time the vehicle is new, the routine maintenance schedule is followed closely and frequent checks are made of fluid levels and high wear items, as suggested throughout this manual, the engine will be kept in relatively good running condition and the need for additional work will be minimized.

More likely than not, however, there will be times when the engine is running poorly due to lack of regular maintenance. This is even more likely if a used vehicle, which has not received regular and frequent maintenance checks, is purchased. In such cases, an engine tune-up will be needed outside of the regular routine maintenance intervals.

The first step in any tune-up or diagnostic procedure to help correct a poor running engine is a cylinder compression check. A compression check (see Chapter 2 Part B) will help determine the condition of internal engine components and should be used as a guide for tune-up and repair procedures. If, for instance, a compression check indicates serious internal engine wear, a conventional tune-up will not improve the performance of the engine and would be a waste of time and money. Because of its importance, the compression check should be done by someone with the right equipment and the knowledge to use it properly.

The following procedures are those most often needed to bring a generally poor running engine back into a proper state of tune.

Minor tune-up

Check all engine related fluids
Clean and check the battery (Section 8)
Check and adjust the drivebelts (Section 26)
Replace the spark plugs (Section 43)
Check the cylinder compression (Chapter 2)
Inspect the distributor cap and rotor (Section 44)
Inspect the spark plug and coil wires (Section 44)
Replace the air filter (Section 29)
Check and adjust the idle speed (Section 35)
Check and adjust the ignition timing (Section 45)
Replace the fuel filter (Section 30)
Check the PCV system (Section 42)
Adjust the valve clearances (Section 15)
Check and service the cooling system (Section 41)

Major tune-up

All items listed under Minor tune-up plus

Check the EGR system (Section 46 and Chapter 6)

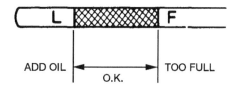

4.4 The oil level should be in the cross-hatched area - if it's below the "L" line, add enough oil to bring the level into the operating range (cross-hatched area)

Check the charging system (Chapter 5)
Check the ignition system (Chapter 5)
Check the fuel system (Section 24 and Chapter 4)
Replace the spark plugs (Section 43)
Replace the spark plug wires, distributor cap and rotor (Section 44)

4   Fluid level checks

Note: *The following are fluid level checks to be done on a 250 mile or weekly basis. Additional fluid level checks can be found in specific maintenance procedures which follow. Regardless of how often the fluid levels are checked, watch for puddles under the vehicle - if leaks are noted, make repairs immediately.*

1   Fluids are an essential part of the lubrication, cooling, brake, clutch and windshield washer systems. Because the fluids gradually become depleted and/or contaminated during normal operation of the vehicle, they must be periodically replenished. See Recommended lubricants and fluids at the beginning of this Chapter before adding fluid to any of the following components. **Note:** *The vehicle must be on level ground when fluid levels are checked.*

Engine oil

Refer to illustrations 4.4 and 4.6

2   The engine oil level is checked with a dipstick that extends through a tube and into the oil pan at the bottom of the engine.

3   The oil level should be checked before the vehicle has been driven, or about 15 minutes after the engine has been shut off. If the oil is checked immediately after driving the vehicle, some of the oil will remain in the upper engine components, resulting in an inaccurate reading on the dipstick.

4   Pull the dipstick from the tube and wipe all the oil from the end with a clean rag or paper towel. Insert the clean dipstick all the way back into the tube, then pull it out again. Note the oil at the end of the dipstick. Add oil as necessary to keep the level between the L and F marks or within the hatched area on the dipstick **(see illustration)**.

5   Don't overfill the engine by adding too much oil since this may result in oil fouled spark plugs, oil leaks or oil seal failures.

6   Oil is added to the engine after removing a threaded cap from the rocker arm cover **(see illustration)**. An oil can spout or funnel may help to reduce spills.

7   Checking the oil level is an important preventive maintenance step. A consistently low oil level indicates oil leakage through damaged seals, defective gaskets or past worn rings or valve guides. If the oil looks milky in color or has water droplets in it, the cylinder head gasket(s) may be blown or the head(s) or block may be cracked. The engine should be checked immediately. The condition of the oil should also be checked. Whenever you check the oil level, slide your thumb and index finger up the dipstick before wiping off the oil. If you see small dirt or metal particles clinging to the dipstick, the oil should be changed (Section 13).

Engine coolant

Refer to illustration 4.8

Warning: *Don't allow antifreeze to come in contact with your skin or painted surfaces of the vehicle. Flush contaminated areas immediately with plenty of water. Don't store new coolant or leave old coolant lying around where it's accessible to children or pets - they're attracted by its sweet taste. Ingestion of even a small amount of coolant can be fatal! Wipe up garage floor and drip pan coolant spills immediately. Keep antifreeze containers covered and repair leaks in your cooling system immediately.*

8   All vehicles covered by this manual are equipped with a pressurized coolant recovery

4.6 Unscrew the oil filter cap to remove it - make sure the area around the opening is clean before unscrewing the cap; this will prevent dirt from entering the engine

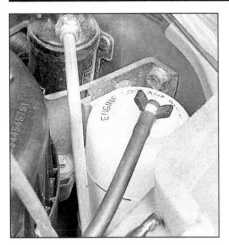

4.8 The coolant reservoir is located in the left front corner of the engine compartment - the level must be maintained between the upper and lower marks

system. A white plastic coolant reservoir located in the engine compartment is connected by a hose to the radiator filler neck **(see illustration)**. If the engine overheats, coolant escapes through a valve in the radiator cap and travels through the hose into the reservoir. As the engine cools, the coolant is automatically drawn back into the cooling system to maintain the correct level.

9   The coolant level in the reservoir should be checked regularly. **Warning:** *Do not remove the radiator cap to check the coolant level when the engine is warm. The level in the reservoir varies with the temperature of the engine. When the engine is cold, the coolant level should be at or slightly above the lower mark on the reservoir. Once the engine has warmed up, the level should be at or near the upper mark. If it isn't, allow the engine to cool, then remove the cap from the reservoir and add a 50/50 mixture of ethylene glycol-based antifreeze and water.*

10   Drive the vehicle and recheck the coolant level. If only a small amount of coolant is required to bring the system up to the proper level, water can be used. However, repeated additions of water will dilute the antifreeze and water solution. In order to maintain the proper ratio of antifreeze and water, always top up the coolant level with the correct mixture. An empty plastic milk jug or bleach bottle makes an excellent container for mixing coolant. Do not use rust inhibitors or additives.

11   If the coolant level drops consistently, there may be a leak in the system. Inspect the radiator, hoses, filler cap, drain plugs and water pump (see Section 9). If no leaks are noted, have the radiator cap pressure tested by a service station.

12   If you have to remove the radiator cap, wait until the engine has cooled, then wrap a thick cloth around the cap and turn it to the first stop. If coolant or steam escapes, let the engine cool down longer, then remove the cap.

13   Check the condition of the coolant as well. It should be relatively clear. If it's brown or rust colored, the system should be drained, flushed and refilled. Even if the coolant appears to be normal, the corrosion inhibitors wear out, so it must be replaced at the specified intervals.

Windshield and rear washer fluid

Refer to illustrations 4.14a and 4.14b

14   Fluid for the windshield washer system is stored in a plastic reservoir located on the passenger's side of the engine compartment **(see illustration)**. If necessary, refer to the underhood component illustration(s) at the beginning of this Chapter to locate the reservoir. On vehicles with rear window washers, the fluid reservoir is located under a cover on the left side of the cargo area **(see illustration)**.

15   In milder climates, plain water can be used in the reservoir, but it should be kept no more than 2/3 full to allow for expansion if the

4.14a The reservoir for the windshield washer fluid is located on the right side of the engine compartment and fluid is added after pulling the top up and out - how often you use the washers will dictate how often you need to check the reservoir

water freezes. In colder climates, use windshield washer system antifreeze, available at any auto parts store, to lower the freezing point of the fluid. Mix the antifreeze with water in accordance with the manufacturer's directions on the container. **Caution:** *Don't use cooling system antifreeze - it will damage the vehicle's paint.*

16   To help prevent icing in cold weather, warm the windshield with the defroster before using the washer.

Battery electrolyte

17   To check the electrolyte level in the battery on conventional batteries, remove all of the cell caps. If the level is low, add distilled water until it's above the plates. Most aftermarket replacement batteries have a splitting-ring indicator in each cell to help you judge when enough water has been added - don't overfill the cells!

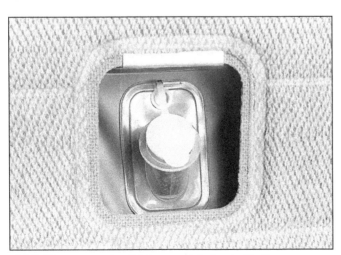

4.14b The rear washer fluid reservoir is located in the rear cargo area, behind a cover

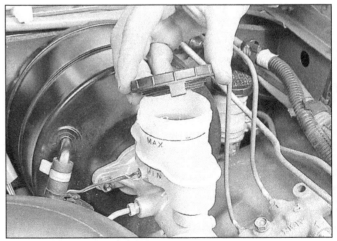

4.19a Keep the brake fluid level near the MAX mark on the reservoir - fluid can be added after unscrewing the cap

4.19b The clutch fluid level must be kept between the upper and lower marks on the fluid reservoir - unscrew the cap to add fluid

Brake and clutch fluid

Refer to illustrations 4.19a and 4.19b

18 The brake master cylinder is mounted on the front of the power booster unit in the engine compartment. The clutch cylinder used on manual transmissions is mounted adjacent to it on the firewall.

19 The fluid inside is readily visible. The level should be between the upper and lower marks on the reservoirs **(see illustrations)**. If a low level is indicated, be sure to wipe the top of the reservoir cover with a clean rag to prevent contamination of the brake and/or clutch system before removing the cover.

20 When adding fluid, pour it carefully into the reservoir to avoid spilling it onto surrounding painted surfaces. Be sure the specified fluid is used, since mixing different types of brake fluid can cause damage to the system. See Recommended lubricants and fluids at the front of this Chapter or your owner's manual. **Warning:** *Brake fluid can harm your eyes and damage painted surfaces, so be very careful when handling or pouring it. Don't use brake fluid that's been standing open or is more than one year old. Brake fluid absorbs moisture from the air. Excess moisture can cause a dangerous loss of brake efficiency.*

21 At this time the fluid and master cylinder can be inspected for contamination. The system should be drained and refilled if deposits, dirt particles or water droplets are seen in the fluid.

22 After filling the reservoir to the proper level, make sure the cover is on tight to prevent fluid leakage.

23 The brake fluid level in the master cylinder will drop slightly as the pads and the brake shoes at each wheel wear down during normal operation. If the master cylinder requires repeated additions to keep it at the proper level, it's an indication of leakage in the brake system, which should be corrected immediately. Check all brake lines and connections (see Section 23 for more information).

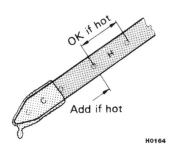

5.6 The automatic transmission fluid level must be between the two upper holes in the dipstick with the fluid at normal operating temperature

24 If, upon checking the master cylinder fluid level, you discover one or both reservoirs empty or nearly empty, the brake system should be bled (Chapter 9).

5 Automatic transmission fluid level check

Refer to illustration 5.6

1 The automatic transmission fluid level should be carefully maintained. Low fluid level can lead to slipping or loss of drive, while overfilling can cause foaming and loss of fluid.

2 With the parking brake set, start the engine, then move the shift lever through all the gear ranges, ending in Park. The fluid level must be checked with the vehicle level and the engine running at idle. **Note:** *Incorrect fluid level readings will result if the vehicle has just been driven at high speeds for an extended period, in hot weather in city traffic, or if it has been pulling a trailer. If any of these conditions apply, wait until the fluid has cooled (about 30 minutes).*

3 With the transmission at normal operating temperature, remove the dipstick from the filler tube. The dipstick is located at the rear of the engine compartment on the

driver's side.

4 Carefully touch the fluid at the end of the dipstick to determine if it's cool (86 to 122-degrees F) or hot (123 to 176-degrees F). Wipe the fluid from the dipstick with a clean rag and push it back into the filler tube until the cap seats.

5 Pull the dipstick out again and note the fluid level.

6 If the fluid felt cool, the level should be within the cold (C) range - between the lower holes **(see illustration)**. If the fluid was hot, the level should be within the hot (H) range.

7 If additional fluid is required, add it directly into the tube using a funnel. It takes about one pint to raise the level from the L mark to the H mark with a hot transmission, so add the fluid a little at a time and keep checking the level until it's correct.

8 The condition of the fluid should also be checked along with the level. If the fluid at the end of the dipstick is a dark reddish-brown color, or if it smells burned, it should be changed. If you are in doubt about the condition of the fluid, purchase some new fluid and compare the two for color and smell.

6 Power steering fluid level check

Refer to illustration 6.4

1 Unlike manual steering, the power steering system relies on fluid which may, over a period of time, require replenishing.

2 The fluid reservoir for the power steering pump is located on the pump body at the front of the engine.

3 For the check, the front wheels should be pointed straight ahead and the engine should be off. The engine and power steering fluid should be cold.

4 The fluid level in the translucent plastic reservoir can be checked without removing the cap **(see illustration)**.

5 If additional fluid is required, pour the specified type directly into the reservoir, using a funnel to prevent spills.

6 If the reservoir requires frequent fluid additions, all power steering hoses, hose connections and the power steering pump should be carefully checked for leaks.

6.4 Unscrew the power steering reservoir cap to add fluid - the level must be kept above the lower mark on the housing

7   Tire and tire pressure checks

Refer to illustrations 7.2, 7.3, 7.4a, 7.4b and 7.8

1   Periodic inspection of the tires may spare you the inconvenience of being stranded with a flat tire. It can also provide you with vital information regarding possible problems in the steering and suspension systems before major damage occurs.

2   The original tires on this vehicle are equipped with wear indicator bars that will appear when tread depth reaches a predetermined limit, usually 1/16-inch, but they don't appear until the tires are worn out. Tread wear can be monitored with a simple, inexpensive device known as a tread depth indicator **(see illustration)**.

3   Note any abnormal tread wear **(see illustration)**. Tread pattern irregularities such as cupping, flat spots and more wear on one side than the other are indications of front end alignment and/or balance problems. If any of these conditions are noted, take the vehicle to a tire shop or service station to correct the problem.

4   Look closely for cuts, punctures and

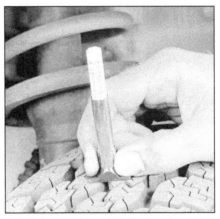

7.2 A tire tread depth indicator should be used to monitor tire wear - they're available at auto parts stores and service stations and cost very little

embedded nails or tacks. Sometimes a tire will hold air pressure for a short time or leak down very slowly after a nail has embedded itself in the tread. If a slow leak persists, check the valve stem core to make sure it's tight **(see illustration)**. Examine the tread for

an object that may have embedded itself in the tire or for a "plug" that may have begun to leak (radial tire punctures are repaired with a plug that's installed in a puncture). If a puncture is suspected, it can be easily verified by spraying a solution of soapy water onto the puncture area **(see illustration)**. The soapy solution will bubble if there's a leak. Unless the puncture is unusually large, a tire shop or service station can usually repair the tire.

5   Carefully inspect the inner sidewall of each tire for evidence of brake fluid leakage. If you see any, inspect the brakes immediately.

6   Correct air pressure adds miles to the lifespan of the tires, improves mileage and enhances overall ride quality. Tire pressure cannot be accurately estimated by looking at a tire, especially if it's a radial. A tire pressure gauge is essential. Keep an accurate gauge in the vehicle. The pressure gauges attached to the nozzles of air hoses at gas stations are often inaccurate.

7   Always check tire pressure when the tires are cold. Cold, in this case, means the vehicle has not been driven over a mile in the three hours preceding a tire pressure check. A pressure rise of four to eight pounds is not uncommon once the tires are warm.

UNDERINFLATION

CUPPING

Cupping may be caused by:
- Underinflation and/or mechanical irregularities such as out-of-balance condition of wheel and/or tire, and bent or damaged wheel.
- Loose or worn steering tie-rod or steering idler arm.
- Loose, damaged or worn front suspension parts.

OVERINFLATION

INCORRECT TOE-IN OR EXTREME CAMBER

FEATHERING DUE TO MISALIGNMENT

7.3 This chart will help you determine the condition of the tires, the probable cause(s) of abnormal wear and the corrective action necessary

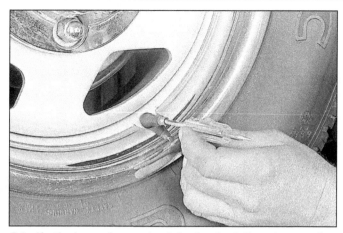

7.4a If a tire loses air on a steady basis, check the valve core first to make sure it's snug (special inexpensive wrenches are commonly available at auto parts stores)

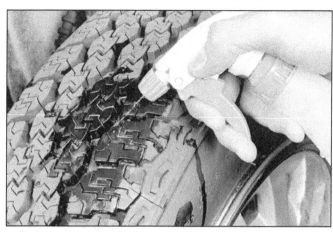

7.4b If the valve core is tight, raise the corner of the vehicle with the low tire and spray a soapy water solution onto the tread as the tire is turned - slow leaks will cause small bubbles to appear

8   Unscrew the valve cap protruding from the wheel or hubcap and push the gauge firmly onto the valve stem **(see illustration)**. Note the reading on the gauge and compare the figure to the recommended tire pressure shown on the placard on the glove compartment door. Be sure to reinstall the valve cap to keep dirt and moisture out of the valve stem mechanism. Check all four tires and, if necessary, add enough air to bring them up to the recommended pressure.

9   Don't forget to keep the spare tire inflated to the specified pressure (refer to your owner's manual or the tire sidewall).

8   Battery check and maintenance

Refer to illustrations 8.1, 8.6, 8.7a, 8.7b and 8.7c
Warning: *Certain precautions must be fol-*

lowed when checking and servicing the battery. Hydrogen gas, which is highly flammable, is always present in the battery cells, so keep lighted tobacco and all other open flames and sparks away from the battery. The electrolyte inside the battery is actually dilute sulfuric acid, which will cause injury if splashed on your skin or in your eyes. It will also ruin clothes and painted surfaces. When removing the battery cables, always detach the negative cable first and hook it up last!

1   Battery maintenance is an important procedure which will help ensure that you aren't stranded because of a dead battery. Several tools are required for this procedure **(see illustration)**.

2   When checking/servicing the battery, always turn the engine and all accessories off.

3   A sealed (sometimes called maintenance-free), battery is standard equipment

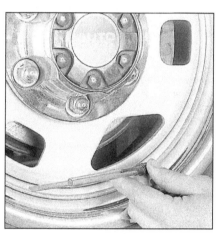

7.8 To extend the life of the tires, check the air pressure at least once a week with an accurate gauge (don't forget the spare!)

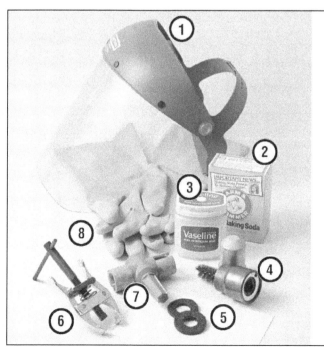

8.1 Tools and materials required for battery maintenance

1   *Face shield/safety goggles - When removing corrosion with a brush, the acidic particles can easily fly up into your eyes*
2   *Baking soda - A solution of baking soda and water can be used to neutralize corrosion*
3   *Petroleum jelly - a layer of this on the battery posts will help prevent corrosion*
4   *Battery post/cable cleaner - This wire brush cleaning tool will remove all traces of corrosion from the battery posts and cable clamps*
5   *Treated felt washers - Placing one of these on each post, directly under the cable clamps, will help prevent corrosion*
6   *Puller - Sometimes the cable clamps are very difficult to pull off the posts, even after the nut/bolt has been completely loosened. This tool pulls the clamp straight up and off the post without damage*
7   *Battery post/cable cleaner - Here is another cleaning tool which is a slightly different version of number 4 above, but it does the same thing*
8   *Rubber gloves - Another safety item to consider when servicing the battery; remember that's acid inside the battery!*

8.6 Removing a cable from the battery post with a wrench - sometimes a special battery pliers is required for this procedure if corrosion has caused deterioration of the nut hex (always remove the ground cable first and hook it up last!)

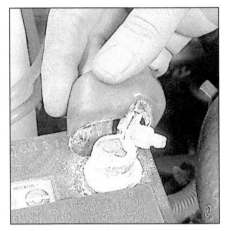

8.7a Battery terminal corrosion usually appears as light, fluffy powder

8.7b When cleaning the cable clamps, all corrosion must be removed (the inside of the clamp is tapered to match the taper on the post, so don't remove too much material)

on some vehicles. The cell caps cannot be removed, no electrolyte checks are required and water cannot be added to the cells. However, if a standard aftermarket battery has been installed, the following maintenance procedure can be used.

4   Remove the caps and check the electrolyte level in each of the battery cells (see Section 4). It must be above the plates. There's usually a split-ring indicator in each cell to indicate the correct level. If the level is low, add distilled water only, then reinstall the cell caps. **Caution:** *Overfilling the cells may cause electrolyte to spill over during periods of heavy charging, causing corrosion and damage to nearby components.*

5   The external condition of the battery should be checked periodically. Look for damage such as a cracked case.

6   Check the tightness of the battery cable bolts **(see illustration)** to ensure good electrical connections. Inspect the entire length of each cable, looking for cracked or abraded

insulation and frayed conductors.

7   If corrosion (visible as white, fluffy deposits) **(see illustration)** is evident, remove the cables from the terminals, clean them with a battery brush and reinstall them **(see illustrations)**. Corrosion can be kept to a minimum by applying a layer of petroleum jelly or grease to the terminals.

8   Make sure the battery carrier is in good condition and the hold-down clamp is tight. If the battery is removed (see Chapter 5 for the removal and installation procedure), make sure that no parts remain in the bottom of the carrier when it's reinstalled. When reinstalling the hold-down clamp, don't overtighten the nuts.

9   Corrosion on the carrier, battery case and surrounding areas can be removed with a solution of water and baking soda. Apply the mixture with a small brush, let it work, then rinse it off with plenty of clean water.

10   Any metal parts of the vehicle damaged by corrosion should be coated with a zinc-based primer, then painted.

11   Additional information on the battery, charging and jump starting can be found in the front of this manual and in Chapter 5.

Check for a chafed area that could fail prematurely.

Check for a soft area indicating the hose has deteriorated inside.

Overtightening the clamp on a hardened hose will damage the hose and cause a leak.

Check each hose for swelling and oil-soaked ends. Cracks and breaks can be located by squeezing the hose.

9.4 Hoses, like drivebelts, have a habit of failing at the worst possible time - to prevent the inconvenience of a blown radiator or heater hose, inspect them carefully as shown here

---

9   Cooling system check

Refer to illustration 9.4

1   Many major engine failures can be attributed to a faulty cooling system. If the vehicle is equipped with an automatic transmission, the cooling system also cools the transmission fluid, prolonging transmission life.

2   The cooling system should be checked with the engine cold. Do this before the vehicle is driven for the day or after it has been shut off for at least three hours.

3   Remove the radiator cap by turning it counterclockwise until it reaches a stop. If you hear a hissing sound (indicating there's still pressure in the system), wait until it stops. Now press down on the cap with the palm of your hand and continue turning until

8.7c Regardless of the type of tool used to clean the battery posts, a clean, shiny surface should be the result

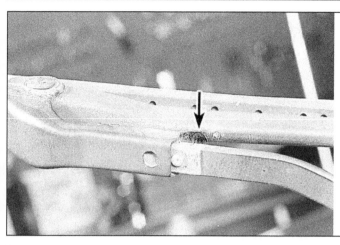

10.1 Depress the tab (arrow) and pull the blade assembly off the wiper arm

it can be removed. Thoroughly clean the cap, inside and out, with clean water. Also clean the filler neck on the radiator. All traces of corrosion should be removed. The coolant inside the radiator should be relatively transparent. If it's rust colored, the system should be drained and refilled (Section 41). If the coolant level is not up to the top, add additional antifreeze/coolant mixture (see Section 4).

4   Carefully check the large upper and lower radiator hoses along with the smaller diameter heater hoses which run from the engine to the firewall. Inspect each hose along its entire length, replacing any hose that's cracked, swollen or deteriorated. Cracks may become more apparent if the hose is squeezed **(see illustration)**. Regardless of condition, it's a good idea to replace hoses with new ones every two years. Make sure that all hose connections are tight. A leak in the cooling system will usually show up as white or rust colored deposits on the areas adjoining the leak. If wire-type clamps are used at the ends of the hoses, it may be a good idea to replace them with more secure screw-type clamps.

5   Use compressed air or a soft brush to remove bugs, leaves, etc. from the front of the radiator or air conditioning condenser. Be careful not to damage the delicate cooling fins or cut yourself on them.

6   Every other inspection, or at the first indication of cooling system problems, have the cap and system pressure tested. If you don't have a pressure tester, most gas stations and repair shops will do this for a minimal charge.

10   Wiper blade inspection and replacement

Refer to illustration 10.1

1   The wiper blade is removed by depressing the release tab at the center of the wiper arm and pulling the blade off the arm **(see illustration)**.

2   The rubber element is part of the blade, so the entire assembly must be replaced.

3   If it's necessary to replace the wiper

arm, remove the cover and nut securing the wiper arm to the linkage and detach the arm.

4   When installing the wiper arm, be sure it's lined up with the other arm and positioned correctly on the shaft.

11   Underhood hose check and replacement

General

Caution: *Replacement of air conditioning hoses must be left to a dealer service department or air conditioning shop that has the equipment to depressurize the system safely. Never remove air conditioning components or hoses until the system has been depressurized.*

1   High temperatures in the engine compartment can cause the deterioration of the rubber and plastic hoses used for engine, accessory and emission systems operation. Periodic inspection should be made for cracks, loose clamps, material hardening and leaks. Information specific to the cooling system hoses can be found in Section 9.

2   Some, but not all, hoses are secured to the fittings with clamps. Where clamps are used, check to be sure they haven't lost their tension, allowing the hose to leak. If clamps aren't used, make sure the hose has not expanded and/or hardened where it slips over the fitting, allowing it to leak.

Vacuum hoses

3   It's quite common for vacuum hoses, especially those in the emissions system, to be color coded or identified by colored stripes molded into them. Various systems require hoses with different wall thicknesses, collapse resistance and temperature resistance. When replacing hoses, be sure the new ones are made of the same material.

4   Often the only effective way to check a hose is to remove it completely from the vehicle. If more than one hose is removed, be sure to label the hoses and fittings to ensure correct installation.

5   When checking vacuum hoses, be sure to include any plastic T-fittings in the check.

Inspect the fittings for cracks and the hose where it fits over the fitting for distortion, which could cause leakage.

6   A small piece of vacuum hose (1/4-inch inside diameter) can be used as a stethoscope to detect vacuum leaks. Hold one end of the hose to your ear and probe around vacuum hoses and fittings, listening for the "hissing" sound characteristic of a vacuum leak. **Warning:** *When probing with the vacuum hose stethoscope, be very careful not to come into contact with moving engine components such as the drivebelts, cooling fan, etc.*

Fuel hose

Warning: *There are certain precautions which must be taken when inspecting or servicing fuel system components. Work in a well ventilated area and don't allow open flames (cigarettes, appliance pilot lights, etc.) or bare light bulbs near the work area. Mop up any spills immediately and don't store fuel soaked rags where they could ignite. On vehicles equipped with fuel injection, the fuel system is under pressure, so if any fuel lines are to be disconnected, the pressure in the system must be relieved first (see Chapter 4 for more information).*

7   Check all rubber fuel lines for deterioration and chafing. Check carefully for cracks in areas where the hose bends and where it's attached to fittings.

8   High quality fuel line, usually identified by the word Fluroelastomer printed on the hose, should be used for fuel line replacement. **Warning:** *Never, under any circumstances, use unreinforced vacuum line, clear plastic tubing or water hose for fuel lines!*

9   Spring-type clamps are commonly used on fuel lines. They often lose their tension over a period of time, and can be "sprung" during removal. Replace all spring-type clamps with screw clamps whenever a hose is replaced.

Metal lines

10   Sections of metal line are often used for fuel line between the fuel pump and carburetor or fuel injection unit. Check carefully to be sure the line has not been bent or crimped and look for cracks.

11   If a section of metal fuel line must be replaced, only seamless steel tubing should be used, since copper and aluminum tubing don't have the strength necessary to withstand normal engine vibration.

12   Check the metal brake lines where they enter the master cylinder and brake proportioning unit (if used) for cracks in the lines and loose fittings. Any sign of brake fluid leakage means an immediate thorough inspection of the brake system should be done.

12   Accelerator linkage check and lubrication

Refer to illustration 12.1

1   At the specified intervals, inspect the

12.1 Check the accelerator cable and linkage for free movement

everything necessary for the job. The correct oil for your application can be found in Recommended lubricants and fluids at the beginning of this Chapter.

7   With the engine oil warm (warm engine oil will drain better and more built-up sludge will be removed with it), raise and support the vehicle. Make sure it's safely supported!

8   Move all necessary tools, rags and newspapers under the vehicle. Set the drain pan under the drain plug. Keep in mind that the oil will initially flow from the pan with some force; position the pan accordingly.

9   Being careful not to touch any of the hot exhaust components, use a wrench to remove the drain plug near the bottom of the oil pan **(see illustration)**. Depending on how hot the oil is, you may want to wear gloves while unscrewing the plug the final few turns.

10   Allow the old oil to drain into the pan. It may be necessary to move the pan as the oil flow slows to a trickle.

11   After all the oil has drained, wipe off the drain plug with a clean rag. Small metal particles may cling to the plug and would immediately contaminate the new oil.

12   Clean the area around the drain plug opening and reinstall the plug. Tighten the plug securely with the wrench. If a torque wrench is available, use it to tighten the plug.

13   Move the drain pan into position under the oil filter.

14   Use the filter wrench to loosen the oil filter **(see illustration)**. Chain or metal band filter wrenches may distort the filter canister, but it doesn't matter since the filter will be discarded anyway.

15   Completely unscrew the old filter. Be careful; it's full of oil. Empty the oil inside the filter into the drain pan.

16   Compare the old filter with the new one to make sure they're the same type.

accelerator linkage and check it for free movement to make sure it is not binding **(see illustration)**.

2   Lubricate the linkage with a few drops of oil.

---

13   Engine oil and filter change

Refer to illustrations 13.3, 13.9, 13.14 and 13.18

1   Frequent oil changes are the most important preventive maintenance procedures that can be done by the home mechanic. As engine oil ages, it becomes diluted and contaminated, which leads to premature engine wear.

2   Although some sources recommend oil filter changes every other oil change, the minimal cost of an oil filter and the fact that it's easy to install dictate that a new filter be used every time the oil is changed.

3   Gather all necessary tools and materials before beginning this procedure **(see illustration)**.

4   You should have plenty of clean rags and newspapers handy to mop up any spills. Access to the underside of the vehicle is greatly improved if the vehicle can be lifted on a hoist, driven onto ramps or supported by jackstands. **Warning:** *Do not work under a vehicle which is supported only by a bumper, hydraulic or scissors-type jack!*

5   If this is your first oil change, get under the vehicle and familiarize yourself with the locations of the oil drain plug and the oil filter. The engine and exhaust components will be warm during the actual work, so note how they're situated to avoid touching them when working under the vehicle.

6   Warm the engine to normal operating temperature. If the new oil or any tools are needed, use the warm-up time to obtain

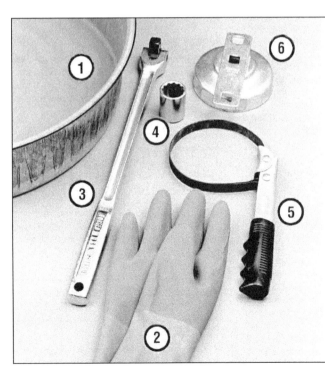

13.3 These tools are required when changing the engine oil and filter

1   **Drain pan** - *It should be fairly shallow in depth, but wide to prevent spills*

2   **Rubber gloves** - *When removing the drain plug and filter, you will get oil on your hands (the gloves will prevent burns)*

3   **Breaker bar** - *Sometimes the oil drain plug is tight and a long breaker bar is needed to loosen it*

4   **Socket** - *To be used with the breaker bar or a ratchet (must be the correct size to fit the drain plug - 6-point preferred*

5   **Filter wrench** - *This is a metal band-type wrench, which requires clearance around the filter to be effective*

6   **Filter wrench** - *This type fits on the bottom of the filter and can be turned with a ratchet or breaker bar (different size wrenches are available for different types of filters)*

13.9 The engine oil drain plug is located at the rear of the oil pan (four-cylinder engine shown) - it's usually very tight, so use a box-wrench or six-point socket to avoid rounding off the hex

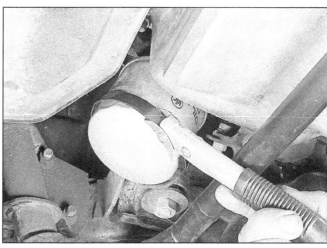

13.14 The oil filter is usually on very tight as well and will require a special wrench for removal - DO NOT use the wrench to tighten the new filter!

17   Use a clean rag to remove all oil, dirt and sludge from the area where the oil filter mounts to the engine. Check the old filter to make sure the rubber gasket isn't stuck to the engine. If the gasket is stuck to the engine, remove it.

18   Apply a light coat of clean oil to the rubber gasket on the new oil filter **(see illustration)**.

19   Attach the new filter to the engine, following the tightening directions printed on the filter canister or packing box. Most filter manufacturers recommend against using a filter wrench due to the possibility of overtightening and damage to the seal.

20   Remove all tools, rags, etc. from under the vehicle, being careful not to spill the oil in the drain pan, then lower the vehicle.

21   Move to the engine compartment and locate the oil filler cap.

22   Pour the fresh oil into the filler opening. A funnel may be used.

23   Pour three or four quarts of fresh oil into the engine. Wait a few minutes to allow the oil

to drain into the pan, then check the level on the oil dipstick (see Section 4 if necessary). If the oil level is above the L mark, start the engine and allow the new oil to circulate.

24   Run the engine for only about a minute and then shut it off. Immediately look under the vehicle and check for leaks at the oil pan drain plug and around the oil filter. If either one is leaking, tighten it a little more.

25   With the new oil circulated and the filter now completely full, recheck the level on the dipstick and add more oil as necessary.

26   During the first few trips after an oil change, make it a point to check frequently for leaks and correct oil level.

27   The old oil drained from the engine cannot be reused in its present state and should be disposed of. Oil reclamation centers, auto repair shops and gas stations will normally accept the oil, which can be refined and used again. After the oil has cooled it can be poured into a container (capped plastic jugs or bottles, milk cartons, etc.) for transport to a disposal site.

14   Tire rotation

Refer to illustration 14.2

1   The tires should be rotated at the specified intervals and whenever uneven wear is noticed.

2   Refer to the accompanying illustration for the preferred tire rotation pattern.

3   Refer to the information in Jacking and towing at the front of this manual for the proper procedures to follow when raising the vehicle and changing a tire. If the brakes are to be checked, don't apply the parking brake as stated. Make sure the tires are blocked to prevent the vehicle from rolling as it's raised.

4   Preferably, the entire vehicle should be raised at the same time. This can be done on a hoist or by jacking up each corner and then lowering the vehicle onto jackstands placed under the frame rails. Always use four jackstands and make sure the vehicle is safely supported.

5   After rotation, check and adjust the tire

13.18 Lubricate the rubber gasket with clean oil before installing the filter on the engine

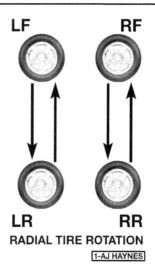

LF   RF

14.2 The recommended tire rotation pattern for these models

LR   RR

RADIAL TIRE ROTATION

1-AJ HAYNES

15.4 Tighten the rocker arm bracket nuts to the specified torque before adjusting the valves

15.9 To make sure the adjusting screw doesn't move when the locknut is tightened, use a box-wrench and have a good grip on the screwdriver

pressures as necessary and be sure to check the lug nut tightness.

6   For additional information on the wheels and tires, refer to Chapter 11.

15 Valve clearance adjustment (four-cylinder engines only)

Refer to illustrations 15.4 and 15.9

1   The valve clearances are checked and adjusted with the engine cold.

2   Remove the air cleaner assembly (Chapter 4).

3   Remove the rocker arm cover (Chapter 2).

4   Tighten the rocker arm bracket nuts to the specified torque **(see illustration)**.

5   Rotate the crankshaft to bring the number one piston to the Top Dead Center (TDC) position on the compression stroke (see Chapter 2, part A for details). The number one cylinder rocker arms (closest to the front of the engine) should be loose (able to move up and down slightly) and the camshaft lobes

should be facing away from the rocker arms.

6   With the crankshaft in this position, the valves on the following cylinders can be checked and adjusted:

 1 Intake and exhaust
 2 Intake
 3 Exhaust

7   First, check/adjust the intake valve clearance. Insert the appropriate size feeler gauge (see the Specifications) between the intake valve stem and the adjusting screw. If the clearance is correct, there should be a slight drag on the feeler gauge as it is inserted between the valve stem and the adjusting screw.

8   If adjustment is required, loosen the adjusting screw locknut. Carefully tighten the adjusting screw until you can feel a slight drag on the feeler gauge as you withdraw it from between the stem and adjusting screw.

9   Hold the adjusting screw with a screwdriver (to keep it from turning) and tighten the locknut **(see illustration)**. Recheck the clearance to make sure it hasn't changed.

10   Repeat the procedure on the exhaust valve on the number 1 cylinder, the intake

valve on number 2 cylinder and the exhaust valve on the number 3 cylinder.

11   Rotate the crankshaft one full turn clockwise until the number 4 piston is at TDC on the compression stroke. The number 4 cylinder rocker arms (closest to the rear of the engine) should be loose, with the camshaft lobes facing away from the rocker arms.

12   With the crankshaft in this position, the valves on the following cylinders can be checked and adjusted:

 2 Exhaust
 3 Intake
 4 Intake and exhaust

13   Install the rocker arm cover and the air cleaner assembly.

16 Chassis lubrication

Refer to illustrations 16.1, 16.3a, 16.3b and 16.3c

1   A grease gun and cartridge filled with the recommended grease are the only items

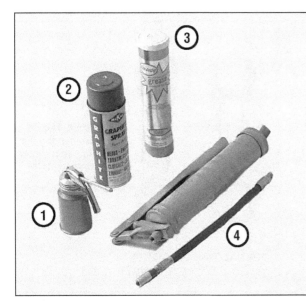

16.1 Materials required for chassis and body lubrication

1   **Engine oil -** *Light engine oil in a can like this can be used for door and hood hinges*

2   **Graphite spay -** *Used to lubricate lock cylinders*

3   **Grease -** *Grease, in a variety of types and weights, is available for use in a grease gun. Check the Specifications for your requirements*

4   **Grease gun -** *A common grease gun, shown here with a detachable hose and nozzle, is needed for chassis lubrication. After use, clean it thoroughly!*

16.3a Upper balljoint grease fitting (arrow)

16.3b Front driveshaft grease fitting (4WD models)

required for chassis lubrication other than some clean rags and equipment needed to raise and support the vehicle safely **(see illustration)**.

2   There are several points on the vehicle's suspension, steering and drivetrain components that must be periodically lubricated with lithium based multi-purpose grease, depending on model and year. Included are the upper and lower suspension balljoints, the swivel joints on the steering linkage and, on 4WD models, the front and rear driveshafts.

3   The grease point for each upper suspension balljoint (if equipped) is on top of the balljoint and is accessible by removing the front wheel and tire **(see illustration)**. The steering linkage swivel joints on some models are designed to be lubricated and the driveshaft universal joints require lubrication as well **(see illustrations)**.

4   For easier access under the vehicle, raise it with a jack and place jackstands under the frame. Make sure the vehicle is safely supported on the stands!

5   If grease fittings aren't already installed, the plugs will have to be removed and fittings screwed into place.

6   Force a little of the grease out of the gun nozzle to remove any dirt, then wipe it clean with a rag.

7   Wipe the grease fitting and push the nozzle firmly over it. Squeeze the trigger on the grease gun to force grease into the component. Both the balljoints and swivel joints should be lubricated until the rubber reservoir is firm to the touch. Don't pump too much grease into the fittings or it could rupture the reservoir. If the grease seeps out around the grease gun nozzle, the fitting is clogged or the nozzle isn't seated all the way. Resecure the gun nozzle to the fitting and try again. If necessary, replace the fitting.

8   Wipe excess grease from the components and the grease fittings.

9   While you're under the vehicle, clean and lubricate the parking brake cable along with the cable guides and levers. This can be done by smearing some of the chassis

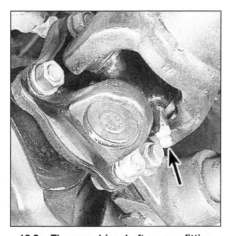

16.3c The rear driveshaft grease fitting (arrow) may be difficult to reach - rotating the driveshaft may improve access

grease onto the cable and its related parts with your fingers.

10   Lower the vehicle to the ground for the remaining body lubrication process.

11   Open the hood and rear gate and smear a little chassis grease on the latch mechanisms. Have an assistant pull the release knob from inside the vehicle as you lubricate the cable at the latch.

12   Lubricate all the hinges (door, hood, hatch) with a few drops of light engine oil to keep them in proper working order.

13   The key lock cylinders can be lubricated with spray-on graphite, which is available at auto parts stores.

---

17   Suspension and steering check

Refer illustrations 17.6a, 17.6b and 17.11

1   Whenever the front of the vehicle is raised for any reason, it's a good idea to visually check the suspension and steering components for wear.

2   Indications of steering or suspension problems include excessive play in the steering wheel before the front wheels react,

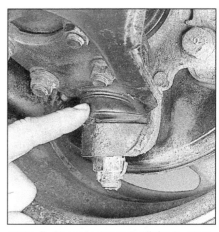

17.6a Push on the lower balljoint boots to check for damage and leaks

excessive swaying around corners or body movement over rough roads and binding at some point as the steering wheel is turned.

3   Before the vehicle is raised for inspection, test the shock absorbers by pushing down aggressively at each corner. If the vehicle doesn't come back to a level position within one or two bounces, the shocks are worn and should be replaced. As this is done listen for squeaks and other noises from the suspension components. Information on shock absorber and suspension components can be found in Chapter 10.

4   Raise the front end of the vehicle and support it on jackstands. Make sure it's safely supported!

5   Crawl under the vehicle and check for loose bolts, broken or disconnected parts and deteriorated rubber bushings on all suspension and steering components. Look for grease or fluid leaking from around the steering gear assembly and shock absorbers. If equipped, check the power steering hoses and connections for leaks.

6   The balljoint boots should be checked at this time **(see illustration)**. This includes not only the upper and lower suspension balljoints, but those connecting the steering

17.6b The grease oozing from this tie-rod end boot indicates that it needs replacing

17.11 Push on the driveaxle boots to check for cracks

linkage parts as well **(see illustration)**. After cleaning around the balljoints, inspect the seals for cracks and damage.

7   Grip the top and bottom of each wheel and try to move it in and out. It won't take a lot of effort to be able to feel any play in the wheel bearings. If the play is noticeable it would be a good idea to adjust it right away or it could confuse further inspections.

8   Grip each side of the wheel and try rocking it laterally. Steady pressure will, of course, turn the steering, but back-and-forth pressure will reveal a loose steering joint. If some play is felt it would be easier to get assistance from someone so while one person rocks the wheel from side to side, the other can look at the joints, bushings and connections in the steering linkage. Generally speaking, there are eight places where the play may occur. The two outer balljoints on the tie-rods are the most likely, followed by the two inner joints on the same rods, where they join to the center rod. Any play in them means replacement of the tie-rod end. Next are two swivel bushings, one at each end of the center gear rod. Finally, check the steering gear arm balljoint and the one on the idler arm which supports the center rod on the side opposite the steering box. This unit is bolted to the side of the frame member and any play calls for replacement of the bushings.

9   To check the steering box, first make sure the bolts holding the steering box to the frame are tight. Then get another person to help examine the mechanism. One should look at, or hold onto, the arm at the bottom of the steering box while the other turns the steering wheel a little from side to side. The amount of lost motion between the steering wheel and the gear arm indicates the degree of wear in the steering box mechanism. This check should be carried out with the wheels first in the straight ahead position and then at nearly full lock on each side. If the play only occurs noticeably in the straight ahead position then the wear is most likely in the worm and/or nut. If it occurs at all positions, then the wear is probably in the sector shaft bearing. Oil leaks from the unit are another indica-

tion of such wear. In either case the steering box will need removal for closer examination and repair.

10   Moving to the vehicle interior, check the play in the steering wheel by turning it slowly in both directions until the wheels can just be felt turning. The steering wheel free play should be less than 1-3/8 inch (35 mm). Excessive play is another indication of wear in the steering gear or linkage. The steering box can be adjusted for wear (see Chapter 10).

11   On 4WD models, inspect the front driveaxle CV joint boots for tears and leakage of grease **(see illustration)**.

12   Following the inspection of the front, a similar inspection should be made of the rear suspension components, again checking for loose bolts, damaged or disconnected parts and deteriorated rubber bushings.

---

18   Exhaust system check

---

1   With the engine cold (at least three hours after the vehicle has been driven), check the complete exhaust system from the manifold to the end of the tailpipe. Be careful around the catalytic converter, which may be hot even after three hours. The inspection should be done with the vehicle on a hoist to permit unrestricted access. If a hoist isn't

available, raise the vehicle and support it securely on jackstands.

2   Check the exhaust pipes and connections for signs of leakage and/or corrosion indicating a potential failure. Make sure that all brackets and hangers are in good condition and tight.

3   Inspect the underside of the body for holes, corrosion, open seams, etc. which may allow exhaust gases to enter the passenger compartment. Seal all body openings with silicone sealant or body putty.

4   Rattles and other noises can often be traced to the exhaust system, especially the hangers, mounts and heat shields. Try to move the pipes, mufflers and catalytic converter. If the components can come in contact with the body or suspension parts, secure the exhaust system with new brackets and hangers.

---

19   Clutch pedal height and freeplay check and adjustment

Refer to illustrations 19.2, 19.3 and 19.4

1   On vehicles equipped with a manual transmission, the clutch pedal height and free play must be correctly adjusted.

2   The height of the clutch pedal is the distance the pedal sits off the floor **(see illustra-**

19.2 The clutch pedal height is measured from the top of the pedal to the floor

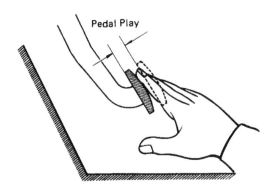

19.3 To check clutch pedal freeplay, measures the distance from the natural resting place of the pedal to the point at which resistance is felt

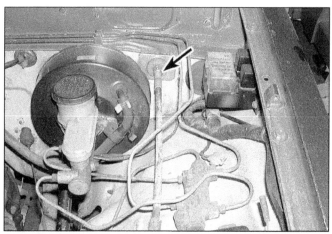

19.4 On cable-operated models, adjust the clutch to achieve the specified play between the nut (arrow) and the rubber cushion

tion). The distance should be as specified. If the pedal height is not within the specified range, loosen the locknut on the clutch stopper bolt or switch located on the pedal bracket and turn the bolt or switch in or out until the pedal height is correct. Retighten the locknut. On some later models, there is provision for adjustment of the clutch master cylinder pushrod. Loosen the locknut and back the stopper bolt or switch out for clearance, then loosen the locknut on the clutch pushrod. Turn the pushrod to achieve the proper height, then retighten the locknut. On cruise control equipped models, turn the clutch switch until the plunger is fully retracted against the clutch pedal arm, then back it out a half turn before tightening the locknut. On non-cruise control equipped models, turn the stopper bolt in until it just contacts the arm and tighten the locknut.

3   The freeplay is the pedal slack, or the distance the pedal can be depressed before it begins to have any effect on the clutch **(see illustration)**. The distance should be as specified. If it isn't, it must be adjusted, as discussed under the appropriate heading below.

Cable actuated clutch

4   Working in the engine compartment, loosen the locknut, pull the outer cable toward the front of the vehicle and turn the adjusting nut in until the washer damper rubber contacts the firewall. Depress and release the clutch pedal several times. Pull the cable forward again, tighten the adjusting nut, then back it off to provide the specified amount of play **(see illustration)**. Tighten the locknut.

Hydraulic clutch

5   On most models, the freeplay is adjusted by turning the clutch switch or stopper bolt. On later models with adjustable clutch master cylinder pushrods, loosen the locknut on the pushrod. Turn the pushrod to achieve the proper freeplay, then retighten the locknut.

20 Manual transmission lubricant level check

Refer to illustration 20.1
Note: *On 1981 through 1987 models, the transmission is integral with the transfer case and the two components share the same lubricant - when you check the transmission lubricant, you're also checking the transfer case lubricant.*

1   Manual transmissions don't have a dipstick. The oil level is checked by removing a plug from the side of the transmission case **(see illustration)**. Locate the plug and use a rag to clean the plug and the area around it. If the vehicle is raised to gain access to the plug, be sure to support it safely on jackstands - DO NOT crawl under the vehicle when it's supported only by a jack!

2   With the engine and transmission cold, remove the plug. If lubricant immediately starts leaking out, thread the plug back into the transmission - the level is correct. If it doesn't, reach inside the hole with your little finger. The level should be even with the bottom of the plug hole.

3   If the transmission needs more lubricant, use a syringe or small pump to add it through the plug hole.

4   Thread the plug back into the transmission and tighten it securely. Drive the vehicle, then check for leaks around the plug.

21 Transfer case lubricant level check (1988 and later 4WD models only)

Note: *On 1981 through 1987 models, the transfer case is integral with the transmission and the two components share the same lubricant (see the previous Section) you're also checking the transfer case lubricant.*

1   Remove the transfer case rock guard (if equipped). The lubricant level is checked by

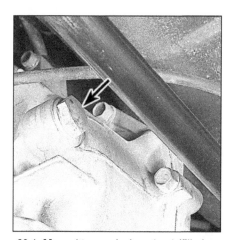

20.1 Manual transmission check/fill plug location (arrow)

removing a plug from the side of the case. If the vehicle is raised to gain access to the plug, be sure to support it safely on jackstands - DO NOT crawl under the vehicle when it's supported only by a jack!

2   With the engine and transfer case cold, remove the plug. If lubricant immediately starts leaking out, thread the plug back into the case - the level is correct. If it doesn't, completely remove the plug and reach inside the hole with your little finger. The level should be even with the bottom of the plug hole.

3   If more lubricant is needed, use a syringe or small pump to add it through the opening.

4   Thread the plug back into the case and tighten it securely. Drive the vehicle, then check for leaks around the plug. Install the rock guard.

22 Differential lubricant level check

Refer to illustration 22.2
1   The differential has a check/fill plug which must be removed to check the lubri-

cant level. If the vehicle is raised to gain access to the plug, be sure to support it safely on jackstands - DO NOT crawl under the vehicle when it's supported only by a jack.

2   Remove the check/fill plug from the differential **(see illustration)**.

3   The lubricant level should be at the bottom of the plug opening. If not, use a syringe to add the recommended lubricant until it just starts to run out of the opening.

4   Install the plug and tighten it securely.

23   Brake check

Refer to illustrations 23.6a, 23.6b and 23.13

Note: *For detailed photographs of the brake system, refer to Chapter 9.*

Warning: *Brake system dust may contain asbestos, which is hazardous to your health. DO NOT blow it out with compressed air and DO NOT inhale it. DO NOT use gasoline or solvents to remove the dust. Use brake system cleaner or denatured alcohol only!*

1   In addition to the specified intervals, the brakes should be inspected every time the wheels are removed or whenever a defect is suspected.

2   To check the brakes, the vehicle must be raised and supported securely on jackstands.

Disc brakes

3   Disc brakes are used on the front, and, on later Trooper II models, rear wheels. Extensive rotor damage can occur if the pads are allowed to wear beyond the specified limit.

4   Raise the vehicle and support it securely on jackstands, then remove the wheels (see Jacking and Towing at the front of the manual if necessary).

5   The disc brake calipers, which contain the pads, are visible with the wheels removed. There's an outer pad and an inner pad in each caliper. All pads should be inspected.

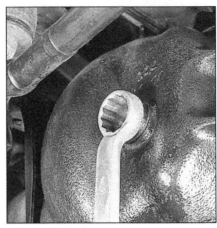

22.2 Use a box-end wrench to remove and install the differential plugs to avoid rounding off the hex

6   Each caliper has openings, which allow you to inspect the pads **(see illustrations)**. If the pad material has worn below the limit listed in this Chapter's Specifications, the pads should be replaced.

7   If you're unsure about the exact thickness of the remaining lining material, remove the pads for further inspection or replacement (refer to Chapter 9).

8   Before installing the wheels, check for leakage and/or damage (cracks, splitting, etc.) around the brake hose connections. Replace the hose or fittings as necessary, referring to Chapter 9.

9   Check the condition of the rotor. Look for score marks, deep scratches and burned spots. If these conditions exist, the hub/rotor assembly should be removed for servicing - Section 40 (2WD models) or Chapter 9 (4WD models).

Drum brakes

10   On rear brakes, remove the drum by removing the retaining screws and pulling it off the axle and brake assembly. If it's stuck, make sure the parking brake is released, then squirt penetrating oil into the joint between

23.6a You'll find an inspection hole like this in each front caliper - placing a steel rule across the hole should enable you to determine the thickness of the remaining lining materials for both inner and outer pads - the lining can also be inspected by looking through each end of the caliper

the hub and drum. Allow the oil to soak in and try to pull the drum off again.

11   If the drum still can't be pulled off, the brake shoes will have to be adjusted. This is done by first disconnecting the parking brake cable from the actuating lever and removing the rubber stopper from the lever. Push the lever against the backing plate.

12   With the drum removed, be careful not to touch any brake dust (see the **Warning** at the beginning of this Section).

13   Note the thickness of the lining material on both the front and rear brake shoes. If the material has worn away to below the specified thickness above the recessed rivets or metal shoe, the shoes should be replaced **(see illustration)**. The shoes should also be replaced if they're cracked, glazed (shiny surface) or contaminated with brake fluid.

14   Make sure that all the brake assembly springs are connected and in good condition.

15   Check the brake components for signs of fluid leakage. Carefully pry back the rubber

23.6b The rear disc brake lining thickness can also be checked through the hole in the caliper (arrow)

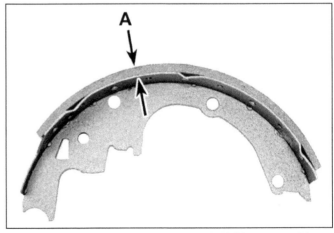

23.13 The rear brake shoe lining thickness (A) is measured from the outer surface of the lining to the metal shoe

cups on the wheel cylinders located at the top of the brake shoes with your finger. Any leakage is an indication that the wheel cylinders should be overhauled immediately (Chapter 10). Also check brake hoses and connections for leakage.

16   Wipe the inside of the drum with a clean rag and brake cleaner or denatured alcohol. Again, be careful not to breathe the asbestos dust.

17   Check the inside of the drum for cracks, score marks, deep scratches and hard spots, which will appear as small discolorations. If imperfections cannot be removed with fine emery cloth, the drum must be taken to a machine shop equipped to turn the drums.

Disc and drum brakes

18   If after the inspection process all parts are in good working condition, reinstall the brake drum.
19   Install the wheels and lower the vehicle.

Parking brake

20   The parking brake is operated by an umbrella type handle next to the steering column and locks the rear brake system. The easiest, and perhaps most obvious method of periodically checking the operation of the parking brake assembly is to park the vehicle on a steep hill with the parking brake set and the transmission in Neutral. If the parking brake cannot prevent the vehicle from rolling within 6 to 10 clicks, it's in need of adjustment (see Chapter 10).

24   Fuel system check

Warning: *Gasoline is extremely flammable, so take extra precautions when you work on any part of the fuel system. Don't smoke or allow open flames or bare light bulbs near the work area, and don't work in a garage where a natural gas-type appliance (such as a water heater or clothes dryer) with a pilot light is present. If you spill any fuel on your skin, rinse it off immediately with soap and water. When you perform any kind of work on the fuel tank, wear safety glasses and have a Class B type fire extinguisher on hand. No components should be disconnected until the pressure has been relieved (see Chapter 4).*

1   On most models, the fuel tank is located at the rear of the vehicle.
2   The fuel system should be checked with the vehicle raised on a hoist so the components underneath the vehicle are readily visible and accessible.
3   If the smell of gasoline is noticed while driving or after the vehicle has been in the sun, the system should be thoroughly inspected immediately.
4   Remove the gas tank cap and check for damage, corrosion and an unbroken sealing imprint on the gasket. Replace the cap with a new one if necessary.
5   With the vehicle raised, check the gas tank and filler neck for punctures, cracks and

other damage. The connection between the filler neck and the tank is especially critical. Sometimes a rubber filler neck will leak due to loose clamps or deteriorated rubber, problems a home mechanic can usually rectify. **Warning:** *Do not, under any circumstances, try to repair a fuel tank yourself (except rubber components). A welding torch or any open flame can easily cause the fuel vapors to explode if the proper precautions are not taken!*

6   Carefully check all rubber hoses and metal lines leading away from the fuel tank. Look for loose connections, deteriorated hoses, crimped lines and other damage. Follow the lines to the front of the vehicle, carefully inspecting them all the way. Repair or replace damaged sections as necessary.
7   If a fuel odor is still evident after the inspection, refer to Section 33 and check the evaporative emissions system.

25   Thermostatic air cleaner check (carbureted models)

1   Carbureted models are equipped with a thermostatically controlled air cleaner, which draws air to the carburetor from different locations depending on engine temperature.
2   This is a simple visual check. However, if access is tight, a small mirror may have to be used.
3   Open the hood and find the air control valve on the air cleaner assembly. It's located inside the long snorkel portion of the metal air cleaner housing.
4   If there's a flexible air duct attached to the end of the snorkel, disconnect it so you can look through the end of the snorkel and see the air control valve inside. A mirror may be needed if you can't safely look directly into the end of the snorkel.
5   The check should be done when the engine and outside air are cold. Start the engine and watch the air control valve, which should move up and close off the snorkel air passage. With the valve closed, air can't enter through the end of the snorkel, but instead enters the air cleaner through the hot air duct attached to the exhaust manifold.
6   As the engine warms up to operating temperature, the valve should open to allow air through the snorkel end. Depending on outside air temperature, this may take 10 to 15 minutes. To speed up the check you can reconnect the snorkel air duct, drive the vehicle and then check the position of the valve.
7   If the thermostatic air cleaner isn't operating properly, see Chapter 6 for more information.

26   Drivebelt check, adjustment and replacement

1   The drivebelts, or V-belts as they are often called, are located at the front of the engine and play an important role in the over-

all operation of the engine and accessories. Due to their function and material makeup, the belts are prone to failure after a period of time and should be inspected and adjusted periodically to prevent major engine damage. On four-cylinder engines V-belts drive the various components while on V6 models a single "serpentine" drivebelt is used.

Four-cylinder engines

Refer to illustrations 26.3, 26.4 and 26.6
2   The number of belts used on a particular vehicle depends on the accessories installed. Drivebelts are used to turn the alternator, power steering pump, water pump and air conditioning compressor. Depending on the pulley arrangement, more than one of the components may be driven by a single belt.
3   With the engine off, locate the drivebelts at the front of the engine. Using your fingers (and a flashlight, if necessary), move along the belts checking for cracks and separation of the belt plies. Also check for fraying and glazing, which gives the belt a shiny appearance **(see illustration)**. Both sides of each belt should be inspected, which means you'll

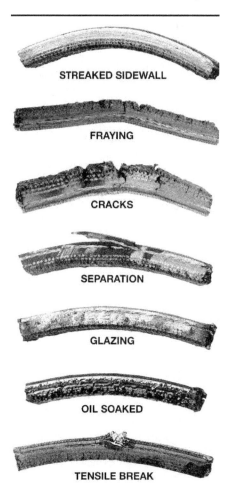

STREAKED SIDEWALL

FRAYING

CRACKS

SEPARATION

GLAZING

OIL SOAKED

TENSILE BREAK

26.3 Here are some of the more common problems associated with drivebelts (check the belts very carefully to prevent an untimely breakdown)

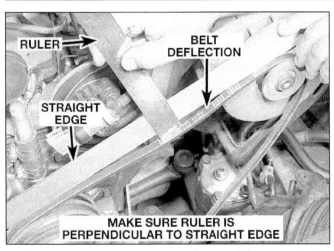

26.4 Measuring drivebelt deflection with a straightedge and ruler

26.6 On air conditioned four-cylinder models, turn the adjusting bolt (arrow) on the idler pulley to adjust drivebelt tension

have to twist each belt to check the underside. Check the pulleys for nicks, cracks, distortion and corrosion.

4   The tension of each belt is checked by pushing on it at a distance halfway between the pulleys. Push firmly with your thumb and see how much the belt moves (deflects) **(see illustration)**. A rule of thumb is that if the distance from pulley center-to-pulley center is between 7 and 11 inches, the belt should deflect 1/4-inch. If the belt travels between pulleys spaced 12-to-16 inches apart, the belt should deflect 1/2-inch.

5   If adjustment is needed, either to make the belt tighter or looser, it's done by moving the belt-driven accessory on the bracket.

6   Each component usually has an adjusting bolt and a pivot bolt. Both bolts must be loosened slightly to enable you to move the component. To adjust the air conditioner compressor belt, loosen the locknut and turn the adjusting bolt on the idler pulley to change the belt tension **(see illustration)**.

7   On components other than the air conditioning compressor, after the two bolts have been loosened, move the component away from the engine to tighten the belt or toward the engine to loosen the belt. Hold the accessory in position and check the belt tension. If it's correct, tighten the two bolts until just snug, then recheck the tension. If the tension is correct, tighten the bolts.

8   You may have to use some sort of pry bar to move the accessory while the belt is adjusted. If this must be done to gain the proper leverage, be very careful not to damage the component being moved or the part being pried against.

9   To replace a belt, follow the above procedures for drivebelt adjustment but slip the belt off the pulleys and remove it. Since belts tend to wear out more or less at the same time, it's a good idea to replace all of them at the same time. Mark each belt and the corresponding pulley grooves so the replacement belts can be installed properly.

10   Take the old belts with you when purchasing new ones in order to make a direct comparison for length, width and design.

11   Adjust the belts as described earlier in this Section.

V6 engines

12   A single drivebelt, or serpentine belt as it is more often called, is located at the front of the engine and drives the alternator, power steering pump, water pump and air conditioning compressor. Belt tension does not need to be periodically adjusted so long as the belt is in good condition.

13   With the engine off, open the hood and use your fingers (and a flashlight, if necessary), to move along the belt checking for cracks and separation of the belt plies. Also check for fraying and glazing, which gives the belt a shiny appearance. Both sides of the belt should be inspected, which means you will have to twist the belt to check the underside.

14   Check the ribs on the underside of the belt. They should all be the same depth, with none of the surface uneven.

15   To replace the belt, place a wrench or socket over the tensioner bolt and rotate the tensioner counterclockwise to release belt tension. The tensioner will swing down when it's released.

16   Remove the belt from the auxiliary components and carefully release the tensioner.

17   Route the new belt over the various pulleys, again rotating the tensioner to allow the belt to be installed, then release the belt tensioner.

27   Seatbelt check

1   Check the seatbelts, buckles, latch plates and guide loops for any obvious damage or signs of wear.

2   Make sure the seatbelt reminder light comes on when the key is turned on.

3   The seatbelts are designed to lock up during a sudden stop or impact, yet allow free movement during normal driving. The retractors should hold the belt against your chest while driving and rewind the belt when the buckle is unlatched.

4   If any of the above checks reveal problems with the seatbelt system, replace parts as necessary.

28   Power steering fluid replacement

1   Place a container under the power steering fluid reservoir.

2   Disconnect and plug the power steering hoses from the reservoir and allow the fluid in the reservoir to drain into the container.

3   Connect the hoses and fill the reservoir with the specified fluid.

4   Bleed the power steering system as described in Chapter 10.

29   Air filter replacement

1   At the specified intervals, the air filter should be replaced with a new one. A thorough program of preventive maintenance would also call for the filter to be inspected periodically between changes, especially if the vehicle is often driven in dusty conditions.

2   The air filter is located inside the air cleaner housing, which is mounted on top of the carburetor or at the front of the engine compartment next to the battery on fuel injected models.

Carbureted models

3   Remove the wing nut(s) that holds the top plate to the air cleaner body, release the clips and lift it off.

4   Lift the air filter out of the housing. If it's covered with dirt, it should be replaced.

5   Wipe the inside of the air cleaner housing with a rag.

6   Place the old filter (if in good condition) or the new filter (if replacement is necessary) into the air cleaner housing.

7   Reinstall the top plate on the air cleaner and tighten the wing nut(s), then snap the clips into place.

29.8 Detach the air cleaner housing cover clips (arrows) (fuel injected models)

29.9 With the hoses still attached, pull the cover up, then lift the element out of the housing

Fuel injected models

Refer to illustrations 29.8 and 29.9

8   Detach the three filter cover retaining clips **(see illustration)**.

9   Lift the cover up for access to the filter element **(see illustration)**.

10   Lower the new filter into the housing, seat the cover and snap the retaining clips into place.

---

30   Fuel filter replacement

Warning: *Gasoline is extremely flammable, so take extra precautions when you work on any part of the fuel system. Don't smoke or allow open flames or bare light bulbs near the work area, and don't work in a garage where a natural gas-type appliance (such as a water heater or clothes dryer) with a pilot light is present. If you spill any fuel on your skin, rinse it off immediately with soap and water. When you perform any kind of work on the fuel tank, wear safety glasses and have a Class B type fire extinguisher on hand.*

1   This job should be done with the engine cold (after sitting at least three hours). Place a metal container, rags or newspapers under the filter to catch spilled fuel.

Carbureted models

2   Raise the vehicle and support it securely on jackstands. The fuel filter is located under the vehicle on the right side frame rail, adjacent to the fuel tank. **Warning:** *Before attempting to remove the fuel filter, disconnect the negative cable from the battery and position it out of the way so it can't accidentally contact the battery post.*

3   To replace the filter, loosen the clamps and slide them down the hoses, past the fittings on the filter.

4   Carefully twist and pull on the hoses to separate them from the filter. If the hoses are in bad shape, now would be a good time to replace them with new ones.

5   Connect the filter to the hoses and

30.7 Detach the hoses (arrows), then unbolt the fuel filter from the bracket (fuel injected models)

tighten the clamps securely. If spring-type clamps were originally installed, it would be a good idea to replace them with screw-type clamps. Make sure that the arrow on the filter housing faces toward the engine. Start the engine and check carefully for leaks at the filter hose connections.

Fuel injected models

Refer to illustration 30.7

6   Depressurize the fuel system (Chapter 4).

7   To replace the filter, loosen the clamps and slide them down the hoses, past the fittings on the filter **(see illustration)**.

8   Note how the filter is installed (the rounded end facing the fuel pump) so the new filter doesn't get installed backwards. Carefully twist and pull on the hoses to separate them from the filter. If the hoses are in bad shape, now would be a good time to replace them with new ones.

9   Remove the filter from the bracket.

10   Install the new filter in the bracket. Make sure the filter is properly oriented - the rounded end should face the fuel pump.

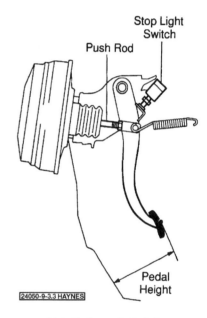

31.1 Brake pedal height adjustment details

11   Connect the hoses to the new filter and install the clamps - if spring-type clamps were originally installed, it would be a good idea to replace them with screw-type clamps.

12   Start the engine and check carefully for leaks at the filter hose connections.

---

31   Brake pedal height and freeplay check and adjustment

Refer to illustration 31.1

1   Brake pedal height is the distance the pedal sits away from the toe pan **(see illustration)**. The distance should be as specified (see the Specifications). If the pedal height is not within the specified range, loosen the locknut on the stop lamp switch located in the bracket to the rear of the brake pedal and turn the switch in or out until the pedal height

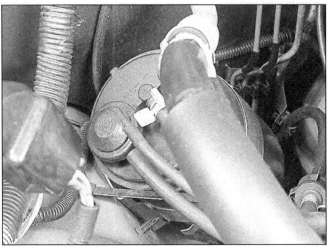

33.2 The charcoal canister is attached to the firewall on the passenger side of the engine compartment - check the canister housing and hoses for damage

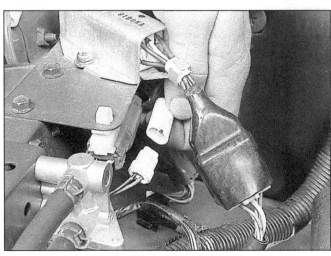

35.7 Unplug the pressure regulator VSV electrical connector before checking the idle speed

is correct. Retighten the locknut.

2   Brake pedal freeplay is the distance the pedal can be depressed before it begins to have any effect on the brakes. Measure the freeplay with the engine off after stepping on the brake pedal five times. The freeplay should be as specified. If it isn't, loosen the locknut on the brake booster push rod, to which the brake pedal is attached. Turn the rod until the freeplay is correct, then retighten the locknut.

32   Distributor advance mechanism check and lubrication

1   Remove the distributor cap and rotor.
2   Use a small screwdriver to check that the distributor advance mechanism moves smoothly and easily and that the springs aren't damaged or stretched.
3   Lubricate the mechanism with a couple of drops of light oil and install the rotor and cap.

33   Evaporative emissions system check

Refer to illustration 33.2
1   The function of the evaporative emissions system is to draw fuel vapors from the fuel tank, store them in a charcoal canister and burn them during normal engine operation.
2   The most common symptom of a fault in the evaporative emissions system is a strong fuel odor in the engine compartment. If a fuel odor is detected, inspect the charcoal canister, located in the engine compartment. Check the canister and all hoses for damage and deterioration **(see illustration)**.
3   The canister is held to the firewall by a spring clip, secured around the outside of the canister body. The canister is removed by marking and disconnecting the hoses, disengaging the clamp and lifting the canister out.

4   The evaporative emissions control system is explained in more detail in Chapter 6.

34   Carburetor choke check

1   The choke operates only when the engine is cold, so this check should be performed before the engine has been started for the day.
2   Open the hood and remove the top plate of the air cleaner assembly. It's held in place by one or two wing nuts at the center and several spring clips around the edge. If any vacuum hoses must be disconnected, tag them to ensure reinstallation in their original positions.
3   Look at the center of the air cleaner housing. You'll notice a flat plate at the carburetor opening.
4   Have an assistant press the throttle pedal to the floor. The plate should close completely. Start the engine while you watch the plate at the carburetor. Don't position your face near the carburetor, as the engine could backfire, causing serious burns! When the engine starts, the choke plate should open slightly.
5   Allow the engine to continue running at an idle speed. As the engine warms up to operating temperature, the plate should slowly open, allowing more air to enter through the top of the carburetor.
6   After a few minutes, the choke plate should be completely open to the vertical position. Blip the throttle to make sure the fast idle cam disengages.
7   You'll notice that engine speed corresponds to the plate opening. With the plate closed, the engine should run at a fast idle speed. As the plate opens and the throttle is moved to disengage the fast idle cam, the engine speed will decrease.
8   With the engine off and the throttle held half-way open, open and close the choke several times. Check the linkage to see if it's

hooked up correctly and make sure it doesn't bind.
9   If the choke or linkage binds, sticks or works sluggishly, clean it with choke cleaner (an aerosol spray available at auto parts stores). If the condition persists after cleaning, replace the troublesome parts.
10   Visually inspect all vacuum hoses to be sure they're securely connected and look for cracks and deterioration. Replace as necessary.
11   If the choke fails to operate normally, but no mechanical causes can be found, check the choke electrical circuits.

35   Idle speed check and adjustment (four-cylinder engines only)

Refer to illustrations 35.7, 35.9a and 35.9b
1   Engine idle speed is the speed at which the engine operates when no throttle pedal pressure is applied. The idle speed is critical to the performance of the engine itself, as well as many engine sub-systems.
2   A hand-held tachometer must be used when adjusting idle speed to get an accurate reading. The exact hook-up for these meters varies with the manufacturer, so follow the particular directions included.
3   Set the parking brake and block the wheels. Be sure the transmission is in Neutral (manual transmission) or Park (automatic transmission).
4   Turn off the air conditioner (if equipped), the headlights and any other accessories during this procedure.
5   On carbureted models, the distributor vacuum, evaporative canister purge and EGR vacuum hoses must be disconnected and plugged. The idle compensator vacuum line must be closed by bending the rubber hose.
6   Start the engine and allow it to reach normal operating temperature.
7   On fuel injected models, make sure the throttle valve is fully closed and unplug the

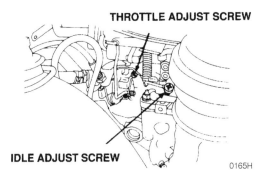

THROTTLE ADJUST SCREW

IDLE ADJUST SCREW

0165H

35.9a On carburetor equipped models, adjust the idle speed by turning the throttle adjust screw

35.9b On fuel injected models, turn the idle adjusting screw located on the throttle body

connector from the pressure regulator Vacuum Switching Valve (VSV) **(see illustration)**.

8 Check the engine idle speed with the tachometer and compare it to the VECI label on the vehicle.

9 If the idle speed is not correct, turn the throttle adjust screw (carbureted models) or the adjusting screw located on the throttle body (fuel injected models) until the idle speed is correct **(see illustrations)**.

10 On air conditioned carbureted models, turn the air conditioning On to Max cold and High blower, then open the throttle about 1/3 to allow the speed up solenoid to reach full travel. Turn the adjusting screw on the Speed Up controller to adjust the idle speed.

36 Transfer case lubricant change (1988 and later 4WD models only)

Refer to illustration 36.5
Note: *On 1981 through 1987 models, the transfer case is integral with the transmission - when you change the transmission lubricant (see the next Section), you're also changing the transfer case lubricant.*

1 Drive the vehicle for at least 15 minutes in 4WD to warm up the oil in the case.

2 Raise the vehicle and support it securely on jackstands.

3 Move a drain pan, rags, newspapers and tools under the vehicle.

4 Remove the check/fill plug (see Section 21).

5 Remove the drain plug from the lower part of the case and allow the old lubricant to drain completely **(see illustration)**.

6 Carefully clean and install the drain plug after the case is completely drained. Tighten the plug to the specified torque.

7 Fill the case with the specified lubricant until it's level with the lower edge of the filler hole.

8 Install the check/fill plug and tighten it securely.

9 Check carefully for leaks around the drain plug after the first few miles of driving.

37 Manual transmission lubricant change

Refer to illustration 37.3
Note: *On 1981 through 1987 models, the transfer case is integral with the transmission - when you change the transmission lubricant, you're also changing the transfer case lubricant.*

1 Drive the vehicle for a few miles to thoroughly warm up the transmission lubricant.

2 Raise the vehicle and support it securely on jackstands.

3 Move a drain pan, rags, newspapers and tools under the vehicle. With the drain pan and newspapers in position under the transmission, loosen the drain plug located in the bottom of the transmission case **(see illustration)**.

4 Once loosened, carefully unscrew it with your fingers until you can remove it from the transmission. Allow all of the oil to drain into the pan. If the plug is too hot to touch, use the wrench to remove it.

5 Clean the drain plug, then reinstall it in the transmission and tighten it to the specified torque.

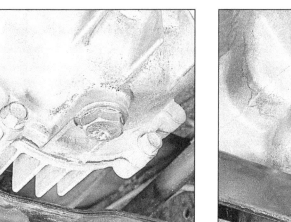

36.5 The transfer case drain plug is located at the bottom of the housing

6 Remove the transmission lubricant check/fill plug (see Section 20). Using a hand pump or syringe, fill the transmission with the correct amount and grade of lubricant, until the level is just at the bottom of the plug hole.

7 Reinstall the check/fill plug and tighten it securely.

38 Differential lubricant change

Note: *The following procedure can be used for the rear differential as well as the front differential used on 4WD vehicles.*

1 Drive the vehicle for several miles to warm up the differential lubricant, then raise the vehicle and support it securely on jackstands.

2 Some differentials can be drained by removing the drain plug, while on others a hand suction pump is required to remove the differential lubricant through the filler hole.

Drain plug equipped models

3 Move a drain pan, rags, newspapers and a 1/2-inch drive breaker bar or ratchet with an extension under the vehicle.

4 With the drain pan under the differential,

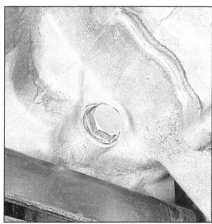

37.3 Manual transmission drain plug (4WD model shown)

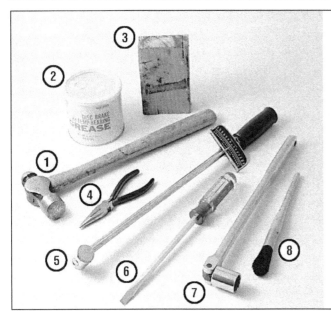

40.1 Tools and materials needed for front wheel bearing maintenance

1   **Hammer** - *A common hammer will do just fine*
2   **Grease** - *High-temperature grease which is formulated specially for front wheel bearings should be used*
3   **Wood block** - *If you have a scrap piece of 2x4, it can be used to drive the new seal into the hub*
4   **Needle-nose pliers** - *Used to straighten and remove the cotter pin in the spindle*
5   **Torque wrench** - *This is very important in this procedure; if the bearing is too tight, the wheel won't turn freely - if it's too loose, the wheel will "wobble" on the spindle. Either way, it could mean extensive damage*
6   **Screwdriver** - *Used to remove the seal from the hub (a long screwdriver would be preferred)*
7   **Socket/breaker bar** - *Needed to loosen the nut on the spindle if it's extremely tight*
8   **Brush** - *Together with some clean solvent, this will be used to remove old grease from the hub and spindle*

use the breaker bar or ratchet and extension to loosen the drain plug. It's the lower of the two plugs.
5   Once loosened, carefully unscrew it with your fingers until you can remove it from the case.
6   Allow all of the lubricant to drain into the pan, then replace the drain plug and tighten it to the specified torque.
7   Feel with your hands along the bottom of the drain pan for any metal bits that may have come out with the lubricant. If there are any, it's a sign of excessive wear, indicating that the internal components should be carefully inspected in the near future.

Models not equipped with a drain plug

8   Move a drain pan, rags, newspapers, a suction pump and a wrench (for the check/fill plug) under the vehicle.
9   Remove the check/fill plug (see Section 22) and insert the suction line from the suction pump into the plug hole. Place the other line in the drain pan.
10   Operate the suction pump until all lubricant is pumped into the pan, then remove the pump.

All models

11   Using a hand pump, syringe or funnel, fill the differential with the correct amount and grade of lubricant until the level is just at the bottom of the fill plug hole.
12   Reinstall the plug and tighten it securely.
13   Lower the vehicle. Check for leaks after the first few miles of driving.

---

39   Automatic transmission fluid and filter change

1   At the specified time intervals, the transmission fluid should be drained and replaced.

Since the fluid should be hot when it's drained, drive the vehicle for 15 or 20 minutes before proceeding.
2   Before beginning work, purchase the specified transmission fluid and filter. Some early models use a fluid strainer instead of a filter, which does not require replacement, only cleaning, so it is important before starting the job to determine whether your model requires a replacement filter.
3   Other tools necessary for this job include jackstands to support the vehicle in a raised position, a drain pan capable of holding at least eight pints, newspapers and clean rags.
4   Raise the vehicle and support it securely on jackstands.
5   Move the drain pan and necessary tools under the vehicle, being careful not to touch any of the hot exhaust components.
6   Place the pan under the drain plug in the transmission pan and remove the plug. Be sure the drain pan is in position, as fluid will come out with some force. Once the fluid is drained, reinstall the drain plug securely.
7   Remove the transmission pan bolts, detach the dipstick tube clip, then carefully pry the pan loose with a screwdriver and remove it.
8   Remove the filter/strainer retaining bolts and lower the filter/strainer from the transmission. Be careful when lowering the filter as it contains residual fluid.
9   If your model is equipped with a fluid strainer, wash it thoroughly in clean transmission fluid. Place the clean strainer or the new filter in position and install the bolts. Tighten the bolts securely.
10   Carefully clean the gasket surface of the transmission to remove all traces of the old gasket and sealant.
11   Drain any remaining fluid from the transmission pan, clean it with solvent and dry it with compressed air. Clean the pan magnet and install it in the pan so it will be directly

below the filter.
12   Apply a thin layer of RTV sealant to the transmission case side of the new gasket.
13   Make sure the gasket surface on the transmission pan is clean, then apply a thin layer of RTV sealant to it and position the new gasket on the pan. Put the pan in place against the transmission, install the bolts and, working around the pan, tighten each bolt a little at a time until the final torque figure is reached.
14   Lower the vehicle and add new automatic transmission fluid through the filler tube (Section 5). The amount should be equal to the amount of fluid that was drained (you don't want to overfill it).
15   With the transmission in Park and the parking brake set, run the engine at a fast idle, but don't race it.
16   Move the gear selector through each range and back to Park, then check the fluid level (Section 5). Add more fluid as required.
17   Check under the vehicle for leaks during the first few miles of driving.

---

40   Front wheel bearing check, repack and adjustment (2WD models only)

Refer to illustrations 40.1 and 40.6
Note: *For 4WD vehicles, see Chapter 8.*
1   In most cases the front wheel bearings will not need servicing until the brake pads are changed. However, the bearings should be checked whenever the front of the vehicle is raised for any reason. Several items, including a torque wrench and special grease, are required for this procedure **(see illustration)**.
2   With the vehicle securely supported on jackstands, spin each wheel and check for noise, rolling resistance and freeplay.
3   Grasp the top of each tire with one hand and the bottom with the other. Move the

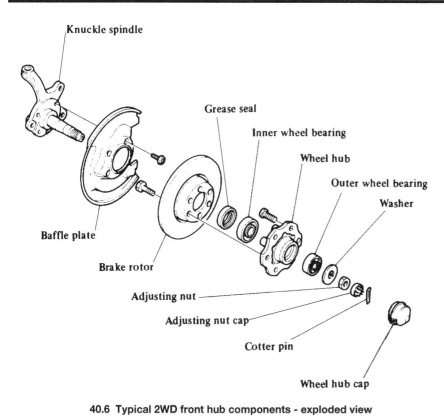

Knuckle spindle

Grease seal

Inner wheel bearing

Wheel hub

Outer wheel bearing

Washer

Baffle plate

Brake rotor

Adjusting nut

Adjusting nut cap

Cotter pin

Wheel hub cap

40.6 Typical 2WD front hub components - exploded view

wheel in-and-out on the spindle. If there's any noticeable movement, the bearings should be checked and then repacked with grease or replaced if necessary.

4 Remove the wheel(s).

5 Remove the brake caliper (Chapter 9) and hang it out of the way on a piece of wire.

6 Pry the hub cap out of the hub using a screwdriver or hammer and chisel **(see illustration)**.

7 Straighten the bent ends of the cotter pin, then pull the cotter pin out of the adjusting nut cap. Discard the cotter pin and use a new one during reassembly.

8 Remove the adjusting nut and washer from the end of the spindle.

9 Pull the hub out slightly, then push it back into its original position. This should force the outer wheel bearing off the spindle enough so it can be removed.

10 Pull the hub off the spindle.

11 Use a screwdriver to pry the grease seal out of the rear of the hub **(see illustration)**. As this is done, note how the seal is installed.

12 Remove the inner wheel bearing from the hub.

13 Use solvent to remove all traces of the old grease from the bearings, hub and spindle. A small brush may prove helpful; however make sure no bristles from the brush embed themselves inside the bearing rollers. Allow the parts to air dry.

14 Carefully inspect the bearings for cracks, heat discoloration, worn rollers, etc. Check the bearing races inside the hub for wear and damage. If the bearing races are

defective, the hubs should be taken to a machine shop with the facilities to remove the old races and press new ones in. Note that the bearings and races come as matched sets and old bearings should never be installed on new races.

15 Use high-temperature front wheel bearing grease to pack the bearings. Work the grease completely into the bearings, forcing it between the rollers, cone and cage from the back side.

16 Apply a thin coat of grease to the spindle at the outer bearing seat, inner bearing seat, shoulder and seal seat.

17 Put a small quantity of grease behind each bearing race inside the hub. Using your finger, form a dam at these points to provide for extra grease and to keep thinned grease from flowing out of the bearing.

18 Place the grease-packed inner bearing into the rear of the hub and put a little more grease outside of the bearing. Place 1/2 oz of grease inside the end of the hub cap.

19 Place a new seal over the inner bearing and tap the seal evenly into place with a hammer and block of wood until it's flush with the hub.

20 Carefully place the hub assembly onto the spindle and push the grease-packed outer bearing into position.

21 Install the washer and adjusting nut. Tighten the nut to about 22 ft-lbs.

22 Spin the hub in a forward direction to seat the bearings and remove any grease or burrs which could cause excessive bearing play later.

23 Attach a spring scale to one of the wheel studs at right angles to the hub and measure the amount of pull necessary to rotate the hub. This is the preload. Compare the preload reading to the Specifications at the front of this Chapter.

24 Tighten or loosen the adjusting nut to achieve the specified preload.

25 Install the adjusting nut cap and see if the hole in the spindle is aligned with the slot in the cap. If it's not, remove the nut cap, rotate it slightly and reinstall it until the slot lines up. Install a new cotter pin.

26 Bend the ends of the cotter pin until they're flat against the nut. Cut off any extra length which could interfere with the dust cap.

27 Install the hub cap, tapping it into place with a hammer.

28 Install the caliper (see Chapter 9).

29 Install the tire/wheel assembly on the hub and tighten the lug nuts.

30 Grasp the top and bottom of the tire and check the bearings in the manner described earlier in this Section.

31 Lower the vehicle.

41 Cooling system servicing (draining, flushing and refilling)

Warning: *Antifreeze is a corrosive and poisonous solution, so be careful not to spill any of the coolant mixture on the vehicle's paint or your skin. If you do, rinse it off immediately with plenty of clean water. Consult local authorities regarding proper disposal of antifreeze before draining the cooling system. In many areas, reclamation centers have been established to collect used oil and coolant mixtures.*

1 Periodically, the cooling system should be drained, flushed and refilled to replenish the antifreeze mixture and prevent formation of rust and corrosion, which can impair the performance of the cooling system and cause engine damage. When the cooling system is serviced, all hoses and the radiator cap should be checked and replaced if necessary.

2 Apply the parking brake and block the wheels. If the vehicle has just been driven, wait several hours to allow the engine to cool down before beginning this procedure.

3 Once the engine is completely cool, remove the radiator cap. Place the heater temperature control in the maximum heat position.

4 Move a large container under the radiator drain to catch the coolant, then unscrew the drain plug (a pair of pliers may be required to turn it).

5 After the coolant stops flowing out of the radiator, move the container under the engine block drain plug(s) (on four-cylinder engines there is one plug on the side of the block; on V6 engines there are two plugs, located on both sides of the block). Remove the plug(s) and allow the coolant in the block to drain.

6   While the coolant is draining, check the condition of the radiator hoses, heater hoses and clamps (refer to Section 9 if necessary).

7   Replace any damaged clamps or hoses.

8   Once the system is completely drained, flush the radiator with fresh water from a garden hose until it runs clear at the drain. The flushing action of the water will remove sediments from the radiator but will not remove rust and scale from the engine and cooling tube surfaces.

9   These deposits can be removed with a chemical cleaner. Follow the procedure outlined in the manufacturer's instructions. If the radiator is severely corroded, damaged or leaking, it should be removed (Chapter 3) and taken to a radiator repair shop.

10   Remove the overflow hose from the coolant recovery reservoir. Drain the reservoir and flush it with clean water, then reconnect the hose.

11   Reinstall and tighten the radiator drain plug. Install and tighten the block drain plug(s).

12   Slowly add new coolant (a 50/50 mixture of water and antifreeze) to the radiator until it's full. Add coolant to the reservoir up to the lower mark.

13   Leave the radiator cap off and run the engine in a well-ventilated area until the thermostat opens (coolant will begin flowing through the radiator and the upper radiator hose will become hot).

14   Turn the engine off and let it cool. Add more coolant mixture to bring the level back up to the lip on the radiator filler neck.

15   Squeeze the upper radiator hose to expel air, then add more coolant mixture if necessary. Replace the radiator cap.

16   Start the engine, allow it to reach normal operating temperature and check for leaks.

42 PCV system check

Refer to illustration 42.6

1   The Positive Crankcase Ventilation (PCV) system directs blowby gases from the crankcase back into the intake manifold so they can be burned in the engine.

2   Rough idling or high idle speed and

42.6 On some later models, the PCV valve screws into the rocker arm cover, so a wrench is needed for removal

stalling are symptoms of faults in the PCV system.

Four cylinder engine

3   The system on these models consists of a hose leading from the rocker arm cover to the intake manifold and a fresh air hose between the air cleaner assembly and the rocker arm cover. Gases from the crankcase are carried by the hose through an orifice (early models) or PCV valve (later models) into the intake manifold.

4   Check the system hoses for cracks, leaks and clogging. Clean the hoses if they are clogged and replace any which are damaged.

5   On early models, pull the orifice out of the end of the hose and inspect it for clogging, cleaning as necessary.

6   On later models using a PCV valve, remove the valve **(see illustration)**. Blow through the valve from the rocker arm cover side to make sure that air passes through easily. Blow through the valve from the manifold side to verify that air will not pass through easily. If it does, replace the valve with a new one.

7   More information on the PCV system can be found in Chapter 6.

V6 engine

8   The system on these models works similarly to the one used on four-cylinder engines, except a PCV valve is used on all models (orifices are not used).

9   Check the hoses for damage and clogging.

10   With the engine idling, pull the PCV valve out of the rocker arm cover and place your thumb over the end to check for vacuum. If there is no vacuum, the valve, hoses or manifold port are clogged.

11   Turn the engine off, remove the PCV valve, shake it back and forth and listen for a rattle. If it does not rattle, replace the valve with a new one.

12   More information on the PCV system can be found in Chapter 6.

43 Spark plug replacement

Refer to illustrations 43.2, 43.5a, 43.5b, 43.6 and 43.10

1   Replace the spark plugs with new ones at the intervals recommended in the *Maintenance schedule*.

2   In most cases, the tools necessary for spark plug replacement include a spark plug socket which fits onto a ratchet (spark plug sockets are padded inside to prevent damage to the porcelain insulators on the new plugs), various extensions and a gap gauge to check and adjust the gaps on the new plugs **(see illustration)**. A special plug wire removal tool is available for separating the wire boots from the spark plugs, but it isn't absolutely necessary. A torque wrench should be used to tighten the new plugs.

3   The best approach when replacing the spark plugs is to purchase the new ones in advance, adjust them to the proper gap and replace them one at a time. When buying the new spark plugs, be sure to obtain the correct plug type for your particular engine. This information can be found on the Emission Control Information label located under the hood and in the factory owner's manual. If differences exist between the plug specified on the emissions label and in the owner's

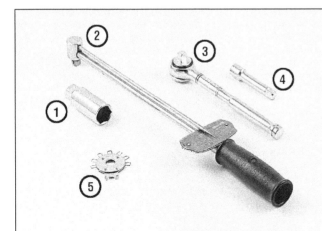

43.2 Tools required for changing spark plugs

1   *Spark plug socket - This will have special padding inside to protect the spark plug's porcelain insulator*

2   *Torque wrench - Although not mandatory, using this tool is the best way to ensure the plugs are tightened properly*

3   *Ratchet - Standard hand tool to fit the spark plug socket*

4   *Extension - Depending on model and accessories, you may need special extensions and universal joints to reach one or more of the plugs*

5   *Spark plug gap gauge - This gauge for checking the gap comes in a variety of styles. Make sure the gap for your engine is included*

43.5a Spark plug manufacturers recommend using a wire-type gauge when checking the gap - if the wire does not slide between the electrodes with a slight drag, adjustment is required

43.5b To change the gap, bend the side electrode only, as indicated by the arrows, and be very careful not to crack or chip the porcelain insulator surrounding the center electrode

manual, assume the emissions label is correct.

4   Allow the engine to cool completely before attempting to remove any of the plugs. While you're waiting for the engine to cool, check the new plugs for defects and adjust the gaps.

5   The gap is checked by inserting the proper thickness gauge between the electrodes at the tip of the plug **(see illustration)**. The gap between the electrodes should be the same as the one specified on the Emissions Control Information label. The wire should just slide between the electrodes with a slight amount of drag. If the gap is incorrect, use the adjuster on the gauge body to bend the curved side electrode slightly until the proper gap is obtained **(see illustration)**. If the side electrode is not exactly over the center electrode, bend it with the adjuster until it is. Check for cracks in the porcelain insulator (if any are found, the plug shouldn't be used).

6   With the engine cool, remove the spark plug wire from one spark plug. Pull only on the boot at the end of the wire - don't pull on the wire. A plug wire removal tool should be used if available **(see illustration)**.

7   If compressed air is available, use it to blow any dirt or foreign material away from the spark plug hole. A common bicycle pump will also work. The idea here is to eliminate the possibility of debris falling into the cylinder as the spark plug is removed.

8   Place the spark plug socket over the plug and remove it from the engine by turning it in a counterclockwise direction.

9   Compare the spark plug to those shown in the photos on the inside back cover to get an indication of the general running condition of the engine.

10   Thread one of the new plugs into the hole until you can no longer turn it with your fingers, then tighten it with a torque wrench (if available) or the ratchet. It might be a good idea to slip a short length of rubber hose over the end of the plug to use as a tool to thread it into place **(see illustration)**. The hose will grip the plug well enough to turn it, but will start to slip if the plug begins to cross-thread in the hole - this will prevent damaged threads and the accompanying repair costs.

11   Before pushing the spark plug wire onto the end of the plug, inspect it following the procedures outlined in Section 44.

12   Attach the plug wire to the new spark plug, again using a twisting motion on the boot until it's seated on the spark plug.

13   Repeat the procedure for the remaining spark plugs, replacing them one at a time to prevent mixing up the spark plug wires.

44 Spark plug wire, distributor cap and rotor check and replacement

Refer to illustrations 44.11 and 44.12

1   The spark plug wires should be checked whenever new spark plugs are installed.

2   Begin this procedure by making a visual check of the spark plug wires while the engine is running. In a darkened garage (make sure there is ventilation) start the engine and observe each plug wire. Be careful not to come into contact with any moving engine parts. If there is a break in the wire, you will see arcing or a small spark at the damaged area. If arcing is noticed, make a note to obtain new wires, then allow the engine to cool and check the distributor cap and rotor.

3   The spark plug wires should be inspected one at a time to prevent mixing up the order, which is essential for proper engine operation. Each original plug wire should be numbered to help identify its location. If the

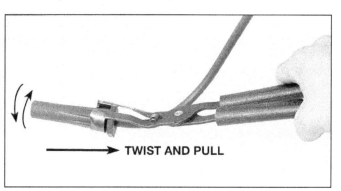

43.6 When removing the spark plug wires, pull only on the boot and twist it back and forth

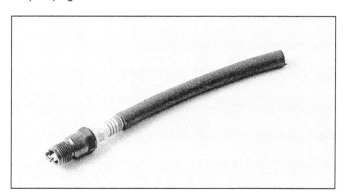

43.10 A length of 3/16-inch ID rubber hose will save time and prevent damaged threads when installing the spark plugs

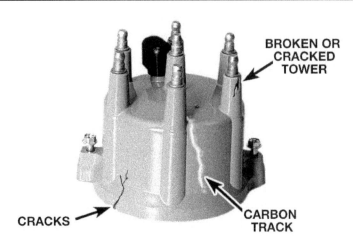

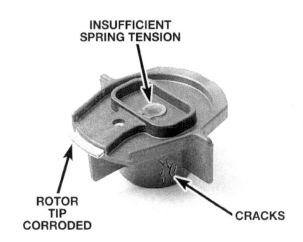

44.12 The ignition rotor should be checked for wear and corrosion as indicated here (if in doubt about its condition, buy a new one)

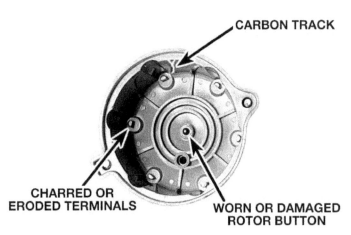

44.11 Shown here are some of the common defects to look for when inspecting the distributor cap (if in doubt about its condition, install a new one)

number is illegible, a piece of tape can be marked with the correct number and wrapped around the plug wire.

4   Disconnect the plug wire from the spark plug. A removal tool can be used for this purpose or you can grasp the rubber boot, twist the boot half a turn and pull the boot free. Do not pull on the wire itself **(see illustration 43.6)**.

5   Check inside the boot for corrosion, which will look like a white crusty powder.

6   Push the wire and boot back onto the end of the spark plug. It should fit tightly onto the end of the plug. If it doesn't, remove the wire and use pliers to carefully crimp the metal connector inside the wire boot until the fit is snug.

7   Using a clean rag, wipe the entire length of the wire to remove built-up dirt and grease. Once the wire is clean, check for burns, cracks and other damage. Do not bend the wire sharply, because the conductor within the wire might break.

8   Disconnect the wire from the distributor. Again, pull only on the rubber boot. Check for corrosion and a tight fit. Press the wire back into the distributor.

9   Inspect the remaining spark plug wires, making sure that each one is securely fastened at the distributor and spark plug when the check is complete.

10   If new spark plug wires are required, purchase a set for your specific engine model. Pre-cut wire sets with the boots already installed are available. Remove and replace the wires one at a time to avoid mixups in the firing order.

11   Detach the distributor cap by prying off the two cap retaining clips. Look inside it for cracks, carbon tracks and worn, burned or loose contacts **(see illustration)**.

12   Pull the rotor off the distributor shaft and examine it for cracks and carbon tracks **(see illustration)**. On some models it will be necessary to remove a retaining screw before the rotor can be pulled off. Replace the cap and rotor if any damage or defects are noted.

13   It is common practice to install a new cap and rotor whenever new spark plug wires are installed, but if you wish to continue using the old cap, clean the terminals first.

14   When installing a new cap, remove the wires from the old cap one at a time and attach them to the new cap in the exact same

location - do not simultaneously remove all the wires from the old cap or firing order mixups may occur.

45   Ignition timing check and adjustment

Refer to illustrations 45.1, 45.2a, 45.2b and 45.9

Note: *If the information in this Section differs from the Vehicle Emission Control Information label in the engine compartment, the information on the label should be considered correct.*

1   The proper ignition timing setting for your vehicle is printed on the VECI label located in the engine compartment. Some special tools will be required for this procedure **(see illustration)**.

2   Locate the timing plate on front of the engine, near the crankshaft pulley **(see illustrations)**. The 0 mark is Top Dead Center (TDC). To locate which mark the notch in the pulley must line up with for the timing to be correct, count back from the 0 mark the number of degrees BTDC (Before Top Dead Center) noted on the VECI label.

3   Locate the timing notch in the pulley and mark it with a dab of paint or chalk so it'll be visible under the strobe light. To locate the notch it may be necessary to have an assistant temporarily turn the ignition key to Start in short bursts to turn the crankshaft. **Warning:** *Stay clear of all moving engine components if the engine is turned in this manner!*

4   Connect a tachometer according to the manufacturer's instructions and make sure the idle speed is correct (four-cylinder engines only). Adjust it, if necessary, as described in Section 35.

5   Allow the engine to reach normal operating temperature. Be sure the air conditioner, if equipped, is off. On V6 engines, make sure the CHECK ENGINE light is not on

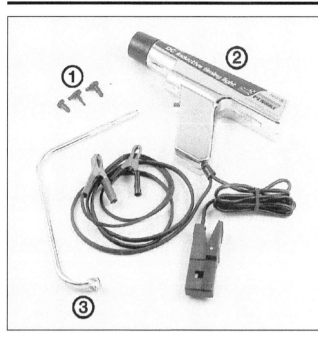

45.1 Tools needed to check and adjust the ignition timing

1   **Vacuum plugs -** *Vacuum hoses will, in most cases, have to be disconnected and plugged. Molded plugs in various shapes and sizes are available for this.*
2   **Inductive pick-up timing light -** *flashes a bright concentrated beam of light when the number one spark plug fires. Connect the leads according to the instructions supplied with the light.*
3   **Distributor wrench -** *On some models, the hold-down bolt for the distributor is difficult to reach and turn with conventional wrenches or sockets. A special wrench like this must be used.*

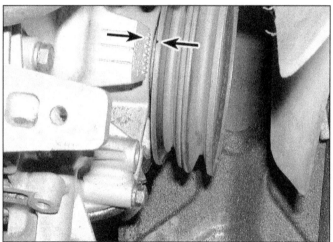

45.2a Typical four-cylinder engine timing marks

45.2b On V6 engines, the timing marks are located on a plate mounted on the engine block

(see Chapter 6) and bypass the Electronic Spark Timing (EST) system by unplugging the EST connector located under the center console in the passenger compartment.

6   With the ignition switch off, connect the pick-up lead of the timing light to the number one spark plug wire. On four-cylinder engines, it's the front one. On V6 engines it's the first spark plug on the right side as viewed from the driver's seat. Use either a jumper lead between the wire and plug or an inductive-type pick up. Don't pierce the wire or attempt to insert a wire between the boot and plug wire. Connect the timing light power leads according to the manufacturer's instructions.

7   Make sure the wiring for the timing light is clear of all moving engine components, then start the engine. Race the engine two or three times, then allow it to idle for a minute.

8   Point the flashing timing light at the tim-

ing marks, again being careful not to come in contact with moving parts. The marks you highlighted should appear stationary. If the marks are in alignment, the timing is correct. If the marks aren't aligned, turn off the engine.

9   Loosen the distributor mounting bolt or nut until the distributor can be rotated **(see illustration)**.

10 Start the engine and slowly rotate the distributor either left or right until the timing marks are aligned.

11 Shut off the engine and tighten the mounting bolt/nut, being careful not to move the distributor.

12 Restart the engine and recheck the timing to make sure the marks are still in alignment.

13 Disconnect the timing light. On V6 engines, reconnect the EST wire harness connector, then clear any trouble codes set

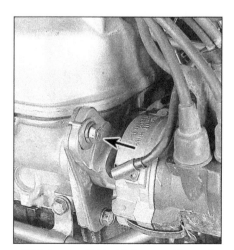

45.9 Loosen the hold-down nut (arrow) and turn the distributor to adjust the ignition timing

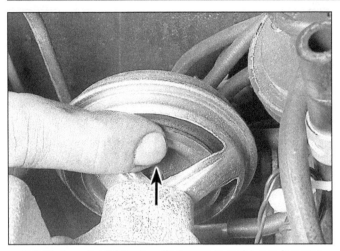

46.2 Push on the EGR valve diaphragm to make sure it moves easily with no binding

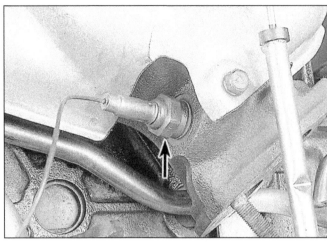

47.2 The oxygen sensor (arrow) threads into the exhaust manifold

during the ignition timing procedure (Chapter 6).

14   Race the engine two or three times, then allow it to run at idle. Recheck the idle speed with the tachometer. If it has changed from the correct setting, readjust it (four-cylinder engines only).

15   Drive the vehicle and listen for "pinging" noises. They'll be noticeable when the engine is hot and under load (climbing a hill, accelerating from a stop). If you hear engine pinging, the ignition timing is too far advanced (Before Top Dead Center). Reconnect the timing light and turn the distributor to move the mark 1 or 2-degrees in the retard direction (counterclockwise). Road test the vehicle again to check for proper operation.

16   To keep "pinging" at a minimum, yet still allow you to operate the vehicle at the specified timing setting, use gasoline of the same octane at all times. Switching fuel brands and octane levels can decrease performance and economy, and possibly damage the engine.

46   Exhaust Gas Recirculation (EGR) system check

Refer to illustration 46.2

1   The EGR valve is located on the intake manifold. Most of the time, when a problem develops in the emissions system, it is due to a stuck or corroded EGR valve.

2   With the engine cold to prevent burns, push on the valve's diaphragm **(see illustration)**. Using moderate pressure, you should be able to move the diaphragm up-and-down within the housing.

3   If the diaphragm does not move or moves only with much effort, replace the EGR valve with a new one. If in doubt about the quality of the valve, compare the free movement of your EGR valve with a new valve.

4   Refer to Chapter 6 for more information on the EGR system.

47   Oxygen sensor replacement

Refer to illustrations 47.2, 47.3a and 47.3b

1   The oxygen (exhaust gas) sensor, used on later models, should be replaced at the specified intervals (every 90,000 miles).

2   The sensor is threaded into the exhaust manifold and can be identified by the wire attached to it **(see illustration)**. Replacement consists of disconnecting the wire harness and unthreading the sensor from the manifold. Tighten the new sensor to the specified torque, then reconnect the wire harness.

3   When the specified mileage has elapsed, a O2 indicator light on the dash will come on. To turn off the indicator light on the Trooper II, slide the reset switch **(see illustration)** on the rear of the instrument cluster. To reset Pick-up models, you will need to remove the instrument cluster. Remove the masking tape from hole "B" **(see illustration)**. Remove the screw from hole "A" and insert it into hole "B".

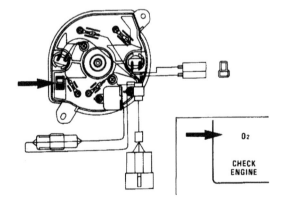

47.3a To turn off the O2 indicator light on an Isuzu Trooper, remove the instrument cluster, locate the reset switch (arrow) on the rear of the instrument cluster and slide the switch to the opposite position

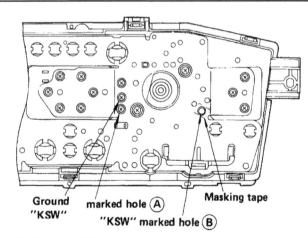

47.3b To turn off the O2 indicator light on an Isuzu Pick-up, remove the instrument cluster, remove the masking tape from hole "B," remove the screw from hole "A" and screw it into hole "B"

Chapter 2 Part A
Four-cylinder engines

Contents

Specifications

General (all engines)

Cylinder numbers (front-to-rear)	1-2-3-4
Firing order	1-3-4-2
Displacement	
G180Z	1800 cc (110.8 cu in)
G200Z	1950 cc (119.0 cu in)
4ZD1	2250 cc (138 cu in)
4ZE1	2559 cc (156.2 cu in)
Valve clearances	See Chapter 1

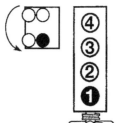

**4ZD1 and 4ZE1
four-cylinder engines**

**G180Z and G200Z
four-cylinder engines**

**Cylinder location and
distributor rotation**

*The blackened terminal shown on the
distributor cap indicates the number
one spark plug wire position*

G180Z and G200Z engines

Camshaft

Journal diameter	1.331 to 1.339 in (33.8 to 34 mm)
Bearing inner diameter	1.3394 to 1.3406 in (34.02 to 34.05 mm)
Bearing oil clearance	0.0016 to 0.0035 in (0.040 to 0.090 mm)
Runout limit	0.0038 in (0.10 mm)
Endplay	0.0020 to 0.0059 in (0.05 to 0.15 mm)
Lobe height	1.431 to 1.451 in (36.25 to 36.85 mm)
Valve spring free length	
Standard	1.893 in (48.1 mm)
Limit	1.787 in (45.4 mm)

Torque specifications

	Ft-lbs
Cylinder head bolts	
Step 1	61
Step 2	72
Cylinder head-to-front cover bolts	3 to 6
Front cover-to-block bolts	
6 mm	3 to 7
8 mm	7 to 12
Chain guide and tensioner bolts	4.3 to 7.2
Timing sprocket bolt	58
Crankshaft drivebelt pulley bolt	87
Manifold bolts/nuts	13 to 18
Rocker arm assembly nuts	16
Rocker arm cover bolts	3.6
Oil pump mounting bolts	13.7
Oil pan mounting bolts	3.6
Flywheel/driveplate-to-crankshaft bolts	72
Motor mount nut	21
Motor mount-to-engine	29

4ZD1 and 4ZE1 engines

Camshaft

Journal diameter	1.3397 to 1.3324 in (34.00 to 33.80 mm)
Endplay	0.002 to 0.0059 in (0.05 to 0.15 mm)
Cam lobe height	
4ZD1	1.4520 to 1.4320 in (36.85 to 36.35 mm)
4ZE1	1.4560 to 1.4320 in (37.00 to 36.35 mm)
Cam lobe wear limit	0.002 in (0.05 mm) or less
Camshaft runout	0.002 in (0.05 mm)
Bearing inside diameter	Not available
Bearing oil clearance	0.0026 to 0.0043 in (0.065 to 0.110 mm)
Valve spring free length	1.8951 to 1.8321 in (48.10 to 46.50 mm)

Torque specifications

	Ft-lbs
Camshaft sprocket bolt	39 to 47
Timing belt tensioner lock bolt	13.7
Crankshaft drivebelt pulley bolt	75 to 97
Cylinder head bolts	
Step 1	57.8
Step 2	65 to 79
Exhaust manifold bolts/nuts	13 to 17
Flywheel/driveplate-to-crankshaft bolts	40 to 46
Rocker arm cover bolts	7
Timing belt cover-to-block bolts	4.4
Intake manifold bolts/nut	12 to 15
Oil pan mounting bolts and nuts	10 to 15
Oil pump mounting bolts	13
Oil pump sprocket nut	55
Rocker arm assembly mounting bolts	14 to 17

1   General information

How to use this Chapter

This Part of Chapter 2 is devoted to in-vehicle repair procedures for the four-cylinder engines. All information concerning engine removal and installation and engine block and cylinder head overhaul can be found in Part C of this Chapter.

The following repair procedures are based on the assumption the engine is installed in the vehicle. If the engine has been removed from the vehicle and mounted on a stand, many of the steps outlined in this Part of Chapter 2 will not apply.

The Specifications included in this Part of Chapter 2 apply only to the procedures contained in this Part. Part C of Chapter 2 contains the Specifications necessary for cylinder head and engine block rebuilding.

Engine code designations

Several different four-cylinder engines were used in these vehicles:
1981 through 1983 (1800cc)..........G180Z
1984 and 1985 (1950cc)................G200Z
1986-on
  2300cc (actually 2250 cc)............4ZD1
  2600cc (actually 2559 cc)............4ZE1

Engine description

The cast iron engine block contains the four cylinder bores and acts as a rigid support for the five main bearing crankshaft. The machined cylinder bores are surrounded by water jackets to dissipate heat and control operating temperature.

The cylinder head is aluminum. Two valves per cylinder are mounted at a slight angle in the cylinder head and are actuated by a rocker arm in direct contact with the camshaft. Double springs are installed on each valve.

The camshaft is driven by a roller chain (G180Z and G200Z) or a belt (4ZD1 and 4ZE1) from the front of the crankshaft. Chain tension is controlled by a tensioner, which is operated by oil and spring pressure or a tension spring and pulley. The rubber shoe type tensioner controls vibration and tension of the chain.

The pistons are a special aluminum casting with struts to control thermal expansion. There are two compression rings and one oil control ring. The piston pin is a hollow steel shaft which is a loose fit in the piston and a press fit in the connecting rod. The pistons are attached to the crankshaft by forged steel connecting rods.

The distributor, which is mounted on the left-hand side of the engine block, is driven by a helical gear mounted on the front of the camshaft. Early engines had the distributor mounted on the right-hand side of the block toward the front. The oil pump, which is mounted low on the front side of the block, is driven by the timing belt.

The crankshaft, which is made of special forged steel, has internal oil passages to provide lubrication to the main and connecting rod bearings. Oil is delivered, via the filter and pressure relief valve, to the main oil gallery. From there it passes to the main bearing journals and then to the connecting rod bearing journals through holes in the crankshaft. Oil throw-off from the connecting rod lower ends, as well as a jet hole drilled through each connecting rod into the lower end, provides lubrication for the pistons and connecting rod upper ends. At the top of the engine, galleries drilled in the camshaft supports provide oil for the five bearings, while two lines that run along the length of the camshaft, deliver oil to provide lubrication for the rocker arm pivots.

2   Repair operations possible with the engine in the vehicle

Many major repair operations can be accomplished without removing the engine from the vehicle.

Clean the engine compartment and the exterior of the engine with some type of pressure washer before any work is done. It'll make the job easier and help keep dirt out of the internal areas of the engine.

Remove the hood to improve access to the engine as repairs are performed (refer to Chapter 11 if necessary).

If vacuum, exhaust, oil or coolant leaks develop, indicating a need for gasket or seal replacement, the repairs can generally be made with the engine in the vehicle. The oil pan gasket, cylinder head gasket, intake and exhaust manifold gaskets, engine front cover gaskets and the crankshaft oil seals are accessible with the engine in place.

Exterior engine components, such as the water pump, the starter motor, the alternator, the distributor and the fuel system components, as well as the intake and exhaust manifolds, can be removed for repair with the engine in place.

Since the cylinder head can be removed without pulling the engine, camshaft and valve component servicing can also be accomplished with the engine in the vehicle.

Replacement of, repairs to or inspection of the timing chain and sprockets and the oil pump are all possible with the engine in place.

In extreme cases caused by a lack of necessary equipment, repair or replacement of piston rings, pistons, connecting rods and rod bearings is possible with the engine in the vehicle. However, this practice is not recommended because of the cleaning and preparation work that must be done to the components involved.

3   Top Dead Center (TDC) for number one piston - locating

Refer to illustration 3.5

1   Top Dead Center (TDC) is the highest point in the cylinder each piston reaches as it travels up-and-down when the crankshaft turns. Each piston reaches TDC on the compression stroke and again on the exhaust stroke, but TDC generally refers to piston position on the compression stroke. The timing marks on the pulley installed on the front of the crankshaft are referenced to the number one piston at TDC on the compression stroke.

2   Positioning the piston(s) at TDC is an essential part of many procedures such as rocker arm removal, camshaft and timing chain/sprocket removal and distributor removal.

3   To bring any piston to TDC, the crankshaft must be turned using one of the methods outlined below. When looking at the front of the engine, normal crankshaft rotation is clockwise. **Warning:** *Before beginning this procedure, be sure to place the transmission in Neutral. Also, detach the coil wire from the center terminal of the distributor cap and ground in on the block with a jumper wire.*

a) *The preferred method is to turn the crankshaft with a large socket and breaker bar attached to the crankshaft pulley hub bolt threaded into the front of the crankshaft.*

b) *A remote starter switch, which may save some time, can also be used. Attach the switch leads to the solenoid terminals. Once the piston is close to TDC, use a socket and breaker bar as described in the previous Paragraph.*

c) *If an assistant is available to turn the ignition switch to the Start position in short bursts, you can get the piston close to TDC without a remote starter switch. Use a socket and breaker bar as described in Paragraph (a) to complete the procedure.*

4   Note the position of the terminal for the number one spark plug wire on the distributor cap. Use a felt-tip pen or chalk to make a mark on the distributor body directly under the terminal. Detach the cap from the distributor and set it aside.

5   Turn the crankshaft (see Step 3 above) until the notch in the crankshaft pulley is

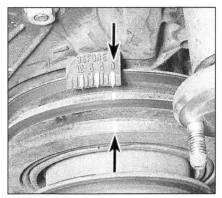

3.5 The number one piston is at TDC on the compression stroke when the notch in the crankshaft pulley is aligned with the 0 on the timing plate (arrows) and the rotor is pointing at the number one spark plug terminal in the distributor cap (G180Z/G200Z engine shown)

4.4 Rocker arm cover mounting bolts (4ZE1 engine)

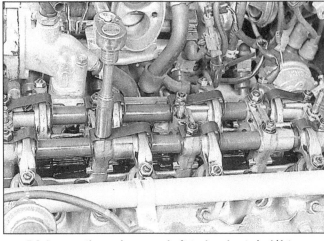

5.2 Loosen the rocker arm shaft-to-head nuts in 1/4-turn increments to avoid distorting the shafts

aligned with the 0 on the timing plate (located at the front of the engine) **(see illustration)**.

6   Look at the distributor rotor - it should be pointing directly at the mark you made on the distributor body. If the rotor is pointing at the terminal for the number four spark plug, the number one piston is at TDC on the exhaust stroke.

7   To get the piston to TDC on the compression stroke, turn the crankshaft one complete turn (360-degrees) clockwise. The rotor should now be pointing at the mark on the distributor. When the rotor is pointing at the number one spark plug wire terminal in the distributor cap and the ignition timing marks are aligned, the number one piston is at TDC on the compression stroke.

8   After the number one piston has been positioned at TDC on the compression stroke, TDC for any of the remaining pistons can be located by turning the crankshaft 180-degrees at a time and following the firing order.

---

4   Rocker arm cover - removal and installation

Refer to illustration 4.4

1   Remove the air cleaner (Chapter 4).

2   Remove the spark plug wires from all spark plugs and remove the wire supports from the rocker arm cover. Don't disconnect the wires from the supports.

3   Disconnect the brake booster vacuum hose and position it out of the way.

4   Remove the rocker arm cover mounting bolts **(see illustration)**.

5   Detach the rocker arm cover. **Caution:** *If the cover is stuck to the head, bump the end with a block of wood and a hammer to jar it loose. If that doesn't work, try to slip a flexible putty knife between the head and cover to break the gasket seal. Don't pry at the cover-to-head joint or damage to the sealing surfaces may occur, leading to oil leaks in the future.*

6   The mating surfaces of the head and rocker arm cover must be perfectly clean

when the cover is installed. Use a gasket scraper to remove all traces of sealant and old gasket material, then clean the mating surfaces with lacquer thinner or acetone. If there's sealant or oil on the mating surfaces when the cover is installed, oil leaks may develop.

7   Apply a continuous bead of sealant to the cover-to-head mating surface of the cover. Be sure to apply it to the inside of the mounting bolt holes.

8   Place the new gasket in position on top of the cylinder head, then place the rocker arm cover on the gasket. While the sealant is still wet, install the mounting bolts and tighten them to the torque listed in this Chapter's specifications.

9   Complete the installation by reversing the removal procedure.

---

5   Rocker arm assembly and camshaft – removal and installation

---

Rocker arm assembly removal

Refer to illustration 5.2

1   Remove the rocker arm cover as

5.5 Mark the timing belt and camshaft sprocket to ensure proper reassembly

described in Section 4.

2   The rocker arm assembly is attached to the cylinder head by several mounting nuts. **Note:** *On later style engines there are two bolts that retain the rocker assembly located near the camshaft sprocket. The rocker arm assembly nuts/bolts should be loosened 1/4-turn at a time, working toward the center from both ends* **(see illustration)**, *to avoid distortion of the shafts by valve spring pressure. Loosen the bolts (later engines only) until they're free, but don't pull them out.*

3   Lift the rocker arm assembly off the cylinder head.

Camshaft removal

All engines

Refer to illustration 5.5

4   Bring the number four piston to TDC on the compression stroke by referring to Section 3, if necessary.

5   If you intend to remove the camshaft only, without disassembling the cylinder head, DO NOT alter the timing sprocket position in relationship to the timing chain or belt! Mark both the timing chain or belt and the camshaft sprocket to preserve the original relationship between the two **(see illustra-**

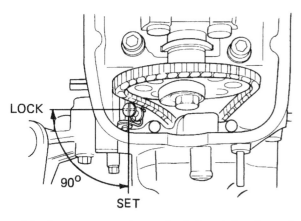

5.6a Release the tension on the timing chain adjuster by depressing and turning the automatic adjuster slide pin 90-degrees in a clockwise direction (early model G180Z and G200Z engines)

5.6b To release tension on the timing chain, pivot the lock lever and push in the adjuster shoe

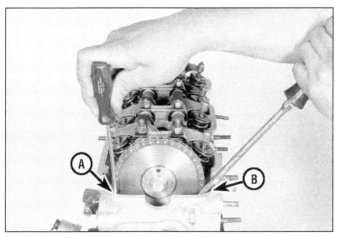

5.6c Using two screwdrivers to retract the timing chain tensioner, pivot the tensioner lock lever with screwdriver A, pry the tensioner foot into the tensioner body with screwdriver B, and then pivot the lever back to hold the tensioner in place

5.14 Position a screwdriver against a bolt head to prevent the camshaft sprocket from turning

tion). The chain or belt must be reinstalled on the camshaft sprocket in the exact same relationship. If a major overhaul is being done, then the sprocket/chain relationship doesn't have to be maintained. Continue the camshaft removal by following the procedure appropriate to the particular engine type.

Engines with timing chain (G180Z and G200Z)

Refer to illustrations 5.6a, 5.6b and 5.6c

6   Release the tension on the timing chain adjuster **(see illustrations)**. After locking the chain adjuster, make sure the chain is loose.

7   Remove the fuel pump (see Chapter 4).

8   Remove the camshaft sprocket bolt. A rod can be positioned through the camshaft sprocket to keep it from turning while the bolt is loosened.

9   Remove the sprocket and fuel pump drive cam from the camshaft. Keep the timing sprocket on the chain damper and tensioner without removing the chain from the sprocket.

10   Remove the rocker arm assembly (see Steps 1 through 3).

11   Lift the camshaft out of the head. Inspection procedures are included in Section 6.

Engines with timing belt (4ZD1 and 4ZE1)

Refer to illustrations 5.14 and 15.15

12   Remove the radiator cooling fan and shroud (see Chapter 3).

13   Remove the timing belt (see Section 17).

14   Remove the camshaft sprocket bolt. A screwdriver can be inserted through one of the sprocket holes and wedged against a bolt head to keep the sprocket from turning while the bolt is loosened **(see illustration)**.

15   Remove the camshaft sprocket and collar **(see illustration)**.

16   Remove the rocker arm assembly (see Steps 1 through 3).

17   Lift the camshaft out of the head. Inspection procedures are included in Section 6.

5.15 The camshaft collar slides off the camshaft

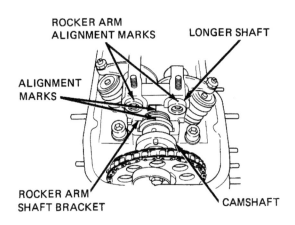

5.21a On engines with a timing chain, install the camshaft with the dowel pin up

5.21b On engines with a timing belt, turn the camshaft until the mark is facing up (arrow)

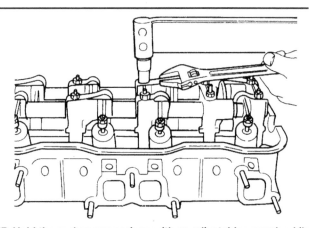

5.27 Hold the rocker arm springs with an adjustable wrench while tightening the nuts to prevent damage to the springs

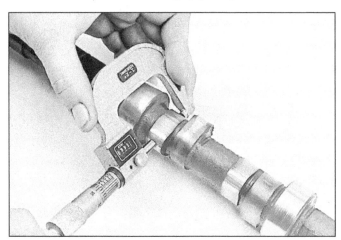

6.3 Measure the camshaft lobe heights . . .

Camshaft installation (all engines)

Refer to illustrations 5.21a and 5.21b

18   Prior to installing the camshaft, apply a coat of moly-base grease or engine assembly lube to the bearing surfaces.

19   Carefully install the camshaft in the head.

20   Reinstall the rocker arm assembly (see Steps 25 through 27).

21   On timing chain-equipped engines, make sure the alignment mark on the number one rocker arm shaft bracket and the mark on the camshaft are aligned **(see illustration)**. On timing belt-equipped engines, make sure the camshaft timing mark is facing up **(see illustration)**. **Note:** *When the mark on the camshaft is facing up, the engine is at TDC, on the compression stroke, for the number four cylinder, on all engines except 2.6L 4ZE1.*

22   Check to make sure the crankshaft pulley timing mark is aligned with the TDC mark on the front cover.

23   Attach the sprocket to the camshaft by aligning it with the pin or mark on the camshaft. Do not remove the chain from the sprocket.

24   Install the sprocket bolt and washer and tighten the bolt to the torque listed in this Chapter's Specifications.

Rocker arm assembly installation

Refer to illustration 5.27

25   Lubricate the rocker arms and shafts with plenty of clean engine oil.

26   Install the rocker arm assemblies with the front marks on the shafts facing up **(see illustration 5.21a) Note:** *The longer shaft must be positioned on the exhaust side and the short shaft on the intake manifold side.*

27   When installing the nuts/bolts, first tighten them until they're finger-tight, starting at the ends and working in toward the center. Continue to tighten them in 1/4-turn increments until they're all at the torque listed in this Chapter's Specifications. Use an adjustable wrench to steady the rocker arm springs when tightening the nuts **(see illustration)**. The remaining installation steps are the reverse of removal.

28   Adjust the valve clearances to the cold specifications as described in Chapter 1.

29   Reinstall the rocker arm cover.

6   Camshaft and rocker arms - inspection

Camshaft

Refer to illustrations 6.3, 6.4 and 6.6

1   Clean the camshaft with solvent, then dry it.

2   Visually inspect the camshaft for wear and/or damage to the distributor drive gear, lobe surfaces, bearing journals and seal contact surfaces. Visually inspect the camshaft bearing surfaces in the cylinder head for scoring and other damage.

3   Measure the camshaft lobe heights **(see illustration)** and compare them to the specifications listed in this Chapter.

4   Measure the camshaft bearing journal diameters **(see illustration)** then measure the inside diameter of the camshaft bearing surfaces in the cylinder head using a telescoping gauge. Subtract the journal measurement from the bearing inside diameter measurement to obtain the camshaft bearing oil clearance. Compare the clearance with the specifications listed in this Chapter.

5   Replace the camshaft if it fails any of the

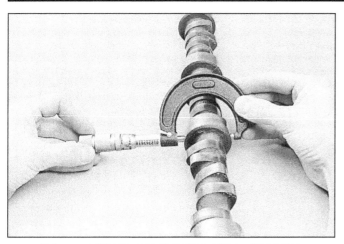

6.4 . . . and the journal diameters with a micrometer

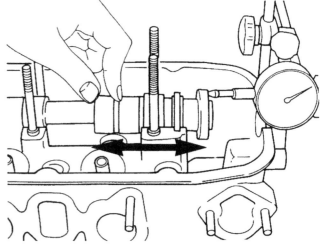

6.6 Use a dial indicator to measure the camshaft endplay

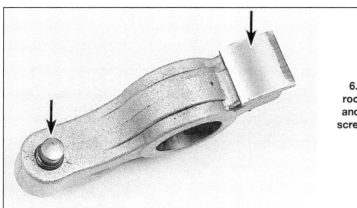

6.9 Check the rocker arm faces and the adjusting screw tips for wear

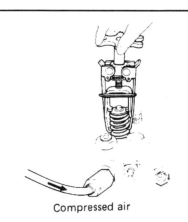

Compressed air

7.4 Compressed air can be used to hold the valves shut as the springs are compressed . . .

above inspections. **Note:** *If the distributor drive gear on the camshaft is worn or damaged, replace the driven gear on the distributor also. If the lobes are worn, replace the rocker arms along with the camshaft. Cylinder head replacement may be necessary if the camshaft bearing surfaces in the head are damaged or excessively worn.*

6   Position the camshaft in the cylinder head and measure the endplay with a dial indicator **(see illustration)**. Compare the results to the specifications listed in this Chapter.

7   If the endplay is excessive, check for damage to the cylinder head and camshaft. Replace the cylinder head if necessary.

Rocker arms

Refer to illustration 6.9

8   Disassemble and inspect the rocker arm assembly. Remove the retaining bolts and slip the rocker arms and springs off the shafts. Keep the parts in order so you can reassemble them in the same locations.

9   Thoroughly clean the parts and inspect them for wear and damage. Check the rocker arm faces that contact the camshaft and the adjusting screw tips **(see illustration)**. Replace any parts that are damaged or excessively worn. Also, make sure the oil holes in the shaft aren't plugged.

7   Valve springs and seals - replacement

Refer to illustrations 7.4 and 7.7
Note: *Broken valve springs and defective valve stem seals can be replaced without removing the cylinder head. Two special tools and a compressed air source are normally required to perform this operation, so read through this Section carefully and rent or buy the tools before beginning the job. If compressed air isn't available, a length of nylon rope can be used to keep the valves from falling into the cylinder during this procedure.*

1   Refer to Section 4 and remove the rocker arm cover. Remove the rocker arm assembly (Section 5).

2   Remove the spark plug from the cylinder with the defective component. If all of the valve stem seals are being replaced, all of the spark plugs should be removed.

3   Turn the crankshaft until the piston in the affected cylinder is at top dead center on the compression stroke (refer to Section 3 for instructions). Since the rocker arm assembly has been removed, and can't hold the camshaft in place, extra care must be taken as the crankshaft is turned. If you're replacing all of the valve stem seals, begin with cylinder number one and work on the valves for one

cylinder at a time. Move from cylinder-to-cylinder following the firing order sequence (1-3-4-2).

4   Thread an adapter into the spark plug hole and connect an air hose from a compressed air source to it **(see illustration)**. Most auto parts stores can supply the air hose adapter. **Note:** *Many cylinder compression gauges utilize a screw-in fitting that may work with your air hose quick-disconnect fitting.*

5   Apply compressed air to the cylinder. The valves should be held in place by the air pressure. If the valve faces or seats are in poor condition, leaks may prevent the air pressure from retaining the valves -refer to the alternative procedure below.

6   If you don't have access to compressed air, an alternative method can be used. Position the piston at a point just before TDC on the compression stroke, then feed a long piece of nylon rope through the spark plug hole until it fills the combustion chamber. Be sure to leave the end of the rope hanging out of the engine so it can be removed easily. Use a large breaker bar and socket to rotate the crankshaft in the normal direction of rotation until slight resistance is felt.

7.7 . . . and the keepers (arrow) are removed

7   Stuff shop rags into the cylinder head holes to prevent parts and tools from falling into the engine, then use a valve spring compressor to compress the spring. Remove the keepers with small needle-nose pliers or a magnet **(see illustration)**.

8   Remove the valve, spring retainer and valve springs, then remove the umbrella type guide seal. **Note:** *If air pressure fails to hold the valve in the closed position during this operation, the valve face or seat is probably damaged. If so, the cylinder head will have to be removed for additional repair operations.*

9   Wrap a rubber band or tape around the top of the valve stem so the valve won't fall into the combustion chamber, then release the air pressure. **Note:** *If a rope was used instead of air pressure, turn the crankshaft slightly in the direction opposite normal rotation.*

10   Inspect the valve stem for damage. Rotate the valve in the guide and check the end for eccentric movement, which would indicate that the valve is bent.

11   Move the valve up-and-down in the guide and make sure it doesn't bind. If the valve stem binds, either the valve is bent or the guide is damaged. In either case, the head will have to be removed for repair.

12   Reapply air pressure to the cylinder to retain the valve in the closed position, then remove the tape or rubber band from the valve stem. If a rope was used instead of air pressure, rotate the crankshaft in the normal direction of rotation until slight resistance is felt.

13   Lubricate the valve stem with engine oil and install a new valve guide oil seal.

14   Measure the valve spring free height and check with the measurements listed in this Chapter's specifications. Replace if the valve spring height exceeds the limit. Install the springs in position over the valve. Make sure the narrow pitch end of the outer spring is against the cylinder head.

15   Install the valve spring retainer. Compress the valve springs and position the keepers in the groove. Apply a small dab of grease to the inside of each keeper to hold it in place if necessary. Remove the pressure from the spring tool and make sure the keepers are seated.

16   Disconnect the air hose and remove the adapter from the spark plug hole. If a rope was used in place of air pressure, pull it out of the cylinder.

17   Install the rocker arm assembly.

18   Install the spark plug(s) and hook up the wire(s).

19   Refer to Section 4 and install the rocker arm cover.

20   Start and run the engine, then check for oil leaks and unusual sounds coming from the rocker arm cover area.

---

8   Intake manifold - removal and installation

Note: *To prevent air leaks and possible damage to the valves, the gasket must be replaced whenever either manifold is removed. The engine must be completely cool when this procedure is done.*

Removal

1   Disconnect the negative battery cable from the battery.

2   Drain the cooling system.

3   Remove the air cleaner.

4   Mark all lines and hoses before removal to ensure correct installation. Disconnect the fuel and vacuum lines from the carburetor or fuel pipe and throttle valve assembly on fuel injected models. Plug the fuel lines immediately to prevent fuel leakage and to keep dirt from entering the lines.

5   Disconnect all hoses from the intake manifold.

6   Unplug all wiring connectors leading to the carburetor or fuel injection system.

7   Disconnect the brake booster hose from the intake manifold.

Carburetor equipped engines

8   Remove the carburetor (see Chapter 4).

9   Remove the EGR pipe clamp bolt at the rear of the cylinder head.

10   Raise the vehicle and support it securely on jackstands.

11   Remove the EGR pipe from the intake and exhaust manifolds.

12   Working under the vehicle, remove the EGR valve and bracket assembly from the lower side of the intake manifold.

13   Remove the jackstands and lower the vehicle.

14   Disconnect the upper radiator hose from the engine. Disconnect the heater hose at the upper side of the manifold. **Note:** *The hoses should be plugged after disconnection.*

15   Loosen them in 1/4-turn increments, then remove all intake manifold mounting nuts.

Fuel injection equipped engines

Refer to illustrations 8.16, 8.20 and 8.21

16   Remove the EGR pipe. Disconnect the EGR pipe from the common chamber **(see illustration 18.5 in Chapter 4)** and the exhaust manifold **(see illustration)**.

17   Carefully mark and remove any vacuum lines and hoses.

18   Remove the common chamber (see Chapter 4).

19   Detach the upper radiator hose from the engine.

20   Remove the fuel rail bolts **(see illustra-**

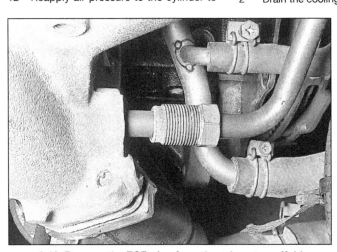

8.16 Remove the EGR pipe from the exhaust manifold

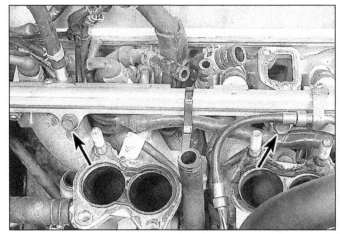

8.20 Remove the bolts that retain the fuel rail to the intake manifold (arrows)

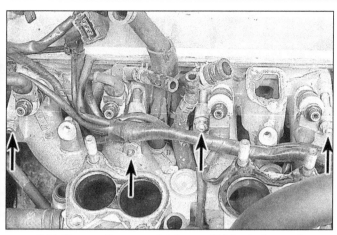

8.21 Remove the upper intake manifold nuts (arrows)

9.3 Remove the exhaust manifold heat shield bolts (arrows) and detach the heat shield

tion) and separate the fuel rail from the injectors.

21 Loosen them in 1/4-turn increments, then remove the upper intake manifold bolts **(see illustration)**. Carefully reach under the intake manifold and remove the retaining nuts.

22 Remove the cylinder head (see Section 10) and lift the manifold and cylinder head off the engine.

Installation

Carburetor equipped engines

23 Before installing the manifolds, stuff clean, lint-free rags into the engine ports and remove all traces of old gasket and sealant with a scraper. Clean the mating surfaces with lacquer thinner or acetone.

24 Begin installation by placing the new gasket in position on the engine and reinstalling the exhaust manifold. Be sure the chamfered side of each washer faces the engine and tighten the mounting bolts only until they support the manifold. The bolts that are common to both manifolds should be finger-tight only!

25 Place the intake manifold in position against the cylinder head and install the remaining bolts, again tightening them only

enough to support the manifold.

26 Once both manifolds are installed, the mounting bolts can be tightened to the specified torque (work from the center out toward the ends and work up to the final torque in three or four steps).

27 The remainder of installation is the reverse of the removal procedure.

28 Fill the radiator with coolant, start the engine and check for leaks. Also check and adjust the idle speed and throttle linkage as described in Chapter 1.

Fuel injection equipped engines

29 Stuff clean, lint-free rags into the engine ports and remove all traces of old gasket and sealant with a scraper. Clean the mating surfaces with lacquer thinner or acetone.

30 Install a new intake manifold gasket on the cylinder head and tighten all intake manifold bolts to the torque listed in this Chapter's specifications.

31 Place the cylinder head and intake manifold assembly on the engine block (see Section 10).

32 The remainder of installation is the reverse of removal.

33 Fill the radiator with coolant, start the engine and check for leaks.

9 Exhaust manifold - removal and installation

Refer to illustrations 9.3, 9.8 and 9.9

1 Disconnect the negative battery cable from the battery.

2 Remove the air cleaner assembly and the hot air hose.

3 Remove the three bolts attaching the exhaust manifold heat shield and detach the heat shield **(see illustration)**.

4 On carburetor equipped engines, remove the EGR pipe clip from the upper portion of the transmission and disconnect the EGR pipe from the exhaust manifold.

5 Disconnect the oxygen sensor wire at the connector (see Chapter 1).

6 Remove the nuts and detach the exhaust pipe from the manifold.

7 Remove the spark plug wires from the exhaust side spark plugs for access to the exhaust manifold bolts/nuts.

8 Remove the bolts/nuts that retain the exhaust manifold to the cylinder head **(see illustration)**.

9 Detach the exhaust manifold from the head **(see illustration)**.

10 Before installing the manifold, remove all

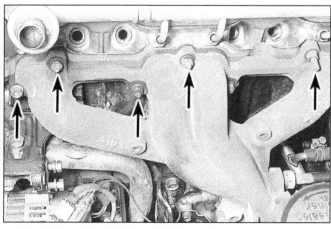

9.8 Remove the exhaust manifold mounting nuts (arrows) . . .

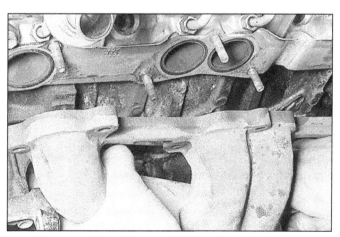

9.9 . . . and detach the manifold from the cylinder head

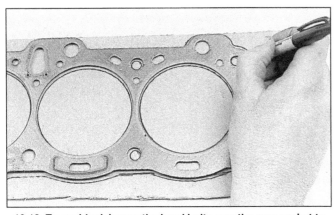

10.13 To avoid mixing up the head bolts, use the new gasket to transfer the hole pattern to a piece of cardboard, punch holes to accept the bolts and push each bolt through the matching hole in the cardboard

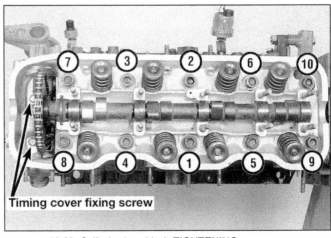

10.23 Cylinder head bolt TIGHTENING sequence

traces of the old gasket with a scraper. Clean the mating surfaces with lacquer thinner or acetone.

11   Place a new exhaust manifold gasket in position on the cylinder head, then hold the manifold in place and install the mounting bolts/nuts finger-tight.

12   Tighten the mounting bolts/nuts, in three or four steps, to the specified torque. Work from the center out toward the ends to prevent distortion of the manifold.

13   Install the remaining components. The remainder of installation is the reverse of the removal procedure.

14   Start the engine and check for exhaust leaks between the manifold and cylinder head and between the manifold and exhaust pipe.

10   Cylinder head - removal and installation

Removal

Refer to illustration 10.13

Note: *On engines equipped with fuel injection, remove the cylinder head and intake manifold as an assembly. The exhaust manifold can be removed along with the cylinder head or separated prior to removal as an option.*

1   Remove the intake and exhaust manifolds as described in Sections 8 and 9.

2   Drain the cooling system (see Chapter 1).

3   Remove the upper radiator hose.

4   Remove the thermostat housing.

5   Disconnect the heater hoses from the fitting on the rear of the cylinder head and the intake manifold.

6   Disconnect the spark plug wires from the spark plugs and position them out of the way, then remove the spark plugs.

7   Disconnect the accelerator linkage, fuel line, and all wires and vacuum hoses. Be sure to mark each one carefully to ensure proper installation.

8   On early models with a carburetor, remove the fuel pump.

9   Raise the vehicle and support it on jackstands. Disconnect the exhaust pipe at the exhaust manifold. Lower the vehicle.

10   Remove the rocker arm cover (see Section 4).

11   On all engines except 2.6L 4ZE1, bring the number four piston to TDC on its compression stroke (see Section 3). On the 2.6L 4ZE1 engine, turn the engine clockwise and align the camshaft and crankshaft marks **(see illustrations 17.7a and 17.7b)**. Remove the camshaft sprocket - be sure to follow the procedure in Section 5.

12   On engines with an air injection system, disconnect the AIR hose and the check valve on the exhaust manifold.

13   Using a new head gasket, outline the cylinders and bolt pattern on a piece of cardboard **(see illustration)**. Be sure to indicate the front of the engine for reference. Punch holes at the bolt locations. Loosen the ten cylinder head bolts in 1/4-turn increments until they can be removed by hand (start at the ends and work toward the center of the head). Store the bolts in the cardboard holder as they're removed - this will ensure that they're reinstalled in their original locations.

14   Lift the head off the engine. If it's stuck, DO NOT pry between the head and block - damage to the mating surfaces will result! To dislodge the head, position a block of wood against it and strike the wood block with a hammer. The timing chain should be left in place, resting on the wooden wedge.

15   Remove the cylinder head gasket. Place the head on a block of wood to prevent damage.

16   Refer to Part C for cylinder head inspection procedures.

Installation

Refer to illustration 10.23

17   The mating surfaces of the cylinder head and block must be perfectly clean when the head is installed. Use a gasket scraper to remove all traces of carbon and old gasket

material, then clean the mating surfaces with lacquer thinner or acetone. If there's oil on the mating surfaces when the head is installed, the gasket may not seal correctly and leaks could develop. When working on the block, stuff the cylinders with clean shop rags to keep out debris. Use a vacuum cleaner to remove any debris that falls into the cylinders.

18   Check the block and head mating surfaces for nicks, deep scratches and other damage. If damage is slight, it can be removed with a file; if it's excessive, machining may be the only alternative.

19   Use a tap of the correct size to chase the threads in the head bolt holes. Mount each bolt in a vise and run a die down the threads to remove corrosion and restore the threads. Dirt, corrosion, sealant and damaged threads will affect torque readings.

20   Install the gasket over the engine block dowel pins.

21   Make sure the crankshaft is still in the same position, then carefully lower the cylinder head onto the engine, over the dowel pins and the gasket. Be careful not to move the gasket.

22   Install the cylinder head mounting bolts. Make sure they're installed in the correct holes.

23   Tighten the bolts in 1/4-turn increments, following the recommended sequence **(see illustration)**, until the Step 1 specified torque is reached. Repeat the procedure, tightening all bolts to the Step 2 specified torque.

24   Install the camshaft sprocket as described in Section 5.

25   The remainder of the installation procedure is the reverse of the removal procedure.

26   Fill the radiator with coolant, start the engine and check for leaks. Be sure to recheck the coolant level once the engine has warmed up to operating temperature. Also check and adjust the idle speed as described in Chapter 1.

27   Readjust the valve clearances to the cold specifications as described in Chapter 1.

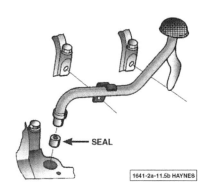

11.5b Oil pump pickup tube details for engines equipped with timing belts (note the location of the seal)

SEAL

1641-2a-11.5b HAYNES

11.5a To remove the oil pump assembly from the engine, remove the pump and pick-up tube mounting bolts

11.6a Use a screwdriver to lock the oil pump sprocket when loosening the nut

11.6b The oil pump bolts can be removed through the sprocket holes

11 Oil pump - removal and installation

Removal

Refer to illustrations 11.5a, 11.5b, 11.6a and 11.6b

1   Position the number one piston at TDC on the compression stroke. Refer to Section 3 if necessary.

2   Remove the distributor cap and mark

the position of the rotor in relation to the distributor body.

3   Raise the front of the vehicle and support it on jackstands. Apply the parking brake and block the rear wheels to keep the vehicle from rolling.

4   Drain the engine oil.

5   On timing chain engines (G180Z and G200Z) remove the oil pan (Section 18) and also the oil pickup tube **(see illustrations)**.

6   On timing belt engines, remove the timing belt (see Section 17) and loosen the oil

pump sprocket nut **(see illustration)**, then remove the pump-to-block bolts **(see illustration)** and withdraw the pump from the block cavity. If necessary, the nut can be removed and the sprocket can be detached from the pump shaft.

7   On timing chain engines, remove the pump mounting bolts and withdraw the oil pump/driveshaft assembly.

8   If the oil pump must be inspected or overhauled, refer to Section 12.

9   Prior to installation, make sure the number one piston is still at TDC.

Installation

Engines with timing chain (G180Z and G200Z)

Refer to illustrations 11.10, 11.11, 11.12a and 11.12b

10   Fill the pump body with engine oil to prime it and align the punch marks on the rotors **(see illustration)**.

11   Install the oil pump assembly by engaging the oil pump drive gear with the pinion gear on the crankshaft so the mark on the drive gear is turned toward the rear and is approximately 20-degrees off the crankshaft centerline in a clockwise direction **(see illustration)**.

12   After the oil pump is installed, check to

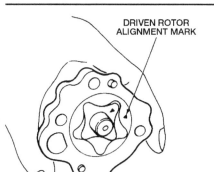

DRIVEN ROTOR
ALIGNMENT MARK

11.10 Align the oil pump drive and driven rotors on timing chain engines

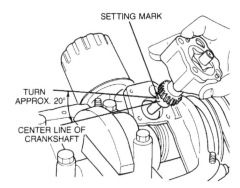

SETTING MARK

TURN
APPROX. 20°

CENTER LINE OF
CRANKSHAFT

11.11 When installing the oil pump on timing chain engines . . .

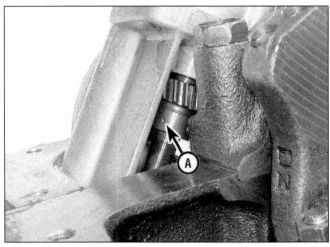

11.12a Check for correct alignment of the oil pump pinion gear by sighting between the rear of the timing cover and the engine block

A   Punch mark

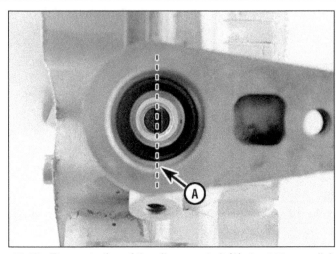

11.12b The centerline of the oil pump shaft (A) should be parallel to the front face of the engine block

make sure the mark on the oil pump drive gear is turned to the rear as viewed from the clearance between the front cover and engine block **(see illustration)** and the slot in the end of the oil pump drive shaft is nearly parallel to the front face of the engine block (offset slightly) as viewed through the distributor hole **(see illustration)**.

13   Install the oil pump cover, carefully aligning it with the dowel pins. Install the oil pump mounting bolts.

14   Install the relief valve assembly and the rubber hose on the cover.

15   Attach the oil pipe to the rubber hose and the engine block.

16   Install the oil pan (see Section 18).

17   The remainder of installation is the reverse of removal.

18   Fill the engine with the proper quantity and grade of engine oil (refer to Chapter 1, if necessary).

19   Finally, check the ignition timing as described in Chapter 1.

Engines with timing belt (4ZD1 and 4ZE1)

20   Lubricate the outer rotor with clean engine oil, then install it in the block cavity, chamfered side OUT.

21   Apply engine oil to the O-ring and insert it into the groove in the oil pump cover.

22   Lubricate the rotor with clean engine oil, then install the pump cover, engaging the inner rotor with the outer rotor in the engine block cavity.

23   Tighten the mounting bolts to the torque listed in this Chapter's Specifications. Tighten them gradually, in 1/4-turn increments, to avoid distorting the cover.

24   Make sure the oil pump shaft turns smoothly. If it doesn't, refer to Section 12.

25   Install the sprocket, apply thread locking cement to the shaft threads, then install the nut. Tighten it to the torque listed in this Chapter's specifications.

12 Oil pump - inspection

If the oil pump is defective or the engine is being overhauled, install a new oil pump - don't attempt to check or repair the original.

13 Crankshaft pulley - removal and installation

Refer to illustration 13.6

1   Disconnect the negative battery cable from the battery.

2   Remove the radiator and shroud (see Chapter 3).

3   Unbolt the lower front cover or skidplate (if equipped).

4   Remove the drivebelts (Chapter 1).

5   On manual transmission equipped models, put the transmission in High gear and apply the parking brake. On automatics, remove the starter and wedge a large screwdriver in the ring gear teeth. Remove the pulley mounting bolt. They're usually very tight, so use a six-point socket and a 1/2-inch drive

13.6 Use a bolt-type puller that applies force to the hub as shown here - if a jaw-type puller that applies force to the outer edge is used, the pulley will be damaged

breaker bar.

6   Use a puller to remove the pulley from the crankshaft **(see illustration)**. Do not use a gear puller that applies force to the outer edge of the pulley!

7   Installation is the reverse of removal. Be sure to apply moly-base grease or clean engine oil to the seal contact surface on the pulley hub before installing the pulley on the crankshaft.

8   Tighten the pulley bolt to the torque listed in this Chapter's specifications.

14 Crankshaft front oil seal - replacement

Refer to illustrations 14.1, 14.2 and 14.3

1   On timing chain engines, remove the crankshaft pulley. On timing belt engines, remove the timing belt (Section 17) and detach the sprocket from the nose of the crankshaft **(see illustration)**.

2   Carefully pry the oil seal out of the front cover or seal housing with a seal removal tool or screwdriver **(see illustration)**. Don't

14.1 The sprocket should slide off the crankshaft

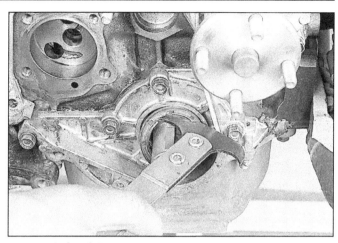

14.2 Carefully pry the old seal out with a seal puller or screwdriver - don't damage the crankshaft

14.3 Use a large socket or piece of pipe and a hammer to install the new seal

15.12 Remove the front cover retaining nuts (arrows) (timing belt engine shown)

scratch the seal bore or damage the crankshaft in the process (if the crankshaft is damaged the new seal will end up leaking).

3   Clean the bore in the cover or housing and coat the outer edge of the new seal with engine oil or multi-purpose grease. Using a socket with an outside diameter slightly smaller than the outside diameter of the seal, carefully drive the new seal into place with a hammer **(see illustration)**. If a socket isn't available, a short section of large diameter pipe will work. Check the seal after installation to be sure the spring didn't pop out.

4   On timing belt engines, reinstall the sprocket.

5   The parts removed to gain access to the seal can now be reinstalled.

6   Run the engine and check for leaks.

15   Front cover - removal and installation

Refer to illustration 15.12

1   Drain the coolant from the engine and

radiator (Chapter 1).

2   Remove the fan and radiator as described in Chapter 3.

3   On timing chain engines, the front cover is sandwiched between the cylinder head and oil pan. Remove the cylinder head (see Section 10).

4   On timing chain engines, remove the distributor as described in Chapter 5.

5   On timing chain engines, remove the water pump as described in Chapter 3.

6   Remove the upper alternator adjusting bolt, then remove the bolts that attach the alternator adjusting arm to the front cover and detach the adjusting arm.

7   First, carefully label them, then disconnect all vacuum and fuel lines attached to the front cover.

8   If equipped, disconnect the air conditioning compressor from the bracket and position it out of the way. Remove the compressor bracket and idler pulley. **Warning:** *Don't disconnect the hoses from the compressor unless the system has been evacuated!*

9   If equipped with power steering, remove

the pump from the bracket and position it out of the way. The hydraulic hoses should not be disconnected from the pump.

10   Remove the coolant fitting from the right side of the front cover.

11   Remove the crankshaft pulley (Section 13).

12   Remove the mounting bolts that attach the front cover to the block and cylinder head **(see illustration)**, then detach the front cover. If it doesn't come off easily, careful tapping with a soft-face hammer will help.

13   Install a new seal in the front cover by referring to Section 14.

14   Place the new gaskets in position on the cover (timing chain engines), then apply RTV sealant.

15   Apply a small amount of grease to the oil seal lip, then place the front cover in position on the block and install the mounting bolts. Tighten them in 1/4-turn increments to the torque listed in this Chapter's specifications.

16   On timing chain engines, align the punch mark on the oil pump drive gear with the oil filter side of the cover. Then align the

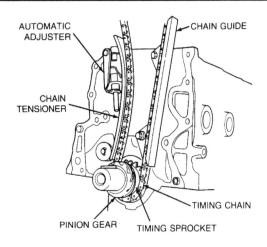

16.2 **Timing chain assembly on G180Z and G200Z engines**

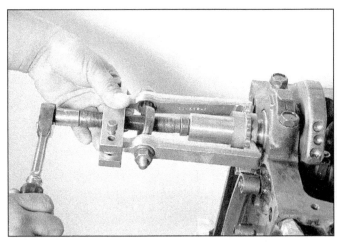

16.5 **Remove the sprocket from the crankshaft with a puller**

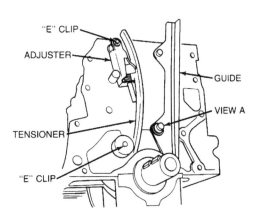

16.8a **Timing chain guide, tensioner and adjuster on G180Z and G200Z engines**

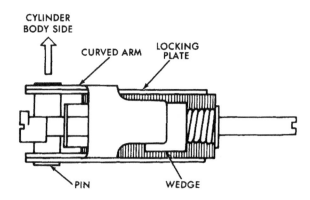

16.8b **Timing chain adjuster components (cross-sectional view)**

center of the dowel pin with the alignment mark on the oil pump case **(see illustrations 11.12a and 11.12b)**.

17   The remainder of installation is the reverse of the removal procedure.

16   Timing chain and sprockets - removal, inspection and installation

Removal and inspection

Refer to illustrations 16.2, 16.5, 16.8a and 16.8b

1   Remove the front cover (Section 15).

2   Remove the bolts that attach the chain tensioner and chain guides to the block, then detach them **(see illustration)**.

3   Remove the camshaft sprocket bolt and detach the sprocket (see Section 5 if necessary).

4   Remove the timing chain from the camshaft and crankshaft sprockets.

5   Remove the pinion gear and timing chain sprocket from the end of the

crankshaft. A puller will most likely be required **(see illustration)**. Note that two Woodruff keys are used for these components.

6   Examine the teeth on both the crankshaft sprocket and the camshaft sprocket for wear. Each tooth forms an inverted V. If worn, the side of each tooth under tension will be slightly concave in shape when compared with the other side of the tooth (i.e. one side of the inverted V will be concave when compared with the other). If the teeth appear to be worn, the sprockets must be replaced with new ones.

7   The chain should be replaced with a new one if the sprockets are worn or if the chain is loose. It's a good idea to replace the chain if the engine is stripped down for a major overhaul. The rollers on a very badly worn chain may be slightly grooved. To avoid future problems, if there's any doubt at all about the chain's condition, replace it with a new one.

8   Remove the E-clip and detach the automatic chain adjuster **(see illustration)**. Check the adjuster for wear and damage. Inspect the adjuster pin, arm, wedge and rack teeth

of the locking plate for wear and damage **(see illustration)**.

9   Examine the chain tensioner and guides. Replace any items that are worn or damaged.

Installation

Refer to illustration 16.11

10   Install the crankshaft sprocket and pinion gear (grooved side out, toward the front cover). Align the grooves in the gears with the keys on the crankshaft and drive them into position with a deep socket.

11   Install the camshaft sprocket with the marked side of the sprocket facing forward and the triangular mark aligned with the chain mark plate **(see illustration)**.

12   Make sure the crankshaft and camshaft sprockets are set at TDC (see Section 3).

13   Install the timing chain by aligning the mark plate on the chain with the mark on the crankshaft sprocket. The side of the chain with the mark plate is on the front side and the side of the chain with the most links between mark plates is on the chain guide side **(see illustration 16.11)**. **Caution:** *Don't turn the crankshaft or camshaft until the timing chain is installed, otherwise the valves will*

16.11 With the camshaft timing sprocket mark at top dead center, position the crankshaft sprocket and chain timing marks (arrow) as shown

17.11 Remove the timing belt cover retaining bolts

17.12a Make sure the camshaft sprocket timing mark is aligned with the upper back cover

17.12b Crankshaft sprocket timing marks - align the crank mark (lower arrow) with the cast-in mark on the front seal retainer (upper arrow)

turn the crankshaft or camshaft until the timing chain is installed, otherwise the valves will contact the piston crowns.

14 Install the chain adjuster assembly and guides on the engine block. Feel the tension on the chain to make sure the system is working properly.

15 Reinstall the front cover, referring to Section 15 as needed.

17 Timing belt and sprockets - removal, inspection installation and adjustment

Removal

Refer to illustrations 17.11, 17.12a, 17.12b, 17.13, 17.14 and 17.16

1 Disconnect the negative cable from the battery.

2 Remove the cooling fan/shroud as described in Chapter 3.

3 Refer to Chapter 3 for removal of the water pump pulley.

4 Position the No. 4 piston at Top Dead Center (Section 3).

5 Remove the drivebelts and spark plugs (see Chapter 1). **Note:** *Some early models have an air pump. If so, remove the air pump drivebelt.*

6 Refer to Section 7 and remove the crankshaft pulley.

7 If equipped, disconnect the air conditioning drivebelt idler pulley.

8 If equipped with power steering, remove the pump from the bracket and position it out of the way (see Chapter 10). The hydraulic hoses should not be disconnected from the pump.

9 Remove the crankshaft pulley (see Section 7).

10 If so equipped, remove the air injection tube that runs in front of the timing belt cover.

11 Unbolt and remove the upper and lower timing belt covers **(see illustration)**. Don't lose the rubber seals.

12 Be sure the camshaft timing mark lines up with the lug on the front housing and the crankshaft sprocket notch lines up with the pointer on the front oil seal housing **(see illustrations)**.

13 If you plan on reinstalling the timing belt, make an arrow on it to indicate the direction of rotation **(see illustration)**.

14 Loosen the tensioner bolt and use a large prybar to push the belt tensioner back against the spring, until there's enough room

17.13 Mark the direction of rotation on the timing belt if the same belt will be used again

17.14 Wedge the tensioner back to allow room for the timing belt to be removed - arrow indicates the tensioner bolt

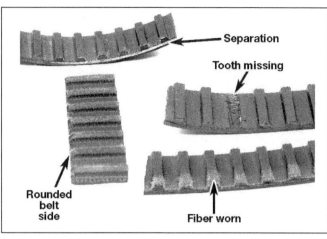

17.16 Inspect the timing belt for cracks and missing teeth; wear on one side of the belt indicates sprocket misalignment problems

to slip the timing belt off **(see illustration)**.

15 Slip the belt off the sprockets and remove it from the engine.

16 Inspect the belt for wear, peeling, cracks, hardness, crimping and signs of oil or other fluid contamination **(see illustration)**. The belt should be replaced if any of these conditions exist or if the specified mileage has elapsed (see Chapter 1). **Caution:** *Unless the engine has very low mileage, it's common practice to replace the timing belt with a new one every time it's removed. Don't reinstall the original belt unless it's in like-new condition. A belt that breaks during engine operation can cause serious valve/piston damage.*

17 Inspect the sprockets for signs of wear or damage. If sprockets need to be replaced, they should be replaced as a set. Refer to Section 7 for removal of the crankshaft sprocket, to Section 9 for removal of the camshaft sprocket, and Section 13 for removal of the oil pump sprocket.

Installation

18 Make sure the crankshaft and camshaft sprocket timing marks are still aligned **(see illustrations 17.12a and 17.12b)**. **Note:**

When the mark on the camshaft is facing up, the engine is at TDC, on the compression stroke, for the number four cylinder.

19 If the old belt is being reinstalled, make sure the arrow is pointed in the proper direction.

20 Slip the belt on, first onto the crankshaft sprocket, then the oil pump sprocket, the camshaft sprocket, and finally the tensioner. Wedge a screwdriver into the pivot area of the tensioner to temporarily release the tension **(see illustration 8.14)** to slip the belt on.

Adjustment

Refer to illustration 17.23

21 Loosen the tensioner lock bolt so only the spring is applying pressure, then tighten the bolt.

22 Install the crankshaft sprocket bolt and use it to turn the crankshaft in the normal direction of rotation (clockwise) through two complete revolutions (720-degrees) so equal tension is applied to both sides of the timing belt, then repeat Step 21.

23 Recheck the timing mark alignment and check the tension of the belt midway between the oil pump and camshaft sprocket

(see illustration). If the tension is incorrect, repeat the adjustment operation.

24 Reinstall the remaining parts in the reverse order of removal.

18 Oil pan - removal and installation

Note: *Early model vehicles are equipped with a one-piece oil pan while later model Troopers are equipped with a two-piece oil pan.*

1 Drain the engine oil.

2 Raise the front of the vehicle and support it on jackstands placed under the frame. Apply the parking brake and block the rear wheels to keep the vehicle from rolling.

3 Remove the splash shield, if equipped.

2WD Vehicles

4 On 2WD models, disconnect the front motor mounts and lift the engine four or five inches, using an overhead chain hoist. **Note:** *In the event there's still not enough clearance to remove the oil pan (depending on the particular model), remove the engine (see Chapter 2, Part C).*

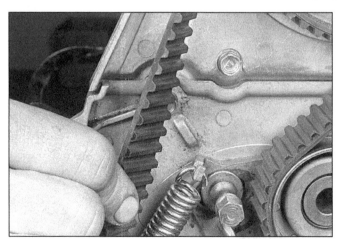

17.23 You should be able to twist the timing belt no further than 90-degrees if it's properly adjusted

18.7 Remove the oil pan mounting bolts (engine removed for clarity)

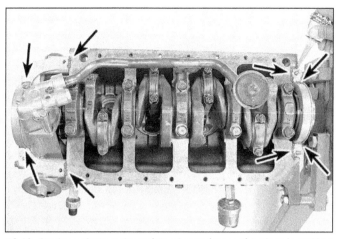

18.10 Apply sealant at the points shown (arrows) before installing the oil pan (1981 through 1991 engines shown, later models similar)

19.2 Wedge a screwdriver in the flywheel teeth to prevent the flywheel from turning as the bolts are loosened

4WD vehicles

Refer to illustration 18.7

5   On 4WD models, remove the front differential as described in Chapter 8, then disconnect the engine and transmission mounts and lift the engine slightly with a hoist. **Warning:** *Do not place any part of your body under the engine/transmission when it's off the mounts! Place wood blocks under the mounts to help support the engine.*

6   Remove the idler arm bolts and lower the cross rod to provide clearance for the oil pan (see Chapter 10).

All vehicles

Refer to illustration 18.10

7   On 1992 and later 4ZD1 engines, remove the bolts, nuts and lockwashers, then remove the two damper plates surrounding the oil pan and detach the oil pan. On all other models, remove the bolts and detach the oil pan. Don't pry between the block and the oil pan or damage to the sealing surface may result and oil leaks could develop. Use a block of wood and a hammer to dislodge the pan if it's stuck.

8   Use a scraper to remove all traces of old gasket material and sealant from the block and oil pan. Clean the gasket sealing surfaces with lacquer thinner or acetone and make sure the bolt holes in the block are clean.

9   Check the oil pan flange for distortion, particularly around the bolt holes. If necessary, place the pan on a block of wood and use a hammer to flatten and restore the gasket surface.

10   Before installing the oil pan, apply a thin coat of RTV sealant to the flange and to the corners of the engine block **(see illustration)**. Attach the new gasket to the pan (make sure the bolt holes are aligned).

11   Position the oil pan against the engine block. On 1992 and later 4ZD1 engines install the two stiffener plates, the mounting bolts, nuts and lockwashers. On all other models, install the mounting bolts. Tighten them to

the specified torque in a criss-cross pattern.

12   Wait at least 30 minutes before filling the engine with oil, then start the engine and check for leaks.

19 Flywheel/driveplate - removal and installation

Refer to illustrations 19.2 and 19.4

1   If the engine is in the vehicle, remove the clutch cover and clutch disc as described in Chapter 8, or the automatic transmission as described in Chapter 7.

2   Flatten the lockplate tabs (if used) and remove the bolts that secure the flywheel/driveplate to the crankshaft rear flange. Be careful, the flywheel is very heavy and shouldn't be dropped. If the crankshaft turns as the bolts are loosened, wedge a screwdriver in the starter ring gear teeth **(see illustration)**.

3   Detach the flywheel/driveplate from the crankshaft flange.

4   Unbolt and remove the engine rear plate **(see illustration)**. Now is a good time to check the engine block rear core plug for leakage.

5   If the teeth on the flywheel/driveplate starter ring gear are badly worn, or if some

are missing, install a new flywheel or driveplate.

6   Refer to Chapter 8 for the flywheel inspection procedure.

7   Before installing the flywheel/driveplate, clean the mating surfaces.

8   If removed, reinstall the engine rear plate.

9   Position the flywheel/driveplate against the crankshaft, using a new spacer, if equipped, and insert the mounting bolts. Use thread locking cement on the bolts. **Note:** *The crankshaft/flywheel bolts will leak oil if the threads are not properly sealed with thread locking cement.*

10   Tighten the bolts in a criss-cross pattern to the torque listed in this Chapter's specifications.

11   The remainder of installation is the reverse of removal.

20 Crankshaft rear oil seal - replacement

Refer to illustration 20.5

1   The rear crankshaft oil seal can be replaced without removing the oil pan or crankshaft.

2   Remove the transmission (Chapter 7).

19.4 Remove the engine rear plate bolts (arrows)

20.5 Use a screwdriver to carefully pry the rear crankshaft oil seal out

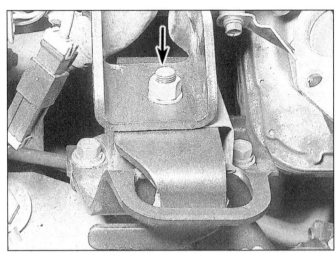

21.3 Remove the nut (arrow) to release the engine mount bracket from the frame mount

3   If equipped with a manual transmission, remove the pressure plate and clutch disc (Chapter 8).

4   Remove the flywheel or driveplate (Section 19).

5   Using a seal removal tool or a screwdriver, carefully pry the seal out of the block **(see illustration)**. Don't scratch or nick the crankshaft in the process!

6   Clean the bore in the block and the seal contact surface on the crankshaft. Check the crankshaft surface for scratches and nicks that could damage the new seal lip and cause oil leaks. If the crankshaft is damaged, the only alternative is a new or different crankshaft.

7   Apply a light coat of engine oil or multipurpose grease to the outer edge of the new seal. Lubricate the seal lip with moly-base grease.

8   The seal lip must face toward the front of the engine. Carefully work the seal lip over the end of the crankshaft and tap the seal in with a hammer and punch until it's seated in the bore.

9   Install the flywheel or driveplate.

10   If equipped with a manual transmission, reinstall the clutch disc and pressure plate.

11   Reinstall the transmission as described in Chapter 7.

21   Engine mounts - replacement

Refer to illustration 21.3

Warning: *Don't position any part of your body under the engine when the engine mounts are unbolted!*

1   Engine mounts are non-adjustable and seldom require service. Periodically they should be inspected for hardness and cracks in the rubber and separation of the rubber from the metal backing.

2   To replace the engine mounts with the engine in the vehicle, use the following procedure.

3   Loosen the nuts and bolts that retain the front mounting insulator to the engine mount bracket and frame. Do this on both sides **(see illustration)**.

4   Next, the weight of the engine must be taken off the engine mounts. This can be done from beneath using a jack and wooden block positioned under the oil pan, or from above by removing the air cleaner and using an engine hoist attached to the two engine brackets. The engine should be raised slowly and carefully, while keeping a constant check on clearances around the engine to prevent anything from binding or breaking. Pay particular attention to areas such as the fan, ignition coil wires, vacuum lines leading to the engine and rubber hoses and ducts.

5   Raise the engine just enough to provide adequate room to remove the mounting insulator.

6   Remove the nuts and bolts retaining the insulator, then lift it out, noting how it's installed.

7   Installation is the reverse of removal, but be sure the insulator is installed in the same position it was in before removal.

Chapter 2 Part B
V6 engine

Contents

Specifications

General

Displacement	171 cu in (2.8 liters) or 189 cu in (3.1 liters)
Cylinder numbers (front-to-rear)	
Left bank (driver's side)	2-4-6
Right bank	1-3-5
Firing order	1-2-3-4-5-6

Camshaft

Lobe lift	
2.8L	
Intake	0.262 in (6.650 mm)
Exhaust	0.273 in (6.940 mm)
3.1L	
Intake	0.230 in (5.858 mm)
Exhaust	0.262 in (6.654 mm)
Journal diameter	
2.8L	1.869 to 1.871 in (47.44 to 47.49 mm)
3.1L	1.868 to 1.882 in (47.44 to 47.79 mm)
Journal-to-bearing (oil) clearance	0.001 to 0.0039 in (0.026 to 0.101 mm)

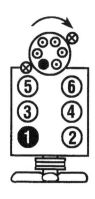

Cylinder locations and distributor rotation

The blackened terminal shown on the distributor cap indicates the Number One spark plug wire position

Torque specifications

	Ft-lbs (unless otherwise indicated)
Camshaft sprocket bolt	17
Camshaft cover (rear) bolts	6 to 9
Crankshaft pulley-to-vibration damper bolts	20 to 30
Cylinder head bolts	40
2.8L	40
3.1L	
Step 1	41
Step 2	Turn an additional 90-degrees (1/4-turn)
Exhaust manifold bolts	25
Flywheel/driveplate bolts	52
Intake manifold bolts	23
Oil pan mounting bolts/nuts	
2.8L	
6 x 1.0 mm	6 to 9
8 x 1.25 mm	14 to 22
3.1L	
Bolts	18
Nuts	7

Torque specifications (continued)

Oil pump 30
Rear main bearing cap bolts... 72
Rocker arm stud .... 43 to 49
Rocker arm cover bolts 72-in-lbs
Tensioner bracket-to-engine block bolts 14
Timing chain cover bolts
   2.8L
      8 x 1.25 mm .... 13 to 18
      10 x 1.5 mm .... 20 to 30
   3.1L 20
Timing chain tensioner bolts... 13 to 18
Vibration damper center bolt 70

1   General information

This Part of Chapter 2 is devoted to in-vehicle repair procedures for the V6 engine. All information concerning engine removal and installation and engine block and cylinder head overhaul can be found in Part C of this Chapter.

The following repair procedures are based on the assumption the engine is installed in the vehicle. If the engine has been removed from the vehicle and mounted on a stand, many of the steps outlined in this Part of Chapter 2 will not apply.

The specifications included in this Part of Chapter 2 apply only to the procedures contained in this Part. Part C of Chapter 2 contains the specifications necessary for cylinder head and engine block rebuilding.

2   Repair operations possible with the engine in the vehicle

Many major repair operations can be accomplished without removing the engine from the vehicle.

Clean the engine compartment and the exterior of the engine with some type of degreaser before any work is done. It will make the job easier and help keep dirt out of the internal areas of the engine.

Depending on the components involved, it may be helpful to remove the hood to improve access to the engine as repairs are performed (refer to Chapter 11 if necessary). Cover the fenders to prevent damage to the paint. Special pads are available, but an old bedspread or blanket will also work.

If vacuum, exhaust, oil or coolant leaks develop, indicating a need for gasket or seal replacement, the repairs can generally be made with the engine in the vehicle. The intake and exhaust manifold gaskets, timing cover gasket, oil pan gasket, crankshaft oil seals and cylinder head gasket are all accessible with the engine in place.

Exterior engine components, such as the intake and exhaust manifolds, the oil pan (and the oil pump), the water pump, the starter motor, the alternator, the distributor and the fuel system components can be removed for repair with the engine in place.

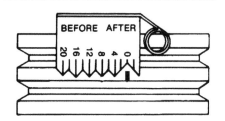

3.1 V6 engine timing marks (typical)

Since the cylinder heads can be removed without pulling the engine, valve component servicing can also be accomplished with the engine in the vehicle. Replacement of the timing chain, sprockets and camshaft is also possible with the engine in the vehicle.

In extreme cases caused by a lack of necessary equipment, repair or replacement of piston rings, pistons, connecting rods and rod bearings is possible with the engine in the vehicle. However, this practice is not recommended because of the cleaning and preparation work that must be done to the components involved.

3   Top Dead Center (TDC) for number one piston - locating

Refer to illustration 3.1

See Chapter 2, Part A, Section 3 for this procedure. The timing plate is attached to the timing chain cover **(see illustration)**. Be sure to use the firing order in the Specifications in this Part of Chapter 2 for the V6 engine.

4   Rocker arm covers - removal and installation

1   Disconnect the negative cable from the battery.
2   Remove the air cleaner assembly (see Chapter 4).
3   Label and disconnect the wires and hoses which would interfere with removal of the rocker arm cover(s).
4   Label and detach the spark plug wires and unclip the wire retainers from the studs.

5   Disconnect the throttle, cruise control and TVS cables and bracket at the TBI unit. If necessary for clearance, remove the air conditioning compressor bracket.
6   Disconnect the PCV valve.
7   Disconnect the vacuum pipe at the manifold. If you're removing a left cover, it may be necessary to disconnect the fuel lines from the throttle body (see Chapter 4).
8   If necessary for clearance, remove the alternator (see Chapter 5). If removing the right cover on a 3.1L engine, remove the ignition coil and bracket (see Chapter 5).
9   Remove the six rocker arm cover mounting bolts and detach the cover. Note that some of the bolts have studs attached to the ends - you'll have to use a deep socket to remove them. If the cover is stuck, use a soft-face hammer or a block of wood and a hammer to dislodge it. If the cover still won't come loose, pry on it carefully with a putty knife, but don't distort the sealing flange surface.
10   Remove all traces of old gasket and sealant with a scraper, then clean the mating surfaces with lacquer thinner or acetone.
11   On models that use RTV sealant in place of a gasket, apply a 1/8-inch bead of sealant to the cover flange. Be sure to apply it to the inside of the bolt holes or oil will leak past the bolt threads. **Caution:** *When applying RTV sealant, keep it out of the bolt holes.*
12   On models that use gaskets, place a small amount of RTV sealant on the seam area where the cylinder head and intake manifold meet before installing the gasket.
13   Install the cover(s) while the RTV is still wet. Tighten the bolts in 1/4-turn increments until the torque listed in this Chapter's Specifications is reached.
14   Reinstall the remaining parts in the reverse order of removal.
15   Run the engine and check for oil leaks.

5   Rocker arms and pushrods - removal, inspection and installation

Removal

Refer to illustration 5.4

1   Refer to Section 3 and detach the rocker arm cover(s) from the cylinder head(s).

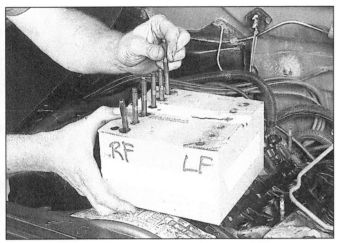

5.4 When removing the pushrods, be sure to store them separately to ensure reinstallation in their original locations

5.10 The ends of the pushrods and the valve stems should be lubricated with moly-base grease prior to installation of the rocker arms

2　Beginning at the front of one cylinder head, remove the rocker arm stud nuts. Store them separately in marked containers to ensure they'll be reinstalled in their original locations. **Note:** *If the pushrods are the only items being removed, loosen each nut just enough to allow the rocker arms to be rotated to the side so the pushrods can be lifted out.*

3　Lift off the rocker arms and pivot balls and store them in the marked containers with the nuts (they must be reinstalled in their original locations).

4　Remove the pushrods and store them separately to make sure they don't get mixed up during installation **(see illustration)**.

Inspection

5　Check each rocker arm for wear, cracks and other damage, especially where the pushrods and valve stems contact the rocker arm faces.

6　Make sure the hole at the pushrod end of each rocker arm is open.

7　Check each rocker arm pivot area for wear, cracks and galling. If the rocker arms are worn or damaged, replace them with new ones and use new pivot balls as well.

8　Inspect the pushrods for cracks and excessive wear at the ends. Roll each pushrod across a piece of plate glass to see if it's bent (if it wobbles, it's bent).

Installation

Refer to illustrations 5.10 and 5.11

9　Lubricate the lower end of each pushrod with clean engine oil or moly-base grease and install them in their original locations. Make sure each pushrod seats completely in the lifter.

10　Apply moly-base grease to the ends of the valve stems and the upper ends of pushrods before positioning the rocker arms over the studs **(see illustration)**.

11　Set the rocker arms in place, then install the pivot balls and nuts. Apply moly-base grease to the pivot balls to prevent damage to the mating surfaces before engine oil pressure builds up **(see illustration)**. Be sure to

install each nut with the flat side against the pivot ball.

12　Adjust the valve lash.

Valve adjustment

Refer to illustration 5.15

13　If the valve train components have been serviced just prior to this procedure, make sure they're completely reassembled.

14　Rotate the crankshaft until the number one piston is at top dead center (TDC) on the compression stroke (see Section 3). To make sure you don't mix up the TDC positions of the number one and four pistons, place your fingers on the number one rocker arms as the timing marks line up at the crankshaft pulley. If the rocker arms aren't moving, the number one piston is at TDC. If they move as the timing marks line up, the number four piston is at TDC.

15　Back off the rocker arm nut until play is felt at the pushrod, then turn it back in until all play is removed. This can be determined by rotating the pushrod while tightening the nut.

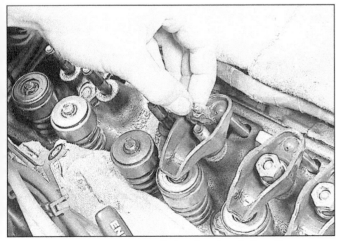

5.11 Moly-base grease applied to the pivot balls will ensure adequate lubrication until oil pressure builds up when the engine is started

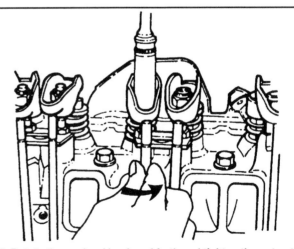

5.15 Rotate the pushrod back and forth and tighten the nut until you feel resistance to movement

**7.12 Use a large screwdriver or pry bar to break the manifold
gasket seal - DO NOT pry between the mating surfaces**

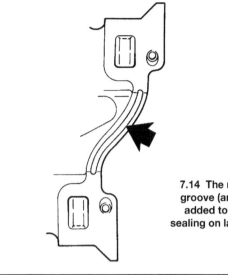

**7.14 The machined
groove (arrow) was
added to improve
sealing on later models**

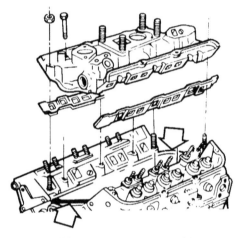

**7.15 Apply a bead of RTV sealant to the block ridges between the
heads (arrows)**

**7.16 The intake manifold gasket must be cut as shown before
installing it so the top can be positioned behind the pushrods**

At the point where the pushrod can no longer
be turned, all lash has been removed **(see
illustration).** Now tighten the nut an additional 3/4-turn.

16   Adjust the number one, five and six
cylinder intake valves and the number one,
two and three cylinder exhaust valves with
the crankshaft in this position, using the
method just described.

17   Rotate the crankshaft until the number
four piston is at TDC on the compression
stroke and adjust the number two, three and
four cylinder intake valves and the number
four, five and six cylinder exhaust valves.

18   Install the rocker arm covers.

6   Valve spring, retainer and seals -
replacement

Refer to Chapter 2, Part A, Section 6 for
this procedure. Remove the rocker arm covers, rocker arms and pushrods; adjust the
valve lash following the procedure in Section
5 of this Part.

7   Intake manifold - removal and
installation

*Refer to illustrations 7.12, 7.14, 7.15, 7.16
and 7.20*

1   Disconnect the negative cable from the
battery.

2   Drain the coolant from the radiator (see
Chapter 1).

3   Remove the wires, hoses, cables and
fuel lines at the TBI unit (see Chapter 4).

4   Label and disconnect all wires and
hoses connected to the intake manifold.

5   If equipped, unbolt the air conditioning
compressor and set it aside without disconnecting the hoses (see Chapter 3).

6   Label and disconnect the spark plug
wires from the plugs, then remove the distributor cap (see Chapter 1).

7   Remove the distributor (see Chapter 5).

8   Detach the EGR vacuum line (see Chapter 6).

9   Detach the evaporative emission hoses.

10   Remove the rocker arm covers (see

Section 4).

11   Disconnect the upper radiator and
heater hoses.

12   Unbolt the intake manifold and lift it off
the engine. If it's stuck, pry carefully against a
protrusion on the manifold **(see illustration).**
Do not pry between the gasket surfaces.

13   Remove all traces of old gasket material
and sealant from the manifold and cylinder
head mating surfaces, then clean them with
lacquer thinner or acetone. Clean the intake
manifold bolt holes in the cylinder head by
chasing them with a tap. Compressed air can
be used to remove the debris from the holes.
Warning: *Wear eye protection when using
compressed air.*

14   Check the underside of the rear of the
intake manifold to see if a machined groove
is present **(see illustration).** This groove was
added to later production models to improve
oil sealing.

15   Apply a 3/16-inch bead of RTV sealant
to each of the ridges between the heads **(see
illustration).**

16   The intake manifold gaskets will have to

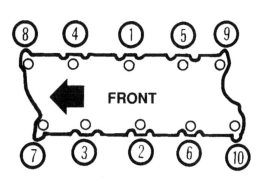

7.20 Intake manifold bolt tightening sequence

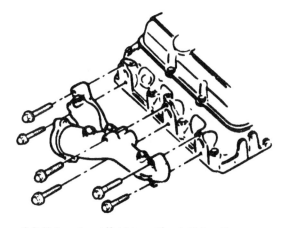

8.8 Exhaust manifold mounting bolt locations

be cut where indicated to position the tops behind the pushrods (see illustration). Cut only those areas necessary to clear the pushrods.

17 Install the new intake manifold gaskets, noting that they're marked Right and Left. Be sure to install them as indicated.

18 Hold the gaskets in place by extending the RTV sealant bead up 1/4-inch onto the gasket ends.

19 Install the intake manifold on the engine and hand tighten the bolts. Make sure the areas between the block ridges and intake manifold are completely sealed.

20 Following the sequence shown (see illustration), tighten the manifold bolts, in several steps, to the torque listed in this Chapter's specifications.

21 Reinstall the remaining parts in the reverse order of removal.

22 Run the engine, adjust the ignition timing and check for oil and vacuum leaks.

8 Exhaust manifolds - removal and installation

Refer to illustration 8.8

Warning: *Allow the engine to cool completely before performing this procedure.*

Both manifolds

1 Detach the cable from the negative battery terminal.

2 Raise the front of the vehicle and support it securely on jackstands. Apply the parking brake and block the rear wheels to keep the vehicle from rolling.

3 Remove the bolts attaching the exhaust pipe to the exhaust manifold, then separate the pipe from the manifold.

4 Remove the four rear manifold bolts.

5 Remove the jackstands and lower the vehicle.

Right manifold

6 Disconnect the air injection pump and alternator brackets from the manifold.

7 Disconnect the spark plug wires from

the spark plugs, labeling them as they're disconnected to simplify installation.

8 Remove the exhaust manifold mounting bolts (see illustration) and separate the manifold from the engine.

Left manifold

9 Remove the heat stove tube.

10 Remove the air cleaner assembly (see Chapter 4), labeling all hoses.

11 Disconnect the hoses leading to the air injection valve.

12 Disconnect and label all wires that will interfere with removal of the manifold.

13 Remove the power steering pump bracket from the cylinder head. Loosen the adjusting bracket bolt and detach the drivebelt from the pulley first. After removing the bracket from the cylinder head, place the pump assembly aside, out of the way. Do not disconnect any hoses and be sure to keep the top of the pump up so no fluid spills.

14 Remove the remaining manifold bolts and separate the manifold and heat shield from the engine.

Both manifolds

15 Installation is the reverse of removal. Be sure to clean the cylinder head and manifold surfaces thoroughly before installing the manifold. Tighten the bolts to the specified torque.

9 Cylinder heads - removal and installation

Refer to illustrations 9.14 and 9.22

Warning: *Allow the engine to cool completely before performing this procedure.*

1 Remove the intake manifold (refer to Section 7).

2 Raise the vehicle and support it securely on jackstands.

3 Locate the engine block drain plugs to the rear of the engine mounts (the plug on the left side is just above the oil filter). Remove the plugs and drain the block.

4 Disconnect the exhaust pipe from the

exhaust manifold.

5 If you're working on the left cylinder head, unbolt and remove the oil dipstick tube assembly from the left side of the engine. If you're working on the right cylinder head, remove the drivebelt, alternator and AIR pump with the mounting bracket from the head. Remove the lifting "eye" from the rear of the head (necessary only if the head is going to be replaced with a new one).

6 Remove the jackstands and lower the vehicle.

7 **Note:** *Steps 8 through 11 should be followed if the head is going to be replaced with a new one. The Steps can be performed either before or after the head has been removed. In the accompanying illustrations, the procedures were performed before the head was removed.*

8 Remove the exhaust manifold (refer to Section 8).

9 Remove the ground strap at the rear of the head and the sensor connector at the front of the head.

10 Detach the power steering pump and bracket from the cylinder head.

11 Remove the air-conditioner compressor bracket from the front of the cylinder head, if equipped.

12 Loosen the rocker arm nuts enough to allow removal of the pushrods, then remove the pushrods (see Section 5).

13 Using a new head gasket, outline the cylinders and bolt pattern on a piece of cardboard (see illustration 10.13 in Part A) Be sure to indicate the front of the engine for reference. Punch holes at the bolt locations. Loosen the ten cylinder head bolts in 1/4-turn increments until they can be removed by hand (start at the ends and work toward the center of the head). Store the bolts in the cardboard holder as they're removed - this will ensure that they're reinstalled in their original locations.

14 Lift the head off the engine. If it's stuck, DO NOT pry between the head and block - damage to the mating surfaces will result! To dislodge the head, use a long screwdriver or pry bar under the cast "ears" to pry up on it

9.14 Be careful not to damage the gasket sealing surface when breaking the head loose with a pry bar or large screwdriver

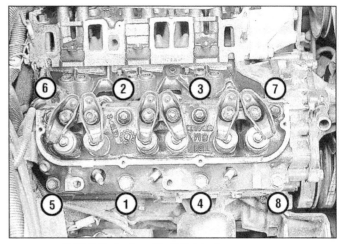

9.22 Cylinder head bolt tightening sequence

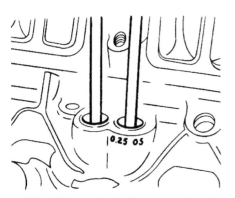

10.7a If the engine is factory equipped with oversize lifters, the lifter boss will be marked with a dab of white paint and will have 0.25 (mm) OS stamped on it

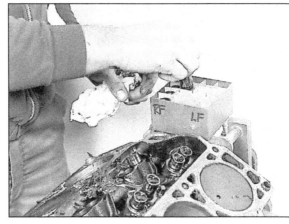

10.7b As the lifters are removed from the engine block, they should be stored separately to ensure reinstallation in their original locations

(see illustration). Don't damage the cylinder head sealing surface.

15   If a new cylinder head is being installed, attach the components previously removed from the old head.

16   The mating surfaces of the cylinder head and block must be perfectly clean when the head is installed. Use a gasket scraper to remove all traces of carbon and old gasket material, then clean the mating surfaces with lacquer thinner or acetone. If there's oil on the mating surfaces when the head is installed, the gasket may not seal correctly and leaks could develop. When working on the block, stuff the cylinders with clean shop rags to keep out debris. Use a vacuum cleaner to remove any debris that falls into the cylinders.

17   Check the block and head mating surfaces for nicks, deep scratches and other damage. If damage is slight, it can be removed with a file; if it's excessive, machining may be the only alternative.

18   Use a tap of the correct size to chase the threads in the head bolt holes. Mount each bolt in a vise and run a die down the threads to remove corrosion and restore the

threads. Dirt, corrosion, sealant and damaged threads will affect torque readings.

19   Install the gasket over the engine block dowel pins with the mark THIS SIDE UP visible.

20   Position the cylinder head over the gasket.

21   Coat the cylinder head bolts with RTV sealant and install them.

22   Tighten the bolts in the recommended sequence (see illustration) to the torque listed in this Chapter's specifications. Work up to the final torque in three steps.

23   Install the pushrods and rocker arms as described in Section 5.

24   The remaining installation steps are the reverse of removal. Before installing the rocker arm covers, adjust the valve lash (refer to Section 5).

---

10 Valve lifters - removal, inspection and installation

1   A noisy valve lifter can be isolated when the engine is idling. Hold a mechanic's stethoscope or a length of hose near the location of each valve while listening at the other end. Another method is to remove the rocker arm cover and, with the engine idling,

place a finger on each of the valve spring retainers, one at a time. If a valve lifter is defective, it will be evident from the shock felt at the retainer as the valve seats.

2   The most likely causes of noisy valve lifters are dirt trapped between the plunger and the lifter body or lack of oil flow, viscosity or pressure. Before condemning the lifters, we recommend checking the oil for fuel contamination, correct level, cleanliness and correct viscosity.

Removal

Refer to illustrations 10.7a and 10.7b

3   Remove the rocker arm cover(s) as described in Section 4.

4   Remove the intake manifold as described in Section 7.

5   Remove the rocker arms and pushrods (Section 5).

6   There are several ways to extract the lifters from the bores. A special tool designed to grip and remove lifters is manufactured by many tool companies and is widely available, but it may not be required in every case. On newer engines without a lot of varnish buildup, the lifters can often be removed with a small magnet or even with your fingers. A machinist's scribe with a bent end can be used to pull the lifters out by positioning the

10.9a If the bottom of any lifter is worn concave, pitted, scratched or galled, replace the entire set with new lifters

10.9b The foot of each lifter should be slightly convex - the side of another lifter can be used as a straightedge to check it; if it appears flat, it's worn and must not be reused

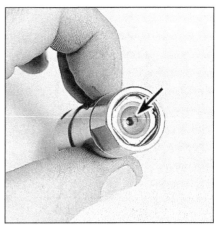

10.9c Check the pushrod seat (arrow) in the top of each lifter for wear

point under the retainer ring in the top of each lifter. **Caution:** *Don't use pliers to remove the lifters unless you intend to replace them with new ones (along with the camshaft). The pliers may damage the precision machined and hardened lifters, rendering them useless.*

7   Before removing the lifters, arrange to store them in a clearly labeled box to ensure they can be reinstalled in their original locations. **Note:** *Some engines may have both standard and 0.25 mm (0.010-inch) oversize lifters installed at the factory. If so, they are marked accordingly* **(see illustration).** Remove the lifters and store them where they won't get dirty **(see illustration).**

Inspection and installation

Refer to illustrations 10.9a, 10.9b and 10.9c
8   Clean the lifters with solvent and dry them thoroughly without mixing them up.
9   Check each lifter wall, pushrod seat and foot for pitting, scuffing, score marks and uneven wear **(see illustration).** Each lifter foot (the surface that rides on the cam lobe) must be slightly convex, although this can be difficult to determine by eye. If the base of the lifter is concave **(see illustration),** the lifters and camshaft must be replaced. If the lifter walls are damaged or worn (which isn't very likely), inspect the lifter bores in the engine block as well. If the pushrod seats **(see illustration)** are worn, check the pushrod ends.
10   If new lifters are being installed, a new camshaft must also be installed. If the camshaft is replaced, then install new lifters as well (see Section 15). Never install used lifters unless the original camshaft is used and the lifters can be installed in their original locations! When installing lifters, make sure they're coated with moly-base grease or engine assembly lube.
11   Soak new lifters in oil to remove trapped air.
12   The remaining installation steps are the reverse of removal.
13   Run the engine and check for oil leaks.

11 Vibration damper - removal and installation

Refer to illustrations 11.5 and 11.8
1   Remove the bolts and separate the radiator shroud from the radiator.
2   Remove the mounting bolts and detach the cooling fan and the radiator shroud. On pick-up models, unbolt the suspension

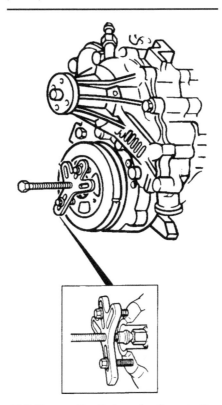

11.5 Use a puller that applies force to the vibration damper hub - be sure the large puller bolt doesn't damage the crankshaft threads

crossmember, if necessary, for clearance.
3   Loosen the adjusting bolts as necessary, then remove the drivebelts, tagging each one as it's removed to simplify installation.
4   Remove the bolts from the crankshaft pulley (a screwdriver can be used to lock the starter ring gear on the flywheel so the crankshaft won't rotate), then detach the pulley. Remove the vibration damper-to-crankshaft bolt.
5   Attach a puller to the damper and draw it off the crankshaft. Be careful not to drop it as it comes free. A common gear puller should not be used - it may separate the outer portion of the damper from the hub. Use only a puller that bolts to the hub **(see illustration).**
6   Before installation, coat the oil seal journal on the damper with moly-base grease.
7   Place the damper in position over the key on the crankshaft. Make sure the damper keyway lines up with the key.
8   Using a damper installation tool, available at most auto part stores, or equivalent), push the damper onto the crankshaft. The special tool **(see illustration)** distributes the pressure evenly around the hub. The bolt that threads into the crankshaft and a large washer can be used if the tool isn't available.
9   Remove the installation tool and install the damper bolt. Tighten the bolt to the specified torque.
10   To install the remaining components, reverse the removal procedure.
11   Adjust the drivebelts (refer to Chapter 1).

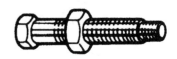

11.8 A special tool is recommended to install the vibration damper

12.3 A special tool is recommended for front crankshaft seal installation

12 Crankshaft front oil seal - replacement

With timing chain cover installed on engine

Refer to illustration 12.3

1   With the vibration damper removed (see Section 11), pry the old seal out of the cover with a large screwdriver. Be very careful not to damage the seal journal on the crankshaft.
2   Place the new seal in position with the open end facing toward the inside of the cover.
3   Drive the seal into the cover until it's seated. A special tool, available at most auto part stores, is recommended for this purpose **(see illustration)**. The tool is designed to exert even pressure around the entire circumference of the seal as it's hammered into place. A section of large-diameter pipe or a large socket can also be used. Be careful not to distort the front cover.
4   Reinstall the remaining parts in the reverse order of removal.

With timing chain cover removed from engine

Refer to illustrations 12.7 and 12.9

5   This method is preferred, as the cover

12.7 Driving the seal out of the front cover

can be supported while the old seal is removed and the new one is installed.
6   Remove the timing chain cover (refer to Section 13).
7   Using a large screwdriver, pry the old seal out of the cover bore. As an alternative, support the cover and drive the seal out from the rear **(see illustration)**. Be careful not to damage the cover.
8   With the front of the cover facing up, place the new seal in position with the open end facing toward the inside of the cover.
9   Using a block of wood and hammer, drive the new seal into the cover until it's completely seated **(see illustration)**.
10   Install the timing chain cover by reversing the removal procedure in Section 13.

13 Timing chain cover - removal and installation

Refer to illustrations 13.6a, 13.6b and 13.8

1   Remove the water pump as described in Chapter 3.

12.9 Installing the new oil seal with a block of wood and a hammer

2   If equipped with air conditioning, remove the compressor from the mounting bracket and secure it out of the way. **Warning:** *Do not disconnect any of the air conditioning system hoses without having the system depressurized by a dealer service department or service station.*
3   Remove the compressor mounting bracket.
4   Remove the vibration damper as described in Section 11.
5   Disconnect the lower radiator hose at the timing chain cover.
6   Remove the timing chain cover mounting bolts and separate the cover from the engine **(see illustrations)**.
7   Clean all oil, dirt and old gasket material off the sealing surfaces of the cover and engine block. Replace the oil seal as described in Section 12.
8   Apply a continuous 3/32-inch bead of anaerobic sealant (Loctite 515 or equivalent) to both mating surfaces of the cover (except the mating surface where the cover engages

13.6a After the water pump (1) is removed, two bolts and two studs secure the timing chain cover (2) to the engine block . . .

13.6b ... and two bolts secure the timing chain cover to the front lip of the oil pan

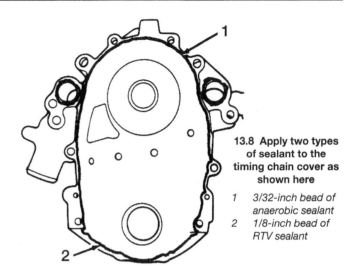

13.8 Apply two types of sealant to the timing chain cover as shown here

1 3/32-inch bead of anaerobic sealant
2 1/8-inch bead of RTV sealant

the oil pan lip). Apply RTV-type sealant to the cover-to-oil pan area. Also apply anaerobic sealant to the areas surrounding the coolant passages **(see illustration)**.

9 Place the timing chain cover in position on the engine block and install the mounting bolts. Tighten the bolts to the specified torque in a criss-cross pattern (to avoid distorting the cover).

10 The remaining installation steps are the reverse of removal.

14 Timing chain and sprockets - inspection, removal and installation

Refer to illustrations 14.8, 14.9, 14.10 and 14.12

1 Disconnect the cable from the negative battery terminal.

2 Remove the vibration damper (see Section 11).

3 Remove the timing chain cover (see Section 13).

4 Before removing the chain and sprockets, visually inspect the teeth on the sprockets for signs of wear and check the chain for looseness.

5 If either or both sprockets show any signs of wear (edges on the teeth of the camshaft sprocket rounded, bright or blue areas on the teeth of either sprocket, chipping, pitting, etc.), they should be replaced with new ones. Wear in these areas is very common. Failure to replace a worn timing chain and sprockets may result in erratic engine performance, loss of power and lowered gas mileage.

6 If any one component (timing chain or either sprocket) requires replacement, the other two components should be replaced as well.

7 If it's determined the components require replacement, proceed as follows.

8 Reinstall the vibration damper mounting bolt and use it to turn the crankshaft clockwise until the marks on the camshaft and crankshaft are in exact alignment **(see illustration)**. At this point the number one and four pistons will be at top dead center with the number four piston in the firing position (verify by checking the position of the rotor in the distributor, which should point to the number four spark plug wire terminal). **Note:** *Do not attempt to remove either sprocket or the timing chain until this is done and do not turn the crankshaft or camshaft after the sprockets/chain are removed.*

9 Remove the three camshaft sprocket mounting bolts **(see illustration)** and detach the camshaft sprocket and timing chain from the front of the engine. You may have to tap the sprocket with a soft-face hammer to dislodge it.

14.8 The timing marks (arrows) on the sprocket should be aligned as shown - a straight line should pass through the center of the camshaft sprocket timing mark, the crankshaft sprocket timing mark and the center of the crankshaft

14.9 A screwdriver will keep the camshaft sprocket from turning while loosening the mounting bolts

14.10　A gear puller will be needed to remove the crankshaft sprocket

14.12　Lubricate the thrust (rear) surface of the camshaft sprocket before installing it

10　If you have to remove the crankshaft sprocket, it can be withdrawn with a gear puller **(see illustration)**.

11　Push the crankshaft sprocket onto the crankshaft with the vibration damper bolt and a large washer or washers.

12　Lubricate the thrust (rear) surface of the camshaft sprocket with moly-base grease or engine assembly lube **(see illustration)**. Install the timing chain over the camshaft sprocket with slack in the chain hanging down.

13　With the timing marks aligned, slip the chain over the crankshaft sprocket and then draw the camshaft sprocket into place with the three bolts. Do not hammer or attempt to drive the camshaft sprocket into place, as it could dislodge the welch plug at the rear of the engine.

14　With the chain and both sprockets in place, check again to make sure the timing marks on the two sprockets are properly aligned. If not, remove the camshaft sprocket and move the chain until the marks align.

15　Lubricate the chain with engine oil and install the remaining components in the reverse order of removal.

15　Camshaft and bearings - removal, inspection and installation

Camshaft lobe lift check

Refer to illustration 15.3

1　To determine the extent of cam lobe wear, the lobe lift should be checked prior to camshaft removal. Refer to Section 4 and remove the rocker arm covers.

2　Position the number one piston at TDC on the compression stroke (see Section 3).

3　Beginning with the number one cylinder valves, mount a dial indicator on the engine and position the plunger against the top surface of the first rocker arm. The plunger should be directly above and in-line with the

15.3　When checking the camshaft lobe lift, the dial indicator plunger must be positioned directly above the pushrod

pushrod **(see illustration)**.

4　Zero the dial indicator, then very slowly turn the crankshaft in the normal direction of rotation until the indicator needle stops and begins to move in the opposite direction. The point at which it stops indicates maximum cam lobe lift.

5　Record this figure for future reference, then reposition the piston at TDC on the compression stroke.

6　Move the dial indicator to the remaining number one cylinder rocker arm and repeat the check. Be sure to record the results for each valve.

7　Repeat the check for the remaining valves. Since each piston must be at TDC on the compression stroke for this procedure, work from cylinder-to-cylinder following the firing order sequence.

8　After the check is complete, compare the results to the Specifications. If camshaft lobe lift is less than specified, cam lobe wear has occurred and a new camshaft should be installed.

Removal

Refer to illustrations 15.17 and 15.18

9　Remove the cable from the negative

battery terminal.

10　Drain the oil (Chapter 1).

11　Remove the radiator (Chapter 3).

12　If equipped with air conditioning, remove the condenser (Chapter 3).

13　Remove the valve lifters (Section 10).

14　Remove the timing chain cover (Section 13).

15　Remove the fuel pump and pushrod (Chapter 4).

16　Remove the timing chain and camshaft sprocket (Section 14).

17　Install two long bolts in the end of the camshaft to use as a handle to pull on and support the camshaft **(see illustration)**.

18　Carefully draw the camshaft out of the engine block. Do this very slowly to avoid damage to the camshaft bearings as the lobes pass over them. Support the camshaft with one hand near the engine block and the other with a wire hook at the other end **(see illustration)**.

Inspection and installation

Refer to illustration 15.20

19　Refer to Chapter 2, Part A, for camshaft and bearing inspection procedures.

20　Prior to installing the camshaft, coat

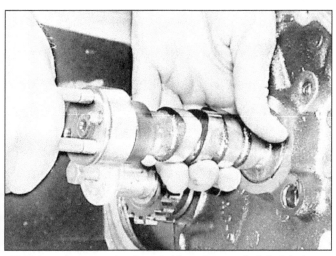

15.17 Long bolts can be threaded into the camshaft bolt holes to provide a handle for removal and installation - support the cam near the block as it's withdrawn

15.18 A length of wire with a hook on it can be used to support the camshaft as you guide it out of the block

each of the lobes and journals with engine assembly lube or moly-base grease **(see illustration)**.

21 Slide the camshaft into the engine block, again taking extra care so you don't damage the bearings.

22 Install the camshaft sprocket and timing chain as described in Section 14.

23 Install the remaining components in the reverse order of removal by referring to the appropriate Chapter or Section.

24 Adjust the valve lash (Section 5).

25 Have the air conditioning system (if equipped) evacuated and recharged.

16 Oil pan - removal and installation

Refer to illustrations 16.19 and 16.20
Note: *The following procedure is based on the assumption the engine is in the vehicle. If its been removed, simply unbolt the oil pan and detach it from the block.*

Removal

1 Disconnect the negative battery cable from the battery, then refer to Chapter 1 and drain the engine oil.

2 Refer to Chapter 5 and remove the starter motor.

3 If necessary for clearance, separate the exhaust pipes from the manifolds.

4 Remove the front skidplate (4WD only).

5 Detach the front crossmember.

6 On Trooper models, remove the catalytic converter bolts.

7 Detach the front driveshaft from the differential pinion flange (4WD only).

8 Remove the braces surrounding the flywheel cover plate.

9 Use paint to mark the pitman arm and shaft for reassembly.

10 Remove the pitman arm from the shaft (4WD only).

11 Detach the idler arm from the shaft (4WD only).

12 Remove the rubber tube from the front axle vent (4WD only).

13 Use jackstands to support the front axle assembly.

14 Remove the bolts from the right and left axle housing isolators and lower the front axle housing (4WD only).

15 Remove the oil pan mounting nuts/bolts.

16 Carefully separate the pan from the block. Don't pry between the block and pan or damage to the sealing surfaces may result and oil leaks could develop. You may have to turn the crankshaft slightly to maneuver the front of the pan past the crankshaft counterweights.

Installation

17 Clean the gasket sealing surfaces with lacquer thinner or acetone. Make sure the bolt holes in the block are clean.

18 Check the oil pan flanges for distortion, particularly around the bolt holes and corners. If necessary, place the pan on a block of wood and use a hammer to flatten and restore the gasket surface.

19 Check the corners of the oil pan for distortion with a straightedge across the cradle opening **(see illustration)**. If distortion is severe, install a new oil pan.

20 Apply RTV sealant to the rear cradle corners **(see illustration)** and use a new pan gasket kit.

21 Carefully position the pan against the block and install the bolts finger-tight. Tighten the nuts/bolts in three steps to the torque listed in this Chapter's specifications. Start at the center of the pan and work out toward the ends in a spiral pattern.

22 The remaining steps are the reverse of

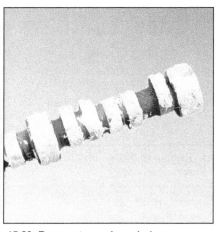

15.20 Be sure to apply moly-base grease or engine assembly lube to the cam lobes and bearing journals before installing the camshaft

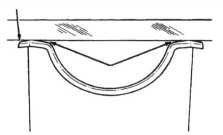

16.19 Using a straightedge, check for distortion at the corners of the oil pan (arrows)

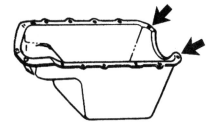

16.20 Apply RTV sealant to the rear cradle corners of the oil pan (arrows)

17.2 Remove the oil pump-to-rear main bearing cap bolt and detach the pump and extension shaft

19.3 Pry the crankshaft rear oil seal out of the bore - be very careful not to scratch he crankshaft seal journal or the edge of the seal bore

removal. **Caution:** *Don't forget to refill the engine with oil before starting it (see Chapter 1).*

23   Start the engine and check carefully for oil leaks at the oil pan.

17 Oil pump - removal and installation

Refer to illustration 17.2

1   Remove the oil pan (refer to Section 16).

2   Remove the pump-to-rear main bearing cap bolt **(see illustration)** and separate the pump and extension shaft from the engine.

3   To install the pump, hold it in position and align the top end of the hexagonal extension shaft with the hexagonal socket in the lower end of the distributor drive gear. The distributor drives the oil pump, so it's essential that the alignment is correct.

4   Install the oil pump-to-rear main bearing cap bolt and tighten it to the torque listed in this Chapter's specifications.

5   Reinstall the oil pan.

18 Flywheel/driveplate - removal and installation

Refer to Chapter 2, Part A, Section 20 for this procedure. Be sure to use the torque specifications in this Part of Chapter 2 for the V6 engine.

19 Crankshaft rear oil seal - replacement

Refer to illustrations 19.3 and 19.4

Note: *Special tools, as noted in the Steps which follow, are required for this procedure. They are available from your dealer or may, in some cases, be rented from an auto parts store or tool rental shop.*

1   Remove the transmission (see Chap-

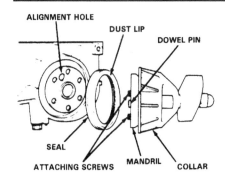

19.4 Crankshaft rear oil seal installation tool

ter 7A or 7B).

2   Remove the flywheel or driveplate (see Section 18).

3   A one-piece seal is utilized, which allows the oil pan to remain in place during this procedure. Pry out the old seal, taking care not to damage the crankshaft seal journal **(see illustration)**. Check the crankshaft for scratches, burrs and nicks on the seal journal.

4   A special seal installation tool, available at most auto part stores, is required to properly seat the seal in the bore without damaging it **(see illustration)**. Lubricate the seal bore, seal lip and seal journal on the crankshaft with engine oil. Slide the seal over the mandril on the tool until the dust lip on the seal bottoms squarely against the collar on the tool.

5   Position the dowel pin on the tool in the alignment hole in the crankshaft and secure the tool to the crankshaft.

6   Turn the T-handle of the tool until the collar pushes the seal into the bore. Make sure the seal is installed squarely and seated completely.

7   To complete the operation, install the flywheel (or driveplate) and transmission, then start the engine and check for leaks.

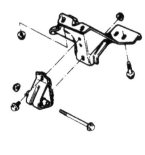

20.1 V6 engine mounts - exploded view

20 Engine mounts - check and replacement

Refer to illustration 20.1

Refer to Chapter 2, Part A, Section 22 for this procedure, but note that the V6 engine mounts are slightly different **(see illustration)** in ways that don't affect the check and replacement procedures.

Chapter 2 Part C
General engine overhaul procedures

Contents

Specifications

G180Z and G200Z four-cylinder engines

General

Oil pressure	56.8 psi at 1400 rpm
Compression pressure (at 350 rpm)	
Standard	171 psi
Minimum	128 psi

Camshaft and rocker arms
See Part A

Cylinder head warpage limit
0.008 in (0.2 mm)

Valves and related components

Valve stem diameter	
Intake	0.3102 in (7.88 mm)
Exhaust	0.3091 in (7.85 mm)
Valve guide inside diameter	Not available
Valve stem-to-guide clearance	
Intake	0.0009 to 0.0022 in (0.023 to 0.056 mm)
Exhaust	0.0015 to 0.0031 in (0.038 to 0.070 mm)
Valve guide height from surface of cylinder head	0.0031 to 0.0047 in (0.08 to 0.12 mm)
Valve spring free length (intake and exhaust)	
Inner	1.7835 in (45.30 mm)
Outer	1.8465 in (46.90 mm)

Engine block

Deck distortion limit	0.016 in (0.40 mm)
Bore	
Diameter	3.425 to 3.466 in (87.000 to 88.040 mm)
Taper/out-of-round limits	Less than 0.008 in (0.20 mm)
Piston-to-bore clearance	0.0018 to 0.0026 in (0.045 to 0.065 mm)

G180Z and G200Z four-cylinder engines (continued)

Pistons and rings

Piston diameter	
Standard	3.4250 to 3.4260 in (87.000 to 87.040 mm)
1st oversize	3.4440 to 3.4450 in (87.500 to 87.520 mm)
2nd oversize	3.4640 to 3.4650 in (88.000 to 88.020 mm)
Piston diameter (G180Z - 1983 only)	
Grade A	3.4169 to 3.4173 in (86.945 to 86.955 mm)
Grade B	3.4173 to 3.4177 in (86.955 to 86.965 mm)
Grade C	3.4177 to 3.4181 in (86.966 to 86.975 mm)
Grade D	3.4181 to 3.4185 in (86.976 to 86.985 mm)
Compression ring side clearance	0.0010 to 0.0024 in (0.025 to 0.060 mm)
Piston ring end gap	
Compression rings	0.0140 to 0.0200 in (0.35 to 0.50 mm)
Oil ring	0.0080 to 0.0350 in (0.20 to 0.90 mm)

Crankshaft, connecting rods and main bearings

Connecting rod end play (side clearance)	0.0137 in (0.35 mm)
Crankshaft end play	
Standard	0.0024 to 0.0094 in (0.06 to 0.24 mm)
Service limit	0.0117 in (0.3 mm)
Crankshaft runout limit	0.0020 in (0.05 mm)
Main journal diameter	2.2016 to 2.2022 in (55.920 to 55.935 mm)
Connecting rod journal diameter	1.9262 to 1.9268 in (48.925 to 48.940 mm)
Journal taper/out-of-round limits	0.002 in (0.05 mm)
Main bearing oil clearance	
Standard	0.008 to 0.0025 in (0.021 to 0.064 mm)
Service limit	0.0046 in (0.12 mm)
Connecting rod bearing oil clearance	0.0080 to 0.0025 in (0.021 to 0.050 mm)

Torque specifications*

	Ft-lbs
Main bearing cap bolts	75
Connecting rod cap nuts	44
Flywheel/driveplate mounting bolts	78

* **Note:** *Refer to Part A for additional torque specifications.*

4ZD1 and 4ZE1 four-cylinder engines

General

Oil pressure (at 3000 rpm)	56.9 psi
Compression pressure	
Standard	171 psi
Minimum	128 psi

Camshaft and rocker arms .... See Part A

Cylinder head warpage limit .... 0.008 in (0.2 mm)

Valves and related components

Valve stem diameter	
Intake	0.3150 to 0.3102 in (8.000 to 7.880 mm)
Exhaust	0.3150 to 0.3091 in (8.000 to 7.850 mm)
Valve face angle	45.5-degrees
Valve seat angle	45-degrees
Valve seat width	
Standard	0.048 to 0.063 in (1.20 to 1.60 mm)
Service limit	0.078 in (2.0 mm)
Valve guide inside diameter	Not available
Valve stem-to-guide clearance	
Standard	
Intake	0.00098 to 0.0022 in (0.023 to 0.056 mm)
Exhaust	0.0015 to 0.0031 in (0.038 to 0.070 mm)
Service limit	
Intake	0.0079 in (0.20 mm)
Exhaust	0.0097 in (0.25 mm)

Valve spring free length
    Standard.. 1.8951 in (48.10 mm)
    Service limit .... 1.8321 in (46.5 mm)
Valve spring out-of-square limit .... 0.0827 in (2.1 mm)

Engine block

Deck distortion limit.. 0.0078 in (0.20 mm)
Cylinder bore diameter
    4ZD1 engine .... 3.5182 to 3.5200 in (89.300 to 89.340 mm)
    4ZE1 engine .... 3.6450 to 3.6470 in (92.600 to 92.640 mm)
    1986 Trooper only
        Grade A cylinder bore.. 3.5160 to 3.5165 in (89.300 to 89.320 mm)
        Grade C cylinder bore.. 3.5166 to 3.5173 in (89.321 to 89.340 mm)
Cylinder bore taper/out-of-round limits...................................... Less than 0.0006 in (0.015 mm)
Piston-to-bore clearance
    4ZD1 engine .... 0.0008 to 0.0016 in (0.020 to 0.040 mm)
    4ZE1 engine .... 0.0010 to 0.0018 in (0.025 to 0.045 mm)

Pistons and rings

Piston diameter
    4ZD1 engine
        Standard.. 3.5160 to 3.5170 in (89.300 to 89.340 mm)
            1st oversize... 3.5337 to 3.5352 in (89.755 to 89.795 mm)
            2nd oversize.. 3.5533 to 3.5549 in (90.255 to 90.295 mm)
    4ZE1 engine
        Standard .... 3.6460 to 3.6470 in (92.600 to 92.640 mm)
            1st oversize... 3.6640 to 3.6660 in (93.065 to 93.105 mm)
            2nd oversize.. 3.6840 to 3.6850 in (93.565 to 93.605 mm)
    1986 Trooper only
        Grade A piston .... 3.5140 to 3.5148 in (89.255 to 89.275 mm)
        Grade C piston .... 3.5148 to 3.5155 in (89.276 to 89.295 mm)
Piston ring end gap
    Top compression ring .... 0.0120 to 0.0180 in (0.30 to 0.45 mm)
    Second compression ring
        4ZE1 .... 0.0240 to 0.0280 in (0.60 to 0.72 mm)
        4ZD1 .... 0.010 to 0.016 in (0.25 to 0.40 mm)
    Oil ring .... 0.0080 to 0.0280 in (0.20 to 0.70 mm)
Piston ring side clearance
    Top compression ring .... 0.0010 to 0.0024 in (0.025 to 0.060 mm)
    Second compression ring .... 0.0024 to 0.0028 in (0.060 to 0.072 mm)

Crankshaft, connecting rods and main bearings

Connecting rod end play (side clearance)
    Standard.. 0.00012 to 0.00080 in (0.003 to 0.020 mm)
    Service limit .... 0.0016 in (0.40 mm)
Crankshaft end play
    Standard.. 0.0024 to 0.0099 in (0.06 to 0.25 mm)
    Service limit .... 0.012 in (0.3 mm)
Main bearing journal diameter .... 2.2016 to 2.2022 in (55.920 to 55.935 mm)
Connecting rod journal diameter .... 1.9262 to 1.9268 in (48.925 to 48.940 mm)
Journal taper/out-of-round
    Standard.. Less than 0.0012 in (0.03 mm)
    Service limit .... 0.004 in (0.10 mm)
Main bearing oil clearance
    Standard.. 0.0009 to 0.0020 in (0.023 to 0.050 mm)
    Service limit .... 0.0047 in (0.12 mm)
Connecting rod bearing oil clearance
    Standard.. 0.0012 to 0.0024 in (0.030 to 0.060 mm)
    Service limit .... 0.0047 in (0.12 mm)

Flywheel runout limit .... 0.00024 in (0.006 mm)

Torque specifications*

Ft-lbs

Main bearing cap bolts 78
Connecting rod cap nuts 43

*__Note:__ *Refer to Part A for additional torque specifications.*

V6 engine

General

Oil pressure	
At 500 rpm	10 psi
At 2000 rpm	30 to 55 psi
Compression pressure	
Standard	Not available
Maximum difference between cylinders	30 psi

Cylinder head warpage limit

Cylinder head warpage limit ... Less than 0.002 in (0.05 mm) per 6 inches

Valves and related components

Valve stem diameter	Not available
Valve guide inside diameter	Not available
Valve stem-to-guide clearance	0.0010 to 0.0027 in (0.026 to 0.068 mm)
Valve face angle (intake and exhaust)	45-degrees
Valve seat angle (intake and exhaust)	46-degrees
Valve seat	
Runout (intake and exhaust)	0.0020 in (0.05 mm)
Width	
2.8L	
Intake	0.0492 to 0.0591 in (1.25 to 1.50 mm)
Exhaust	0.0630 to 0.0750 in (1.60 to 1.90 mm)
3.1L	
Intake	0.061 to 0.073 in (1.55 to 1.85 mm)
Exhaust	0.067 to 0.079 in (1.70 to 2.0 mm)
Valve spring	
Free length	1.9094 in (48.5 mm)
Installed height	1.5748 in (40 mm)

Engine block

Cylinder bore diameter	
Standard	3.5036 to 3.5067 in (88.992 to 89.070 mm)
Wear limit	0.0008 in (0.020 mm)
Taper/out-of-round limit	Less than 0.0005 in (0.013 mm)

Pistons and rings

Piston diameter	Not available
Piston-to-bore clearance	
2.8L	0.0007 to 0.0017 in (0.017 to 0.043 mm)
3.1L	0.0009 to 0.0022 in (0.0235 to 0.0565 mm)
Piston ring side clearance	
2.8L	
Top compression ring	0.0012 to 0.0028 in (0.030 to 0.070 mm)
Second compression ring	0.0016 to 0.0037 in (0.040 to 0.095 mm)
3.1L (top and second ring)	0.002 to 0.0035 in (0.05 to 0.09 mm)
Piston ring end gap	
2.8L	
Compression rings	0.0098 to 0.1969 in (0.25 to 0.50 mm)
Oil control ring	0.0079 to 0.0299 in (0.20 to 0.76 mm)
3.1L	
Top compression ring	0.01 to 0.02 in (0.25 to 0.50 mm)
Second compression ring	0.02 to 0.028 in (0.5 to 0.71 mm)
Oil control ring	0.010 to 0.030 in (0.25 to 0.75 mm)

Crankshaft, connecting rods and main bearings

Connecting rod end play (side clearance)	0.0063 to 0.0267 in (0.16 to 0.64 mm)
Crankshaft end play	0.0024 to 0.0083 in (0.06 to 0.21 mm)
Crankshaft runout	Less than 0.0002 in (0.005 mm)
Main journal diameter	
3 Dots	2.6473 to 2.6476 in (67.241 to 67.249 mm)
2 Dots	2.6476 to 2.6480 in (67.249 to 67.257 mm)
1 Dot	2.6480 to 2.6482 in (67.257 to 67.265 mm)
Connecting rod journal diameter	
2 Dots	1.9983 to 1.9989 in (50.758 to 50.771 mm)
1 Dot	1.9989 to 1.9994 in (50.771 to 50.784 mm)
Journal taper/out-of-round limits	Less than 0.0002 in (0.005 mm)
Main bearing oil clearance	0.0012 to 0.0027 in (0.032 to 0.081 mm)
Connecting rod bearing oil clearance	0.0011 to 0.0037 in (0.028 to 0.095 mm)

Torque specifications*

	Ft-lbs
Main bearing cap bolts	72
Connecting rod cap nuts	39
Rear oil seal retainer bolts	4

* Note: *Refer to Part B for additional torque specifications.*

1 General information

Included in this portion of Chapter 2 are the general overhaul procedures for the cylinder head(s) and internal engine components. The information ranges from advice concerning preparation for an overhaul and the purchase of replacement parts to detailed, step-by-step procedures covering removal and installation of internal engine components and the inspection of parts. The following procedures have been written based on the assumption the engine has been removed from the vehicle. For information concerning in-vehicle engine repair, as well as removal and installation of the external components necessary for the overhaul, see Part A or B of this Chapter and Section 7 of this Part.

The Specifications included here in Part C are only those necessary for the inspection and overhaul procedures which follow. Refer to Parts A and B for additional Specifications.

2 Cylinder compression check

Refer to illustration 2.4

1   A compression check will tell you what mechanical condition the upper end (pistons, rings, valves, head gaskets) of your engine is in. Specifically, it can tell you if the compression is down due to leakage caused by worn piston rings, defective valves and seats or a blown head gasket. **Note:** *The engine must be at normal operating temperature and the battery must be fully charged for this check. Also, if the engine is equipped with a carburetor, the choke valve must be all the way open to get an accurate compression reading (if the engine's warm, the choke should be open).*

2   Begin by cleaning the area around the spark plugs before you remove them (compressed air should be used, if available, otherwise a small brush or even a bicycle tire pump will work). The idea is to prevent dirt from getting into the cylinders as the compression check is being done. Remove all of the spark plugs from the engine (Chapter 1).

3   Block the throttle wide open. Detach the coil wire from the distributor cap and ground it on the engine block. Use a heavy jumper wire with alligator clips at both ends to ensure a good ground.

4   With the compression gauge in the number one spark plug hole **(see illustration)**, depress the accelerator pedal all the way to the floor to open the throttle valve. Crank the engine over at least four compression strokes and watch the gauge. The compression should build up quickly in a healthy

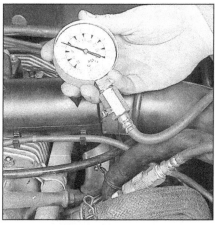

2.4 A compression gauge with a threaded fitting for the spark plug hole is preferred over the type that requires hand pressure to maintain the seal - be sure to open the throttle and choke valves as far as possible during the compression check!

engine. Low compression on the first stroke, followed by gradually increasing pressure on successive strokes, indicates worn piston rings. A low compression reading on the first stroke, which doesn't build up during successive strokes, indicates leaking valves or a blown head gasket (a cracked head could also be the cause). Record the highest gauge reading obtained.

5   Repeat the procedure for the remaining cylinders and compare the results to the Specifications.

6   Add some engine oil (about three squirts from a plunger-type oil can) to each cylinder, through the spark plug hole, and repeat the test.

7   If the compression increases after the oil is added, the piston rings are definitely worn. If the compression doesn't increase significantly, the leakage is occurring at the valves or head gasket. Leakage past the valves may be caused by burned valve seats and/or faces or warped, cracked or bent valves.

8   If two adjacent cylinders have equally low compression, there's a strong possibility that the head gasket between them is blown. The appearance of coolant in the combustion chambers or the crankcase would verify this condition.

9   If the compression is unusually high, the combustion chambers are probably coated with carbon deposits. If that's the case, the cylinder head(s) should be removed and decarbonized.

10   If compression is way down or varies greatly between cylinders, it would be a good

idea to have a leak-down test performed by an automotive repair shop. This test will pinpoint exactly where the leakage is occurring and how severe it is.

3 Engine removal - methods and precautions

If you've decided the engine must be removed for overhaul or major repair work, several preliminary steps should be taken.

Locating a place to work is extremely important. Adequate work space, along with storage space for the vehicle, will be needed. If a shop or garage isn't available, at the very least a flat, level, clean work surface made of concrete or asphalt is required.

Cleaning the engine compartment and engine before beginning the removal procedure will help keep tools clean and organized.

An engine hoist or A-frame will also be needed. Make sure the equipment is rated in excess of the combined weight of the engine and transmission. Safety is of primary importance, considering the potential hazards involved in lifting the engine out of the vehicle.

If the engine is being removed by a novice, a helper should be available. Advice and aid from someone more experienced would also be helpful. There are many instances when one person cannot simultaneously perform all of the operations required when lifting the engine out of the vehicle.

Plan the operation ahead of time. Arrange for or obtain all of the tools and equipment you'll need prior to beginning the job. Some of the equipment necessary to perform engine removal and installation safely and with relative ease are (in addition to an engine hoist) a heavy duty floor jack, complete sets of wrenches and sockets as described at the front of this manual, wooden blocks and plenty of rags and cleaning solvent for mopping up spilled oil, coolant and gasoline. If the hoist must be rented, make sure you arrange for it in advance and perform all of the operations possible without it ahead of time. This will save you money and time.

Plan for the vehicle to be out of use for quite a while. A machine shop will have to perform some of the work the do-it-yourselfer can't accomplish without special equipment. They often have a busy schedule, so it would be a good idea to consult them before removing the engine to accurately estimate the amount of time required to rebuild or repair components that need work.

Always be extremely careful when

removing and installing the engine. Serious injury can result from careless actions. Plan ahead, take your time and a job of this nature, although major, can be accomplished successfully.

4   Engine - removal and installation

Refer to illustration 4.40

Warning: *The engine is very heavy, so equipment designed specifically for lifting engines should be used. Never position any part of your body under the engine when it's supported by a hoist - it could shift or fall and cause serious injury or death! Also, the air conditioning system is under high pressure; loosening fittings will cause a sudden discharge of refrigerant, which could cause injuries. Have the system discharged by a service station before disconnecting any hoses or lines.*

1   Read through the Section entitled Engine removal - methods and precautions before beginning any work. On 2WD models, the transmission must be removed along with the engine. On 4WD models, the transmission should be removed before removing the engine (refer to Chapter 7 for the transmission removal procedure). **Note:** *On 1988 and later 2WD pick-ups, the transmission is removed separately, prior to removing the engine, just like 4WD vehicles.*

2   Remove the hood (see Chapter 11) and store it in a safe place where it won't be damaged.

3   Remove the air cleaner assembly (see Chapter 4).

4   Disconnect the cables from the battery (negative first, then positive).

5   Drain the coolant from the radiator and engine block (refer to Chapter 1 if necessary).

6   Drain the engine oil.

7   Drain the oil from the transmission. Refer to Chapter 1 if necessary.

8   Remove the upper and lower radiator hoses.

9   Disconnect all wires and vacuum hoses running between the engine and components attached to the body. **Note:** *Before disconnecting or removing wires or hoses, mark them with pieces of tape to identify their installed locations. This will eliminate possible problems and confusion during the installation procedure.*

10   Disconnect the fuel hoses from the mechanical fuel pump, if equipped, and plug them to prevent fuel leakage and the entry of dirt.

11   On fuel injected vehicles, relieve the fuel pressure (Chapter 4), then disconnect the fuel line from the fuel rail and return line and plug it.

12   If the vehicle is equipped with air conditioning, remove the compressor from the bracket and secure it out of the way. **Caution:** *Don't disconnect the hoses from the compressor unless the system has been discharged.*

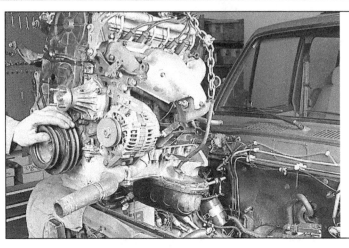

4.40 Be sure the chain is securely attached to the engine brackets and the hoist before lifting the engine out of the vehicle

13   If the vehicle is equipped with power steering, remove the power steering pump and secure it out of the way. Leave the hoses connected to it.

14   Disconnect the ignition coil wire from the distributor cap.

15   Detach the throttle cable from the carburetor or throttle valve on the common chamber (see Chapter 4).

16   Disconnect the brake booster vacuum hose and position it out of the way.

17   Disconnect the heater hoses where they attach to the engine.

18   On carburetor-equipped engines, disconnect the rubber hose at the vacuum switch.

19   Raise the front of the vehicle and support it on jackstands. Apply the parking brake and block the rear wheels.

20   Disconnect the wires from the alternator and starter. On vehicles with a V6 engine, remove the starter (Chapter 5).

21   If the vehicle is equipped with an automatic transmission, disconnect the oil cooler hoses leading from the oil cooler located at the base of the radiator. Allow the fluid to drain into a container.

22   Remove the radiator and shroud. Refer to Chapter 3 if necessary.

23   Detach the speedometer cable from the transmission. Disconnect all wires attached to the transmission.

24   On 4WD models, remove the primary and front driveshafts as described in Chapter 8.

25   On 4WD models, remove the four mounting bolts that attach the front differential to the crossmember, then remove the forward mounting bolt and lower the front differential so it rests on the crossmember.

26   On 2WD models, remove the driveshaft as described in Chapter 8.

27   Disconnect the front exhaust pipe from the exhaust manifold.

28   If the vehicle is equipped with a hydraulic clutch, detach the clutch operating cylinder from the transmission. Refer to Chapter 8 if necessary. If equipped with a clutch cable, take the tension off the cable by loosening the adjusting nut.

29   Disconnect the parking brake cable and

position it out of the way.

30   Disconnect any remaining lines, hoses or wires leading to the engine. Mark them first to avoid confusion during installation.

31   On 2WD models, disconnect the transmission (see Chapter 7) from the chassis but leave the bellhousing bolts attached to the engine so the transmission can be lifted out with the engine.

32   On 4WD models, remove the transmission from the vehicle (see Chapter 7).

33   On 4WD models with an automatic transmission, unbolt the torque converter from the driveplate (see Chapter 7).

34   Attach an engine hoist to the two brackets mounted on the engine. They're located at the corners of the head. Make sure the chain is attached securely with hooks or high-strength nuts, bolts and washers. The hook on the hoist should be over the center of the engine with the two lengths of chain the same length so the engine can be lifted straight up.

35   Raise the engine just enough to remove all slack from the chains.

36   On 4WD models, remove the bolts that attach the rear mounting member for the front differential to the frame.

37   Remove the left side engine mounting stopper plate (G180Z and G200Z engines - chain-driven camshaft).

38   On 2WD models, remove the bolts that attach the transmission mounting bracket to the body, then remove the bolts that attach the bracket to the transmission. Remove the damper, if equipped.

39   Remove the nuts that attach the front engine mount bracket to the chassis.

40   On 4WD models, lift the engine slightly, then move it forward to separate it from the transmission. Check carefully to make sure everything has been disconnected. On 2WD models, tilt the engine/transmission assembly down at the rear to clear the radiator support as you lift it up and out of the engine compartment **(see illustration).**

41   To separate the engine from an automatic transmission on 2WD models:

a) Remove the dust cover from the lower half of the torque converter housing.

b) Remove the torque converter-to-drive-

plate bolts. Access to the bolts can be gained (one at a time), by turning the crankshaft until each bolt comes into view through the lower half of the converter housing.

c) Withdraw the automatic transmission, leaving the driveplate bolted to the crankshaft rear flange.

42   To separate the engine from a manual transmission on 2WD models:

a) Remove the bolts that secure the bellhousing to the engine block.

b) Pull the transmission straight back so its weight doesn't hang up on the input shaft while it's still engaged in the clutch plate.

43   Installation is the reverse of removal.

44   Tighten all fasteners to the torque listed in this Chapter's specifications.

45   Before starting the engine, make sure the oil and coolant levels are correct.

46   Run the engine and check for leaks and proper operation of all accessories, then install the hood and test drive the vehicle.

5   Engine overhaul - general information

Refer to illustration 5.4

It's not always easy to determine when, or if, an engine should be completely overhauled, as a number of factors must be considered.

High mileage isn't necessarily an indication an overhaul is needed, while low mileage doesn't preclude the need for an overhaul. Frequency of servicing is probably the most important consideration. An engine that's had regular and frequent oil and filter changes, as well as other required maintenance, will most likely give many thousands of miles of reliable service. Conversely, a neglected engine may require an overhaul very early in its life.

Excessive oil consumption is an indication that piston rings, valve seals and/or valve guides are in need of attention. Make sure oil leaks aren't responsible before deciding the rings and/or guides are bad. Have a cylinder compression or leakdown test performed by an experienced tune-up mechanic to determine the extent of the work required.

If the engine is making obvious knocking or rumbling noises, the connecting rod and/or main bearings may be at fault. Check the oil pressure with a gauge installed in place of the oil pressure sending unit (sometimes called a switch) **(see illustration)** and compare it to the Specifications. If it's extremely low, the bearings and/or oil pump are probably worn out.

Loss of power, rough running, excessive valve train noise and high fuel consumption rates may also point to the need for an overhaul, especially if they're all present at the same time. If a complete tune-up doesn't remedy the situation, major mechanical work is the only solution.

An engine overhaul involves restoring the internal parts to the specifications of a new engine. During an overhaul, the piston rings are replaced and the cylinder walls are reconditioned (rebored and/or honed). If a rebore is done, new pistons are required. The main and connecting rod bearings are generally replaced with new ones and, if necessary, the crankshaft may be reground to restore the journals. Generally, the valves are serviced as well, since they're usually in less-than-perfect condition at this point. While the engine is being overhauled, other components, such as the distributor, starter and alternator, can be rebuilt as well. The end result should be a like new engine that will give many trouble free miles. **Note:** *Critical cooling system components such as the hoses, drivebelts, thermostat and water pump MUST be replaced with new parts when an engine is overhauled. The radiator should be checked carefully to ensure it isn't clogged or leaking; if in doubt, replace it with a new one. Also, we don't recommend overhauling the oil pump - install a new one if and when the engine is rebuilt.*

Before beginning the engine overhaul, read through the entire procedure to familiarize yourself with the scope and requirements of the job. Overhauling an engine isn't difficult, but it is time consuming. Plan on the vehicle being tied up for a minimum of two weeks, especially if parts must be taken to an automotive machine shop for repair or reconditioning. Check on availability of parts and make sure any necessary special tools and equipment are obtained in advance. Most work can be done with typical hand tools, although a number of precision measuring tools are required for inspecting parts to determine if they must be replaced. Often an automotive machine shop will handle the inspection of parts and offer advice concerning reconditioning and replacement. **Note:** *Always wait until the engine has been completely disassembled and all components, especially the engine block, have been inspected before deciding what service and repair operations must be performed by an automotive machine shop. Since the block's*

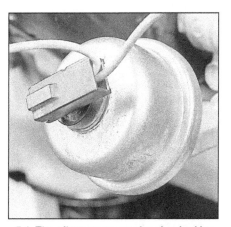

5.4 The oil pressure can be checked by removing the oil pressure switch and installing a pressure gauge in the hole

condition will be the major factor to consider when determining whether to overhaul the original engine or buy a rebuilt one, never purchase parts or have machine work done on other components until the block has been thoroughly inspected. As a general rule, time is the primary cost of an overhaul, so it doesn't pay to install worn or substandard parts.

As a final note, to ensure maximum life and minimum trouble from a rebuilt engine, everything must be assembled with care in a spotlessly clean environment.

6   Engine rebuilding alternatives

The do-it-yourselfer is faced with a number of options when performing an engine overhaul. The decision to replace the engine block, piston/connecting rod assemblies and crankshaft depends on a number of factors, with the number one consideration being the condition of the block. Other considerations are cost, access to machine shop facilities, parts availability, time required to complete the project and the extent of prior mechanical experience on the part of the do-it-yourselfer.

Some of the rebuilding alternatives include:

Individual parts - If the inspection procedures reveal that the engine block and most engine components are in reusable condition, purchasing individual parts may be the most economical alternative. The block, crankshaft and piston/connecting rod assemblies should all be inspected carefully. Even if the block shows little wear, the cylinder bores should be surface honed.

Crankshaft kit - This rebuild package consists of a reground crankshaft and a matched set of pistons and connecting rods. The pistons will already be installed on the connecting rods. Piston rings and the necessary bearings will be included in the kit. These kits are commonly available for standard cylinder bores, as well as for engine blocks which have been bored to a regular oversize.

Short block - A short block consists of an engine block with a crankshaft and piston/connecting rod assemblies already installed. All new bearings are incorporated and all clearances will be correct. The existing camshaft, valve train components, cylinder head(s) and external parts can be bolted to the short block with little or no machine shop work necessary.

Long block - A long block consists of a short block plus an oil pump, oil pan, cylinder head(s), rocker arm cover(s), camshaft and valve train components, timing sprockets and chain/belt. All components are installed with new bearings, seals and gaskets incorporated throughout. The installation of manifolds and external parts is all that's necessary.

Give careful thought to which alternative is best for you and discuss the situation with local automotive machine shops, auto parts dealers and experienced rebuilders before ordering or purchasing replacement parts.

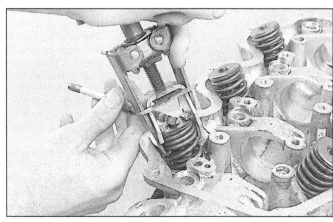

8.2 Have several plastic bags ready (one for each valve) before disassembling the head - label each bag and put the entire contents of each valve assembly in one bag as shown

8.3 Use a valve spring compressor to compress the springs, then remove the keepers from the valve stem with a small magnet

7   Engine overhaul - disassembly sequence

1   It's much easier to disassemble and work on the engine if it's mounted on a portable engine stand. A stand can often be rented quite cheaply from an equipment rental yard. Before the engine is mounted on a stand, the flywheel/driveplate should be removed from the engine.

2   If a stand isn't available, it's possible to disassemble the engine with it blocked up on a sturdy workbench or on the floor. Be extra careful not to tip or drop the engine when working without a stand.

3   If you're going to obtain a rebuilt engine, all external components must come off first, to be transferred to the replacement engine, just as they will if you're doing a complete engine overhaul yourself. These include:

Alternator and brackets
A/C compressor brackets
Power steering pump brackets
Emissions control components
Distributor, spark plug wires and spark
  plugs
Thermostat and housing cover
Water pump
Fuel injection components or carburetor
Intake/exhaust manifolds
Oil filter
Engine mounts
Clutch and flywheel/driveplate
Engine rear plate

Note: *When removing the external components from the engine, pay close attention to details that may be helpful or important during installation. Note the installed position of gaskets, seals, spacers, pins, brackets, washers, bolts and other small items.*

4   If you're obtaining a short block, which consists of the engine block, crankshaft, pistons and connecting rods all assembled, then the cylinder head(s), oil pan and oil pump will have to be removed as well. See Engine rebuilding alternatives for additional information regarding the different possibilities to be considered.

5   If you're planning a complete overhaul, the engine must be disassembled and the internal components removed in the following general order:

Four-cylinder engines

Rocker arm cover
Intake and exhaust manifolds
Engine front cover/timing belt covers
Timing chain/belt and sprockets
Rocker arm assembly and camshaft
Cylinder head
Oil pan
Oil pump/pickup tube
Piston/connecting rod assemblies
Crankshaft and main bearings

V6 engine

Rocker arm covers
Intake and exhaust manifolds
Rocker arms and pushrods
Valve lifters
Cylinder heads
Timing chain cover
Timing chain and sprockets
Camshaft
Oil pan
Oil pump
Piston/connecting rod assemblies
Crankshaft and main bearings

6   Before beginning the disassembly and overhaul procedures, make sure the following items are available:

Common hand tools
Small cardboard boxes or plastic bags for
  storing parts
Gasket scraper
Ridge reamer
Vibration damper puller
Micrometers
Telescoping gauges
Dial indicator set
Valve spring compressor
Cylinder surfacing hone
Piston ring groove cleaning tool
Electric drill
Tap and die set
Wire brushes
Oil gallery brushes
Cleaning solvent

8   Cylinder head - disassembly

Refer to illustrations 8.2 and 8.3
Note: *New and rebuilt cylinder heads are usually available for most engines at dealerships and auto parts stores. Due to the fact that some specialized tools are necessary for the disassembly and inspection procedures, and replacement parts may not be readily available, it may be more practical and economical for the home mechanic to purchase replacement head(s) rather than taking the time to disassemble, inspect and recondition the original(s).*

1   Cylinder head disassembly involves removal of the intake and exhaust valves and related components. It's assumed the rocker arms and pivot balls (V6 engine) or rocker arm shaft assembly and camshaft (four-cylinder engine) have been removed from the head.

2   Before the valves are removed, arrange to label and store them, along with their related components, so they can be kept separate and reinstalled in the same valve guides they're removed from **(see illustration)**.

3   Compress the springs on the first valve with a spring compressor and remove the keepers **(see illustration)**. Carefully release the valve spring compressor and remove the retainer, the springs and the spring seat. Next, remove the oil seal from the guide, then pull the valve out of the head. If the valve binds in the guide (won't pull through), push it back into the head and deburr the area around the keeper groove with a fine file or whetstone.

4   Repeat the procedure for the remaining valves. Remember to keep all parts for each valve together so they can be reinstalled in the same locations.

5   Once the valves and related components have been removed and stored in an organized manner, the head should be thoroughly cleaned and inspected. If a complete engine overhaul is being done, finish the engine disassembly procedures before beginning the cylinder head cleaning and inspection process.

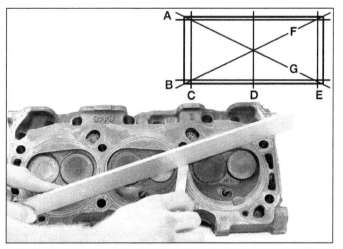

9.12 Check the cylinder head gasket surface for warpage by trying to slip a feeler gauge under the precision straightedge (see the Specifications for the maximum warpage allowed and use a feeler gauge of that thickness)

9.14 A dial indicator can be used to determine the valve stem-to-guide clearance (move the valve stem as indicated by the arrows)

9 Cylinder head - cleaning and inspection

Refer to illustrations 9.12, 9.14, 9.17, 9.18 and 9.19

1 Thorough cleaning of the cylinder head(s) and related valve train components, followed by a detailed inspection, will enable you to decide how much valve service work must be done during the engine overhaul.

Cleaning

2 Scrape all traces of old gasket material and sealing compound off the head gasket, intake manifold and exhaust manifold sealing surfaces. Be very careful not to gouge the cylinder head. Special gasket removal solvents that soften gaskets and make removal much easier are available at auto parts stores.
3 Remove all built up scale from the coolant passages.
4 Run a stiff wire brush through the various holes to remove deposits that may have formed in them.
5 Run an appropriate size tap into each of the threaded holes to remove corrosion and thread sealant that may be present. If compressed air is available, use it to clear the holes of debris produced by this operation. **Warning:** *Wear eye protection when using compressed air!*
6 Clean the exhaust and intake manifold stud threads with a wire brush.
7 Clean the cylinder head with solvent and dry it thoroughly. Compressed air will speed the drying process and ensure that all holes and recessed areas are clean. **Note:** *Decarbonizing chemicals are available and may prove very useful when cleaning cylinder heads and valve train components. They're very caustic and should be used with caution. Be sure to follow the instructions on the container.*
8 Clean the rocker arms with solvent and

dry them thoroughly (don't mix them up during the cleaning process). Compressed air will speed the drying process and can be used to clean out the oil passages.
9 Clean all the valve springs, keepers and retainers with solvent and dry them thoroughly. Do the components from one valve at a time to avoid mixing up the parts.
10 Scrape off any heavy deposits that may have formed on the valves, then use a motorized wire brush to remove deposits from the valve heads and stems. Again, make sure the valves don't get mixed up.

Inspection

Cylinder head
11 Inspect the head very carefully for cracks, evidence of coolant leakage and other damage. If cracks are found, a new cylinder head should be obtained.
12 Using a precision straightedge and feeler gauge, check the head gasket mating surface for warpage **(see illustration)**. If the warpage exceeds the specified limit, the head can be resurfaced at an automotive machine shop. **Note:** *If the V6 engine heads are resurfaced, the intake manifold flanges may also require machining.*
13 Examine the valve seats in each of the combustion chambers. If they're pitted, cracked or burned, the head will require valve service that's beyond the scope of the home mechanic.
14 Check the valve stem deflection parallel to the rocker arms with a dial indicator attached securely to the head **(see illustration)**. The valve must be in the guide and approximately 1/16-inch off the seat. The total valve stem movement indicated by the gauge needle must be noted. If it exceeds the specified limit, the valve stem-to-guide clearance should be checked by an automotive machine shop (the cost should be minimal).
15 Refer to Part A of Chapter 2 and inspect the camshaft bearing surfaces in the cylinder head.

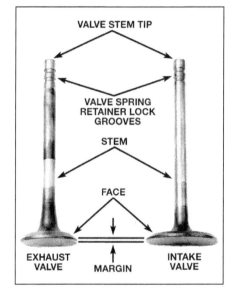

9.17 The margin width on each valve must be as specified (if no margin exists, the valve cannot be reused)

Valves
16 Carefully inspect each valve face for uneven wear, deformation, cracks, pits and burned areas. Check the valve stem for scuffing and galling and the neck for cracks. Rotate the valve and check for any obvious indication that it's bent. Look for pits and excessive wear on the end of the stem. The presence of any of these conditions indicates the need for valve service by an automotive machine shop.
17 Measure the margin width on each valve **(see illustration)**. Any valve with a margin narrower than 1/32-inch will have to be replaced with a new one.

Valve components
18 Check each valve spring for wear (on the ends) and pits. Measure the free length

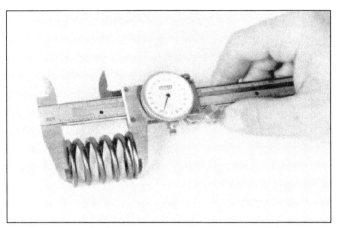

9.18 Measure the free length of each valve spring with a dial or vernier caliper

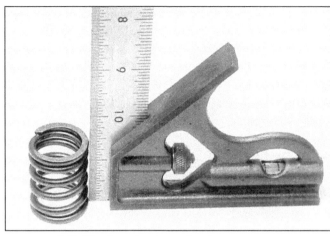

9.19 Check each valve spring for squareness

and compare it to the specifications listed in this Chapter **(see illustration)**. Any springs that are shorter than specified have sagged and should not be reused. The tension of all springs should be checked with a special fixture before deciding they're suitable for use in a rebuilt engine (take the springs to an automotive machine shop for this check).

19   Stand each spring on a flat surface and check it for squareness **(see illustration)**. If any of the springs are distorted or sagged, replace all of them with new parts.

20   Check the spring retainers and keepers for obvious wear and cracks. Any questionable parts should be replaced with new ones, as extensive damage will occur if they fail during engine operation.

Rocker arm components

21   Refer to Part A (four-cylinder engines) or B (V6 engine) of Chapter 2 and inspect the rocker arms and related components.

22   If the inspection process indicates the valve components are in generally poor condition and worn beyond the limits specified, which is usually the case in an engine that's being overhauled, reassemble the valves in the cylinder head and refer to Section 10 for valve servicing recommendations.

23   If the inspection turns up no excessively worn parts, and if the valve faces and seats are in good condition, the valve train components can be reinstalled in the cylinder head without major servicing. Refer to the appropriate Section for the cylinder head reassembly procedure.

10   Valves - servicing

1   Because of the complex nature of the job and the special tools and equipment needed, servicing of the valves, the valve seats and the valve guides, commonly known as a valve job, is best left to a professional.

2   The home mechanic can remove and disassemble the head, do the initial cleaning and inspection, then reassemble and deliver it to a dealer service department or an auto-

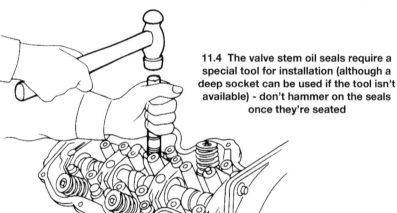

11.4 The valve stem oil seals require a special tool for installation (although a deep socket can be used if the tool isn't available) - don't hammer on the seals once they're seated

motive machine shop for the actual valve servicing.

3   The dealer service department, or automotive machine shop, will remove the valves and springs, recondition or replace the valves and valve seats, recondition or replace the valve guides, check and replace the valve springs, spring retainers and keepers (as necessary), replace the valve seals with new ones, reassemble the valve components and make sure the installed spring height is correct. The cylinder head gasket surface will also be resurfaced if it's warped.

4   After the valve job has been performed by a professional, the head will be in like new condition. When it's returned, be sure to clean it again before installation on the engine to remove any metal particles and abrasive grit that may still be present from the valve service or head resurfacing operations. Use compressed air, if available, to blow out all the oil holes and passages.

11   Cylinder head - reassembly

Refer to illustrations 11.4, 11.5, 11.6, 11.7 and 11.8

1   Regardless of whether or not the head was sent to an automotive repair shop for

valve servicing, make sure it's clean before beginning reassembly.

2   If the head was sent out for valve servicing, the valves and related components will already be in place. Begin the reassembly procedure with Step 8.

3   Install the valve spring seats (where applicable) prior to valve seal installation.

4   Install new seals on each of the valve guides. Intake valve seals require a special tool or an appropriate size deep socket. Gently tap each intake valve seal into place until it's seated on the guide **(see illustration)**. **Caution:** *Don't hammer on the intake valve seals once they're seated or you may damage them. Don't twist or cock the seals during installation or they won't seal properly on the valve stems.*

5   Apply moly-base grease or engine assembly lube to the first valve and install it in the head. Don't damage the new valve guide oil seal. Set the retainer and keepers in place. Check the installed spring height by lifting up on the retainer until the valve is seated. Measure the distance between the top of the spring seat(s) and the underside of the retainer **(see illustration)**. Compare your measurement to the specified installed height. Add shims, if necessary to obtain the specified height.

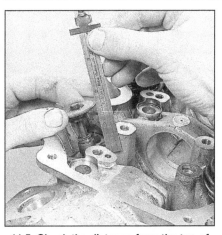

11.5 Check the distance from the top of the spring seat to the under side of the retainer to determine the installed valve spring height

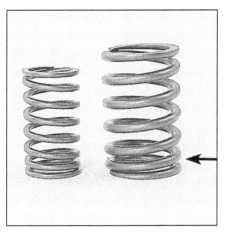

11.6 Make sure outer valve spring (right) is installed with the narrow pitch end (arrow) against the cylinder head

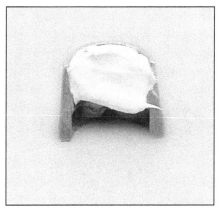

11.7 Keepers don't always stay in place, so apply a small dab of grease to each one as shown here before installation - it'll hold them in place on the valve stem as the spring is released

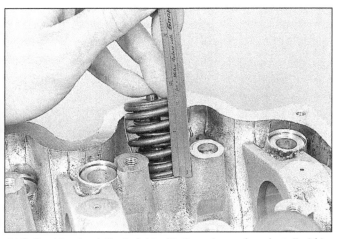

11.8 Double-check the height with the valve springs installed (do this for each valve)

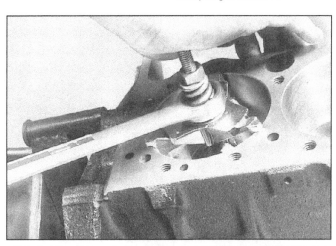

12.1 A ridge reamer is required to remove the ridge from the top of the cylinder - do this before removing the pistons!

6   Once the correct height is established, remove the keepers and retainer and install the valve springs. **Note:** *The outer spring has a graduated pitch. Install it with the narrow pitch end against the cylinder head* **(see illustration)**.

7   Compress the springs and retainer with a valve spring compressor and slip the keepers into place. Release the compressor and make sure the keepers are seated completely in the valve stem groove. If necessary, grease can be used to hold the keepers in place as the compressor is released **(see illustration)**.

8   Double-check the installed valve spring height for each valve **(see illustration)**. If it was correct prior to reassembly, it should still be within the specified limits. If it isn't, you must install more shims until it's correct. **Caution:** *Do not, under any circumstances shim the springs to the point where the installed height is less than specified!*

9   If you're working on a four-cylinder engine, the valves should be adjusted cold (see Chapter 1).

10   Store the head in a clean plastic bag until you're ready to install it.

12   Piston/connecting rod assembly - removal

Refer to illustrations 12.1, 12.3, 12.4 and 12.5
Note: *Prior to removing the piston/connecting rod assemblies, remove the cylinder head(s), the oil pan and the oil pump by referring to the appropriate Sections in Part A or B.*

1   Completely remove the ridge at the top of each cylinder with a ridge reaming tool **(see illustration)**. Follow the manufacturer's instructions provided with the tool. Failure to remove the ridge before attempting to remove the piston/connecting rod assemblies may result in piston breakage.

2   After the cylinder ridges have been removed, turn the engine upside-down so the crankshaft is facing up.

3   Before the connecting rods are removed, check the endplay with feeler gauges. Slide them between the connecting rod and the crankshaft throw until the play is removed **(see illustration)**. The endplay is equal to the thickness of the feeler gauge(s). If the endplay exceeds the service limit, new

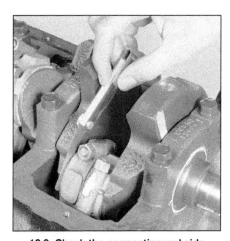

12.3 Check the connecting rod side clearance with a feeler gauge as shown

connecting rods will be required. If new rods (or a new crankshaft) are installed, the endplay may fall under the specified minimum (if it does, the rods will have to be machined to restore it - consult an automotive machine

12.4 The connecting rods and caps should be marked to indicate which cylinder they're installed in - if they aren't, mark them with a center punch to avoid confusion during reassembly

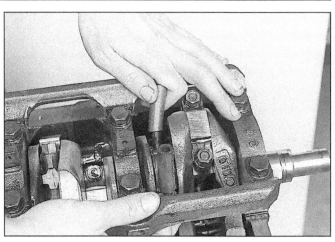

12.5 To prevent damage to the crankshaft journal and cylinder, slip sections of hose over the rod bolts, then carefully drive the piston/connecting rod assembly out through the top of the engine with the end of a wooden hammer handle

13.1 Crankshaft end play can be checked with a dial indicator, as shown here . . .

13.3 . . . or with feeler gauge

13.4 The main bearing caps should be marked so they don't get mixed up - if they aren't, mark them with a center punch (the arrow indicates the front of the engine)

shop for advice if necessary). Repeat the procedure for the remaining connecting rods.

4   Check the connecting rods and caps for identification marks **(see illustration)**. If they aren't plainly marked, use a small center punch to make the appropriate number of indentations on each rod and cap (1 - 4 or 6, depending on the engine type and cylinder they're associated with).

5   Loosen each of the connecting rod cap nuts 1/2-turn at a time until they can be removed by hand. Remove the number one connecting rod cap and bearing insert. Don't drop the bearing insert out of the cap. Slip a short length of plastic or rubber hose over each connecting rod cap bolt to protect the crankshaft journal and cylinder wall as the piston is removed **(see illustration)**. Push the connecting rod/piston assembly out through the top of the engine - use a wooden hammer handle to push on the upper bearing insert in the connecting rod. If resistance is felt, double-check to make sure all of the ridge was removed from the cylinder.

6   Repeat the procedure for the remaining pistons. After removal, reassemble the con-

necting rod caps and bearing inserts in their respective connecting rods and install the cap nuts finger-tight. Leaving the old bearing inserts in place until reassembly will help prevent the connecting rod bearing surfaces from being accidentally nicked or gouged.

---

13   Crankshaft - removal

Refer to illustrations 13.1, 13.3 and 13.4
Note: *The crankshaft can be removed only after the engine has been removed from the vehicle. It's assumed the flywheel/driveplate, vibration damper, timing chain or belt, oil pan, oil pick-up, oil pump, front cover (four-cylinder engine) and piston/connecting rod assemblies have already been removed. The rear main oil seal retainer also must be unbolted and separated from the block before proceeding with crankshaft removal.*

1   Before the crankshaft is removed, check the endplay. Mount a dial indicator with the stem in-line with the crankshaft and just touching one of the crank throws **(see illustration)**.

2   Push the crankshaft all the way to the rear and zero the dial indicator. Next, pry the crankshaft to the front as far as possible and check the reading on the dial indicator. The distance that it moves is the endplay. If it's greater than specified, check the crankshaft thrust surfaces for wear. New main bearings or thrust bearings (four-cylinder engine) should correct the endplay.

3   If a dial indicator isn't available, feeler gauges can be used. Gently pry or push the crankshaft all the way to the front of the engine. Slip feeler gauges between the crankshaft and the front face of the thrust main bearing to determine the clearance **(see illustration)**. The thrust bearing on all engines is number three (counting from the front of the engine).

4   Check the main bearing caps to see if they're marked to indicate their locations. They should be numbered consecutively from the front of the engine to the rear. If they aren't, mark them with number stamping dies

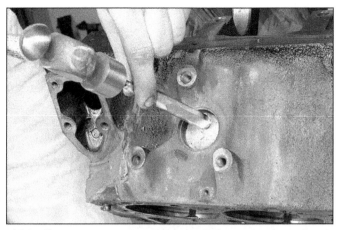

14.1a If necessary, use a punch to mark the connecting rod and bearing cap (one mark for number one, two for number two, etc.)

14.1b Pull the core plugs from the block with pliers

14.2 Use a gasket scraper to remove all traces of old gasket material and sealant from the block surfaces (gasket removal solvents are available and can make removal of stubborn gaskets much easier)

14.7 All bolt holes in the block - particularly the main bearing cap and head bolt holes - should be cleaned and restored with a tap (be sure to remove debris from the holes after this is done)

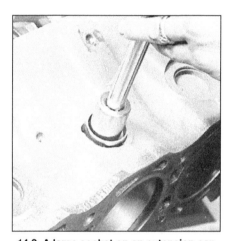

14.9 A large socket on an extension can be used to drive the new core plugs into the bores

or a center punch. Main bearing caps generally have a cast-in arrow, which points to the front of the engine (see illustration). Working from the ends toward the center, loosen the main bearing cap bolts 1/4-turn at a time, until they can be removed by hand.

5   Gently tap the caps with a soft-face hammer, then separate them from the engine block. If necessary, use the bolts as levers to remove the caps.

6   Carefully lift the crankshaft out of the engine. With the bearing inserts in place in the engine block and main bearing caps or cap assembly, install the caps on the engine block and tighten the bolts finger-tight.

14   Engine block - cleaning

Refer to illustrations 14.1a, 14.1b, 14.2, 14.7 and 14.9

Note: *The core plugs (also known as freeze or soft plugs) may be difficult or impossible to retrieve if they're driven into the block coolant passages.*

1   Remove the freeze plugs from the engine block. Knock one side of the plugs into the block with a hammer and punch, and then grasp them with large pliers and pull them back through the holes **(see illustrations)**.

2   Using a gasket scraper, remove all traces of gasket material from the engine block **(see illustration)**. Be very careful not to nick or gouge the gasket sealing surfaces.

3   Remove the main bearing caps and separate the bearing inserts from the caps and the engine block. Tag the bearings, indicating which cylinder they were removed from and whether they were in the cap or the block, then set them aside.

4   If the engine is extremely dirty, it should be taken to an automotive machine shop to be steam cleaned or hot tanked.

5   After the block is returned, clean all oil holes and oil galleries one more time. Brushes specifically designed for this purpose are available at most auto parts stores. Flush the passages with warm water until the water runs clear, dry the block thoroughly and wipe all machined surfaces with a light,

rust preventive oil. If you have access to compressed air, use it to speed the drying process and blow out all oil holes and galleries. **Warning:** *Wear eye protection when using compressed air!*

6   If the block isn't extremely dirty or sludged up, you can do an adequate cleaning job with hot soapy water and a stiff brush. Take plenty of time and do a thorough job. Regardless of the cleaning method used, be sure to clean all oil holes and galleries very thoroughly, dry the block completely and coat all machined surfaces with light oil.

7   The threaded holes in the block must be clean to ensure accurate torque readings during reassembly. Run the proper size tap into each of the holes to remove rust, corrosion, thread sealant or sludge and restore damaged threads **(see illustration)**. If possible, use compressed air to clear the holes of debris produced by this operation. Now is a good time to clean the threads on the head bolts and the main bearing cap bolts as well.

8   Reinstall the main bearing caps or cap assembly and tighten the bolts finger-tight.

9   After coating the sealing surfaces of the new core plugs with RTV sealant, install them in the engine block **(see illustration)**. Make

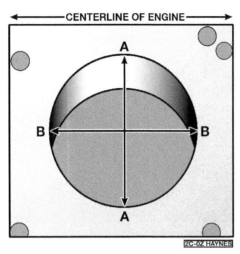

15.4a Use "A" and "B" axes for cylinder measurements as described in text

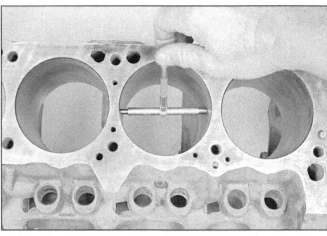

15.4b The ability to "feel" when the telescoping gauge is at the correct point will be developed over time, so work slowly and repeat the check until you're satisfied the bore measurement is accurate

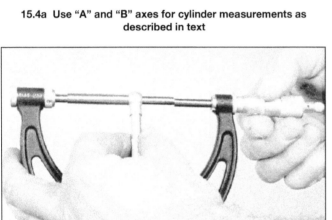

15.4c The gauge is then measured with a micrometer to determine the bore size

15.6 Check the block deck (four-cylinder engines) for distortion with a precision straightedge and feeler gauges

sure they're driven in straight and seated properly or leakage could result. Special tools are available for this purpose, but equally good results can be obtained using a large socket, with an outside diameter that will just slip into the core plug, a 1/2-inch drive extension and a hammer.

10   If the engine isn't going to be reassembled right away, cover it with a large plastic trash bag to keep it clean.

15   Engine block - inspection

Refer to illustrations 15.4a, 15.4b, 15.4c and 15.6

1   Before the block is inspected, it should be cleaned as described in Section 14. Double-check to make sure the ridge at the top of each cylinder has been completely removed.

2   Visually check the block for cracks, rust and corrosion. Look for stripped threads in the threaded holes. It's also a good idea to have the block checked for hidden cracks by an automotive machine shop that has the special equipment to do this type of work. If

defects are found, have the block repaired, if possible, or replaced.

3   Check the cylinder bores for scuffing and scoring.

4   Measure the diameter of each cylinder at the top (just under the ridge area), center and bottom of the cylinder bore, parallel to the crankshaft axis **(see illustrations)**. Next, measure each cylinder's diameter at the same three locations across the crankshaft axis. **Note:** *These measurements should not be made with the bare block mounted on an engine stand - the cylinders will be distorted and the measurements will be inaccurate.* Compare the results to the Specifications. If the cylinder walls are badly scuffed or scored, or if they're out-of-round or tapered beyond the limits given in the Specifications, have the engine block rebored and honed at an automotive machine shop. If a rebore is done, oversize pistons and rings will be required.

5   If the cylinders are in reasonably good condition and not worn to the outside of the limits, and if the piston-to-cylinder clearances can be maintained properly, then they don't

have to be rebored. Honing is all that's necessary (Section 16).

6   Using a precision straightedge and feeler gauge, check the block deck (the surface that mates with the cylinder head) for distortion (four-cylinder engines only) **(see illustration)**. If it's distorted beyond the specified limit, it can be resurfaced by an automotive machine shop.

16   Cylinder honing

Refer to illustrations 16.3a and 16.3b

1   Prior to engine reassembly, the cylinder bores must be honed so the new piston rings will seat correctly and provide the best possible combustion chamber seal. **Note:** *If you don't have the tools or don't want to tackle the honing operation, most automotive machine shops will do it for a reasonable fee.*

2   Before honing the cylinders, install the main bearing caps (without bearing inserts) and tighten the bolts to the specified torque.

3   Two types of cylinder hones are commonly available - the flex hone or "bottle

16.3a A "bottle brush" hone will produce better results if you've never done cylinder honing before

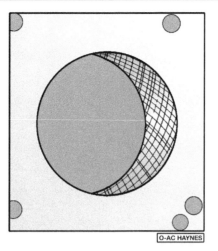

16.3b The cylinder hone should leave a smooth, crosshatch pattern with the lines intersecting at approximately a 60-degree angle

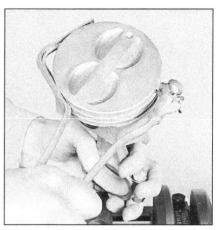

17.4a The piston ring grooves can be cleaned with a special tool, as shown here . . .

17.4b . . . or a section of a broken ring

brush" type and the more traditional surfacing hone with spring-loaded stones. Both will do the job, but for the less experienced mechanic the "bottle brush" hone will probably be easier to use. You'll also need plenty of light oil or honing oil, some rags and an electric drill motor. Proceed as follows:

a) *Mount the hone in the drill motor, compress the stones and slip it into the first cylinder* **(see illustration)**. *Be sure to wear safety goggles or a face shield!*

b) *Lubricate the cylinder with plenty of oil, turn on the drill and move the hone up-and-down in the cylinder at a pace that will produce a fine crosshatch pattern on the cylinder walls. Ideally, the crosshatch lines should intersect at approximately a 60-degree angle* **(see illustration)**. *Be sure to use plenty of lubricant and don't take off any more material than absolutely necessary to produce the desired finish.* **Note:** *Piston ring manufacturers may specify a smaller crosshatch angle than the traditional 60-degrees - read and follow any instructions included with the new rings.*

c) *Don't withdraw the hone from the cylinder while it's running. Instead, shut off the drill and continue moving the hone up-and-down in the cylinder until it comes to a complete stop, then compress the stones and withdraw the hone. If you're using a "bottle brush" type hone, stop the drill motor, then turn the chuck in the normal direction of rotation while withdrawing the hone from the cylinder.*

d) *Wipe the oil out of the cylinder and repeat the procedure for the remaining cylinders.*

4 After the honing job is complete, chamfer the top edges of the cylinder bores with a small file so the rings won't catch when the pistons are installed. Be very careful not to nick the cylinder walls with the end of the file!

5 The entire engine block must be washed again very thoroughly with warm, soapy water to remove all traces of abrasive grit

produced during the honing operation. **Note:** *The bores can be considered clean when a white cloth - dampened with clean engine oil - used to wipe them down doesn't pick up any more honing residue, which will show up as gray areas on the cloth.* Be sure to run a brush through all oil holes and galleries and flush them with running water.

6 After rinsing, dry the block and apply a coat of light rust preventive oil to all machined surfaces. Wrap the block in a plastic trash bag to keep it clean and set it aside until reassembly.

17 Piston/connecting rod assembly - inspection

Refer to illustrations 17.4a, 17.4b, 17.5, 17.10 and 17.11

1 Before the inspection process can be carried out, the piston/connecting rod assemblies must be cleaned and the original piston rings removed from the pistons. **Note:** *Always use new piston rings when the engine is reassembled.*

2 Using a piston ring installation tool, carefully remove the rings from the pistons. Be careful not to nick or gouge the pistons in the process.

3 Scrape all traces of carbon from the top of the piston. A hand-held wire brush or a piece of fine emery cloth can be used once the majority of the deposits have been scraped away. Do not, under any circumstances, use a wire brush mounted in an electric drill to remove deposits from the pistons. The piston material is soft and will be eroded away by the wire brush.

4 Use a piston ring groove cleaning tool to remove carbon deposits from the ring grooves. If a tool isn't available, a piece broken off the old ring will do the job. Be very careful to remove only the carbon deposits - don't remove any metal and don't nick or

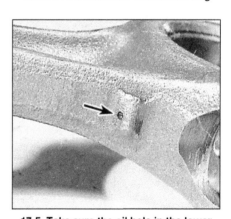

17.5 Take sure the oil hole in the lower end of each connecting rod is clear - if the rods are separated from the pistons, make sure the oil hole is on the correct side when they're reassembled

scratch the sides of the ring grooves **(see illustrations)**.

5 Once the deposits have been removed, clean the piston/rod assemblies with solvent and dry them with compressed air (if available). Make sure the oil return holes in the back sides of the piston ring grooves and the oil hole in the lower end of each rod are clear **(see illustration)**.

6 If the pistons and cylinder walls aren't

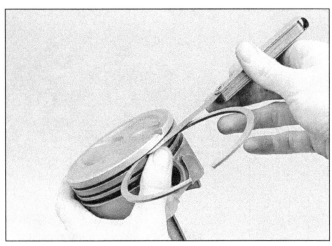

17.10 Check the ring side clearance with a feeler gauge at several points around the groove

17.11 On four-cylinder engines with a timing chain, measure the piston diameter at a 90-degree angle to the piston pin, the specified distance up from the lower edge of the skirt (see text)

damaged or worn excessively, and if the engine block isn't rebored or replaced, new pistons won't be necessary. However, if new pistons are required, be sure to match them to each cylinder by referring to the grade numbers on the block. Normal piston wear appears as even vertical wear on the piston thrust surfaces and slight looseness of the top ring in its groove. New piston rings, on the other hand, should always be used when an engine is rebuilt.

7   Carefully inspect each piston for cracks around the skirt, at the pin bosses and at the ring lands.

8   Look for scoring and scuffing on the thrust faces of the skirt, holes in the piston crown and burned areas at the edge of the crown. If the skirt is scored or scuffed, the engine may have been suffering from overheating and/or abnormal combustion, which caused excessively high operating temperatures. The cooling and lubrication systems should be checked thoroughly. A hole in the piston crown is an indication that abnormal combustion (preignition) was occurring. Burned areas at the edge of the piston crown are usually evidence of spark knock (detonation). If any of the above problems exist, the causes must be corrected or the damage will occur again.

9   Corrosion of the piston, in the form of small pits, indicates coolant is leaking into the combustion chamber and/or the crankcase. Again, the cause must be corrected or the problem may persist in the rebuilt engine.

10   Measure the piston ring side clearance by laying a new piston ring in each ring groove and slipping a feeler gauge in beside it **(see illustration)**. Check the clearance at three or four locations around each groove. Be sure to use the correct ring for each groove; they are different. If the side clearance is greater than specified, new pistons will have to be used.

11   Check the piston-to-bore clearance by measuring the bore (see Section 15) and the

piston diameter. Make sure the pistons and bores are correctly matched. Measure the piston across the skirt, at a 90-degree angle to the piston pin and in-line with it (V6 engine only). If you're working on a four-cylinder engine with a timing chain, take the measurement 1-37/64 inch (40 mm) up from the lower edge of the skirt **(see illustration)**. If you're working on a four-cylinder engine with a timing belt, the measurement should be done 1-21/32 inch (42 mm) down from the top of the piston. Subtract the piston diameter from the bore diameter to obtain the clearance. If it's greater than specified, the block will have to be rebored and new pistons and rings installed.

12   Check the piston-to-rod clearance by twisting the piston and rod in opposite directions. Any noticeable play indicates excessive wear, which must be corrected. The piston/connecting rod assemblies should be taken to an automotive machine shop to have the pistons and rods rebored and new pins installed.

13   If the pistons must be removed from the connecting rods for any reason, they should

be taken to an automotive machine shop. While they're there, have the connecting rods checked for bend and twist, since automotive machine shops have special equipment for this purpose. **Note:** *Unless new pistons and/or connecting rods must be installed, do not disassemble the pistons and connecting rods.*

14   Check the connecting rods for cracks and other damage. Temporarily remove the rod caps, lift out the old bearing inserts, wipe the rod and cap bearing surfaces clean and inspect them for nicks, gouges and scratches. After checking the rods, replace the old bearings, slip the caps into place and tighten the nuts finger-tight.

18   Crankshaft - inspection

Refer to illustrations 18.1, 18.3, 18.4, 18.6 and 18.8

1   Clean the crankshaft with solvent and dry it with compressed air (if available). **Warning:** *Wear eye protection when using compressed air.* Be sure to clean the oil holes

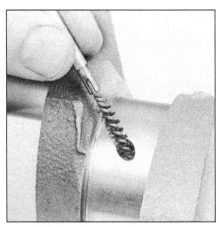

18.1 Clean the crankshaft oil passages with a wire or stiff plastic bristle brush and flush out with solvent

18.3 Rubbing a penny lengthwise on each journal will give you a quick idea of its condition - if copper rubs off the penny and adheres to the crankshaft, the journals should be reground

18.4 Chamfer the oil holes to remove sharp edges that might gouge or scratch the new bearings

18.6 Measure the diameter of each crankshaft journal at several points to detect taper and out-of-round conditions

18.8 If the seals have worn grooves in the crankshaft journals, or if the seal journals are nicked or scratched, the new seal(s) will leak

with a stiff brush **(see illustration)** and flush them with solvent.

2   Check the main and connecting rod bearing journals for uneven wear, scoring, pits and cracks.

3   Rub a penny across each journal several times **(see illustration)**. If a journal picks up copper from the penny, it's too rough and

must be reground.

4   Remove all burrs from the crankshaft oil holes with a stone, file or scraper **(see illustration)**.

5   Check the rest of the crankshaft for cracks and other damage. It should be magnafluxed to reveal hidden cracks - an automotive machine shop will handle the procedure.

6   Using a micrometer, measure the diameter of the main and connecting rod journals and compare the results to this Chapter's Specifications **(see illustration)**. By measuring the diameter at a number of points around each journal's circumference, you'll be able to determine whether or not the journal is out-of-round. Take the measurement at each end of the journal, near the crank throws, to determine if the journal is tapered.

7   If the crankshaft journals are damaged, tapered, out-of-round or worn beyond the limits given in the Specifications, have the crankshaft reground by an automotive machine shop. Be sure to use the correct size bearing inserts if the crankshaft is reconditioned.

8   Check the oil seal journals at each end of the crankshaft for wear and damage. If the seal has worn a groove in the journal, or if it's nicked or scratched **(see illustration)**, the new seal may leak when the engine is reassembled. In some cases, an automotive machine shop may be able to repair the journal by pressing on a thin sleeve. If repair isn't feasible, a new or different crankshaft should be installed.

9   Refer to Section 19 and examine the main and rod bearing inserts.

19 Main and connecting rod bearings - inspection and main bearing selection

Inspection

Refer to illustration 19.1

1   Even though the main and connecting rod bearings should be replaced with new ones during the engine overhaul, the old bearings should be retained for close examination, as they may reveal valuable information about the condition of the engine **(see illustration)**.

2   Bearing failure occurs because of lack of lubrication, the presence of dirt or other

CRATERS OR POCKETS

FATIGUE FAILURE

BRIGHT (POLISHED) SECTIONS

IMPROPER SEATING

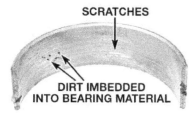

SCRATCHES

DIRT IMBEDDED INTO BEARING MATERIAL

SCRATCHED BY DIRT

OVERLAY WIPED OUT

LACK OF OIL

OVERLAY GONE FROM ENTIRE SURFACE

EXCESSIVE WEAR

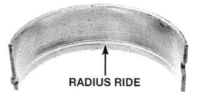

RADIUS RIDE

TAPERED JOURNAL

19.1 Typical bearing failures

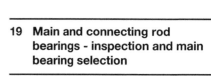

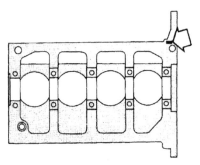

19.9a Locate the cylinder body journal hole diameter size mark (arrow)

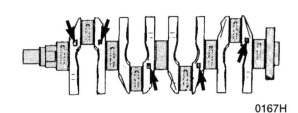

0167H

19.9b Locate the crankshaft journal diameter size mark (arrows)

foreign particles, overloading the engine and corrosion. Regardless of the cause of bearing failure, it must be corrected before the engine is reassembled to prevent it from happening again.

3   When examining the bearings, remove them from the engine block, the main bearing caps, the connecting rods and the rod caps and lay them out on a clean surface in the same general position as their location in the engine. This will enable you to match any bearing problems with the corresponding crankshaft journal.

4   Dirt and other foreign particles get into the engine in a variety of ways. It may be left in the engine during assembly, or it may pass through filters or the PCV system. It may get into the oil, and from there into the bearings. Metal chips from machining operations and normal engine wear are often present. Abrasives are sometimes left in engine components after reconditioning, especially when parts are not thoroughly cleaned using the proper cleaning methods. Whatever the source, these foreign objects often end up embedded in the soft bearing material and are easily recognized. Large particles will not embed in the bearing and will score or gouge the bearing and journal. The best prevention for this cause of bearing failure is to clean all parts thoroughly and keep everything spotlessly clean during engine assembly. Frequent and regular engine oil and filter

changes are also recommended.

5   Lack of lubrication (or lubrication breakdown) has a number of interrelated causes. Excessive heat (which thins the oil), overloading (which squeezes the oil from the bearing face) and oil leakage or throw off (from excessive bearing clearances, worn oil pump or high engine speeds) all contribute to lubrication breakdown. Blocked oil passages, which usually are the result of misaligned oil holes in a bearing shell, will also oil starve a bearing and destroy it. When lack of lubrication is the cause of bearing failure, the bearing material is wiped or extruded from the steel backing of the bearing. Temperatures may increase to the point where the steel backing turns blue from overheating.

6   Driving habits can have a definite effect on bearing life. Full throttle, low speed operation (lugging the engine) puts very high loads on bearings, which tends to squeeze out the oil film. These loads cause the bearings to flex, which produces fine cracks in the bearing face (fatigue failure). Eventually the bearing material will loosen in pieces and tear away from the steel backing. Short trip driving leads to corrosion of bearings because insufficient engine heat is produced to drive off the condensed water and corrosive gases. These products collect in the engine oil, forming acid and sludge. As the oil is carried to the engine bearings, the acid attacks and corrodes the bearing material.

7   Incorrect bearing installation during engine assembly will lead to bearing failure as well. Tight fitting bearings leave insufficient bearing oil clearance and will result in oil starvation. Dirt or foreign particles trapped behind a bearing insert result in high spots on the bearing which lead to failure.

Selection (4ZE1 engine main bearings only)

Refer to illustrations 19.9a, 19.9b and 19.10

8   If the original main bearings are worn or damaged, or if the oil clearances are incorrect (Section 22), the following procedure should be used to select the correct new main bearings for 4ZE1 engine reassembly. However, if the crankshaft has been reground, new undersize bearings must be installed - the following procedure should not be used if undersize bearings are required. The automotive machine shop that reconditions the crankshaft will provide or help you select the correct size bearings. Regardless of how the bearing sizes are determined, use the oil clearance, measured with Plastigage (Section 22), as a guide to ensure the bearings are the right size.

9   Locate the cylinder body journal hole diameter size mark and the crankshaft journal diameter size mark **(see illustrations)**.

10   Locate the particular size marks on the chart **(see illustration)** and determine the

CRANKSHAFT BEARING SELECTION TABLE

mm(in.)

CYLINDER BODY JOURNAL		CRANKSHAFT JOURNAL		CRANKSHAFT BEARING	
① Size Mark	Diameter	② Size mark	Diameter	Bearing Thickness	Size Mark (Upper & Lower)
1	59.990–60.000 (2.3636–2.3640)	–	55.928–55.935 (2.2035–2.2038)	2.015–2.019 (0.0793–0.0795)	Blue
		– –	55.920–55.927 (2.2032–2.2035)	2.011–2.015 (0.0792–0.0793)	Black
2	59.980–59.989 (2.3632–2.3635)	–	55.928–55.935 (2.2035–2.2038)		
		– –	55.920–55.927 (2.2032–2.2035)	2.007–2.011 (0.0790–0.0792)	Brown

0168H

19.10 Crankshaft bearing selection table for 4ZE1 engines

21.3 When checking piston ring end gap, the ring must be square in the cylinder bore - this is done by pushing the ring down with the top of a piston

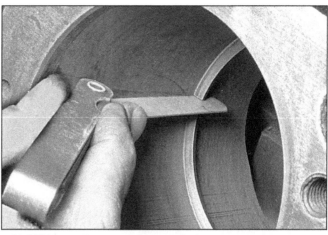

21.4 With the ring square in the cylinder, measure the end gap with a feeler gauge

color code for the crankshaft bearings.

11 Remember, the oil clearance is the final judge when selecting new bearing sizes. If you have any questions or are unsure which bearings to use, get help from an Isuzu dealer parts or service department.

20 Engine overhaul - reassembly sequence

1 Before beginning engine reassembly, make sure you have all the necessary new parts, gaskets and seals as well as the following items on hand:

Common hand tools
A 1/2-inch drive torque wrench
Piston ring installation tool
Piston ring compressor
Short lengths of rubber or plastic hose to fit over connecting rod bolts
Plastigage
Feeler gauges
A fine-tooth file
New engine oil
Engine assembly lube or moly-base grease
RTV-type gasket sealant
Thread locking compound

2 To save time and avoid problems, engine reassembly must be done in the following general order:

Four-cylinder engines

Piston rings
Crankshaft and main bearings
Piston/connecting rod assemblies
Rear main oil seal
Cylinder head (Part A)
Camshaft (Part A)
Rocker arm assembly (Part A)
Timing chain or belt and sprockets (Part A)
Front cover (Part A)
Oil strainer (Part A)
Oil pump (Part A)
Oil pan (Part A)

Intake and exhaust manifolds (Part A)
Rocker arm cover (Part A)
Flywheel/driveplate (Part A)

V6 engine

Piston rings
Crankshaft and main bearings
Piston/connecting rod assemblies
Rear main oil seal
Oil pump (Part B)
Oil pan (Part B)
Cylinder heads (Part B)
Camshafts and lifters (Part B)
Rocker arm assemblies (Part B)
Timing chain (Part B)
Intake and exhaust manifolds (Part B)
Rocker arm covers (Part B)
Engine rear plate
Flywheel/driveplate (Part B)

21 Piston rings - installation

Refer to illustrations 21.3, 21.4, 21.5, 21.9a, 21.9b and 21.12

1 Before installing the new piston rings, the ring end gaps must be checked. It's assumed the piston ring side clearance has been checked and verified correct (Section 17).

2 Lay out the piston/connecting rod assemblies and the new ring sets so the ring sets will be matched with the same piston and cylinder during the end gap measurement and engine assembly.

3 Insert the top (number one) ring into the first cylinder and square it up with the cylinder walls by pushing it in with the top of the piston **(see illustration)**. The ring should be near the bottom of the cylinder, at the lower limit of ring travel.

4 To measure the end gap, slip feeler gauges between the ends of the ring until a gauge equal to the gap width is found **(see illustration)**. The feeler gauge should slide between the ring ends with a slight amount of drag. Compare the measurement to the

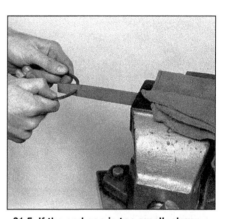

21.5 If the end gap is too small, clamp a file in a vise and file the ring ends (from the outside in only) to enlarge the gap slightly

Specifications. If the gap is larger or smaller than specified, double-check to make sure you have the correct rings before proceeding.

5 If the gap is too small, it must be enlarged or the ring ends may come in contact with each other during engine operation, which can cause serious damage to the engine. The end gap can be increased by filing the ring ends very carefully with a fine file. Mount the file in a vise equipped with soft jaws, slip the ring over the file with the ends contacting the file face and slowly move the ring to remove material from the ends. When performing this operation, file only from the outside in **(see illustration)**.

6 Excess end gap isn't critical unless it's greater than 0.040-inch. Again, double-check to make sure you have the correct rings for your engine.

7 Repeat the procedure for each ring that will be installed in the first cylinder and for each ring in the remaining cylinders. Remember to keep rings, pistons and cylinders matched up.

8 Once the ring end gaps have been checked/corrected, the rings can be installed on the pistons.

9 The oil control ring (lowest one on the piston) is installed first. On most models it's composed of three separate components. Slip the spacer/expander into the groove **(see illustration)**. If an anti-rotation tang is used, make sure it's inserted into the drilled hole in the ring groove. Next, install the lower side rail. Don't use a piston ring installation tool on the oil ring side rails, as they may be damaged. Instead, place one end of the side rail into the groove between the spacer/expander and the ring land, hold it firmly in place and slide a finger around the piston while pushing the rail into the groove **(see illustration)**. Next, install the upper side rail in the same manner.

10 After the three oil ring components have been installed, check to make sure both the upper and lower side rails can be turned smoothly in the ring groove.

11 The number two (middle) ring is installed next. It's stamped with a mark which must face up, toward the top of the piston. **Note:** *Always follow the instructions printed on the ring package or box - different manufacturers may require different approaches. Do not mix up the top and middle rings, as they have different cross sections.*

12 Use a piston ring installation tool and make sure the identification mark is facing the top of the piston, then slip the ring into the middle groove on the piston **(see illustration)**. Don't expand the ring any more than is necessary to slide it over the piston.

13 Install the number one (top) ring in the same manner. Make sure the mark is facing up. Be careful not to confuse the number one and number two rings.

14 Repeat the procedure for the remaining pistons and rings.

22 Crankshaft - installation and main bearing oil clearance check

Refer to illustrations 22.5, 22.6a, 22.6b, 22.11 and 22.15

1 Crankshaft installation is the first step in

21.9a Install the three-piece oil control ring first, one part at a time, beginning with the spacer/expander . . .

engine reassembly. It's assumed at this point that the engine block and crankshaft have been cleaned, inspected and repaired or reconditioned.

2 Position the engine with the bottom facing up.

3 Remove the main bearing cap bolts and lift out the caps. Lay them out in the proper order to ensure correct installation.

4 If they're still in place, remove the original bearing inserts from the block and the main bearing caps. Wipe the bearing surfaces of the block and caps with a clean, lint-free cloth. They must be kept spotlessly clean.

Main bearing oil clearance check

Note: *Don't touch the faces of the new bearing inserts with your fingers. Oil and acids from your skin can etch the bearings.*

5 Clean the back sides of the new main bearing inserts and lay one in each main bearing saddle in the block. If one of the bearing inserts from each set has a large groove in it, make sure the grooved insert is

21.9b . . . followed by the side rails - DO NOT use a piston ring installation tool to install the oil ring rails

installed in the block **(see illustration)**. Lay the other bearing from each set in the corresponding main bearing cap. Make sure the tab on the bearing insert fits into the recess in the block or cap. **Caution:** *The oil holes in the block must line up with the oil holes in the bearing inserts. Do not hammer the bearing into place and don't nick or gouge the bearing faces. No lubrication should be used at this time.*

6 If you're working on a V6 engine, the flanged thrust bearing must be installed in the number three cap and saddle **(see illustration)**. On four-cylinder engines, the thrust bearing must be installed in the number three (center) cap **(see illustration)**.

7 Clean the faces of the bearings in the block and the crankshaft main bearing journals with a clean, lint-free cloth.

8 Check or clean the oil holes in the crankshaft, as any dirt here can go only one way - straight through the new bearings.

9 Once you're certain the crankshaft is clean, carefully lay it in position in the main bearings.

10 Before the crankshaft can be permanently installed, the main bearing oil clear-

21.12 Install the compression rings with a ring expander (make sure the mark is facing up)

22.5 The bearing inserts with the oil groove (arrow) must be installed in the block

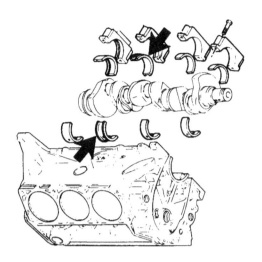

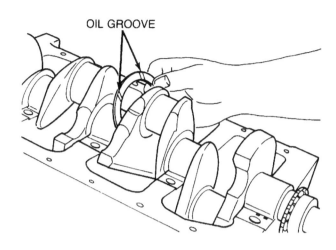

OIL GROOVE

22.6a If you're working on a V6 engine, be sure the thrust bearings (arrows) are installed in the number three cap and saddle (counting from the front of the engine)

22.6b The four-cylinder engine thrust bearings must be installed with the oil grooves against the crankshaft face

ance must be checked.

11   Cut several pieces of the appropriate size Plastigage (they should be slightly shorter than the width of the main bearings) and place one piece on each crankshaft main bearing journal, parallel with the journal axis **(see illustration)**.

12   Clean the faces of the bearings in the caps and install the caps in their original locations (don't mix them up) with the arrows pointing toward the front of the engine. Don't disturb the Plastigage.

13   Starting with the center main and working out toward the ends, tighten the main bearing cap bolts, in three steps, to the torque figure listed in this Chapter's Specifications. Don't rotate the crankshaft at any time during this operation.

14   Remove the bolts and carefully lift off the main bearing caps. Keep them in order. Don't disturb the Plastigage or rotate the crankshaft. If any of the main bearing caps are difficult to remove, tap them gently from side-to-side with a soft-face hammer to loosen them.

15   Compare the width of the crushed Plastigage on each journal to the scale printed on the Plastigage envelope to obtain the main bearing oil clearance **(see illustration)**. Check the Specifications to make sure it's correct.

16   If the clearance is not as specified, the bearing inserts may be the wrong size (which means different ones will be required). Before deciding different inserts are needed, make sure no dirt or oil was between the bearing inserts and the caps or block when the clearance was measured. If the Plastigage was wider at one end than the other, the journal may be tapered (refer to Section 18).

17   Carefully scrape all traces of the Plastigage material off the main bearing journals and/or the bearing faces. Use your fingernail

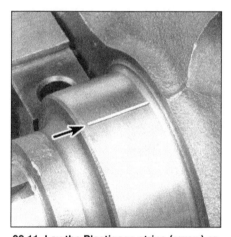

22.11 Lay the Plastigage strips (arrow) on the main bearing journals, parallel to the crankshaft centerline

or the edge of a credit card - don't nick or scratch the bearing faces.

Final crankshaft installation

18   Carefully lift the crankshaft out of the engine.

19   Clean the bearing faces in the block, then apply a thin, uniform layer of moly-base grease or engine assembly lube to each of the bearing surfaces. Be sure to coat the thrust faces as well as the journal face of the thrust bearing.

20   Make sure the crankshaft journals are clean, then lay the crankshaft back in place in the block.

21   Clean the faces of the bearings in the caps, then apply lubricant to them.

22   Install the caps in their original locations with the arrows pointing toward the front of the engine.

23   Install the bolts.

24   Tighten all except the thrust bearing cap bolts to the specified torque (work from the

22.15 Compare the width of the crushed Plastigage to the scale on the envelope to determine the main bearing oil clearance (always take the measurement at the widest point of the Plastigage); be sure to use the correct scale - inch and metric ones are included

center out and approach the final torque in three steps).

25   Tighten the thrust bearing cap bolts to 10-to-12 ft-lbs.

26   Tap the ends of the crankshaft forward and backward with a lead or brass hammer to line up the main bearing and crankshaft thrust surfaces.

27   Retighten all main bearing cap bolts to the torque specified in this Chapter, starting with the center main and working out toward the ends.

28   Rotate the crankshaft a number of times by hand to check for any obvious binding.

29   The final step is to check the crankshaft endplay with feeler gauges or a dial indicator as described in Section 13. The endplay should be correct if the crankshaft thrust faces aren't worn or damaged and new bear-

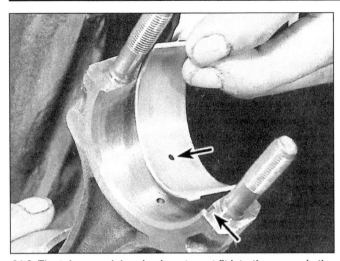

24.3 The tab on each bearing insert must fit into the recess in the rod or cap and the oil holes must line up (arrows)

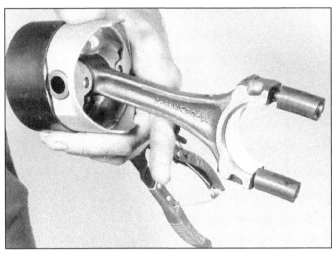

24.6 Slip sections of rubber hose over the rod bolts, then compress the rings with a ring compressor - leave the bottom of the piston sticking out so it'll slip into the cylinder

ings have been installed.
30   Refer to Section 23 and install the new rear main oil seal.

23   Crankshaft rear oil seal - installation

Four-cylinder engines

1   Clean the bore in the block and rear main bearing cap and the seal contact surface on the crankshaft. Check the crankshaft surface for scratches and nicks that could damage the new seal lip and cause oil leaks. If the crankshaft is damaged, the only alternative is a new or different crankshaft.
2   Apply a light coat of engine oil or multi-purpose grease to the outer edge of the new seal. Lubricate the seal lip with moly-base grease.
3   Carefully tap the new seal into place with a special installation tool, available at most auto part stores. The open (spring) side of the seal must face toward the front of the engine. If the special tool isn't available, carefully work the seal lip over the end of the crankshaft and tap the seal in with a hammer and punch until it's seated in the bore.

V6 engine

4   A special seal installation tool, available at most auto part stores, is required to properly seat the seal in the bore without damaging it **(see illustration 19.4 in Part B)**. Lubricate the seal bore, seal lip and seal surface on the crankshaft with engine oil. Slide the seal over the mandril on the tool until the dust lip on the seal bottoms squarely against the collar of the tool.
5   Position the dowel pin on the tool in the dowel pin hole in the crankshaft and secure the tool to the crankshaft.
6   Turn the T-handle of the tool until the collar pushes the seal completely into the bore. Make sure the seal is installed squarely.

24   Piston/connecting rod assembly - installation and rod bearing oil clearance check

Refer to illustrations 24.3, 24.6, 24.9, 24.11, 24.13 and 24.17
1   Before installing the piston/connecting rod assemblies, the cylinder walls must be perfectly clean, the top edge of each cylinder must be chamfered, and the crankshaft must be in place.
2   Remove the cap from the end of the number one connecting rod (check the marks made during removal). Remove the original bearing inserts and wipe the bearing surfaces of the connecting rod and cap with a clean, lint-free cloth. They must be kept spotlessly clean.

Connecting rod bearing oil clearance check

Note: *Don't touch the faces of the new bearing inserts with your fingers. Oil and acids from your skin can etch the bearings.*
3   Clean the back side of the new upper bearing insert, then lay it in place in the connecting rod. Make sure the tab on the bearing fits into the recess in the rod **(see illustration)**. Don't hammer the bearing insert into place and be very careful not to nick or gouge the bearing face. Don't lubricate the bearing at this time.
4   Clean the back side of the other bearing insert and install it in the rod cap. Again, make sure the tab on the bearing fits into the recess in the cap, and don't apply any lubricant. It's critically important that the mating surfaces of the bearing and connecting rod are perfectly clean and oil free when they're assembled.
5   Position the piston ring gaps at 120-degree intervals around the piston.
6   Slip a section of plastic or rubber hose over each connecting rod cap bolt **(see illustration)**.

24.9 The mark or arrow on each piston must point towards the front of the engine

7   Lubricate the piston and rings with clean engine oil and attach a piston ring compressor to the piston. Leave the skirt protruding at least 1/4-inch to guide the piston into the cylinder. The rings must be compressed until they're flush with the piston.
8   Rotate the crankshaft until the number one connecting rod journal is at BDC (bottom dead center) and apply a coat of engine oil to the cylinder walls.
9   With the mark or arrow on top of the piston facing the front of the engine **(see illustration)**, gently insert the piston/connecting rod assembly into the number one cylinder bore and rest the bottom edge of the ring compressor on the engine block.
10   Tap the top edge of the ring compressor to make sure it's contacting the block around its entire circumference.
11   Gently tap on the top of the piston with the end of a wooden or plastic hammer handle **(see illustration)** while guiding the end of the connecting rod into place on the crankshaft journal. The piston rings may try to pop out of the ring compressor just before entering the cylinder bore, so keep some

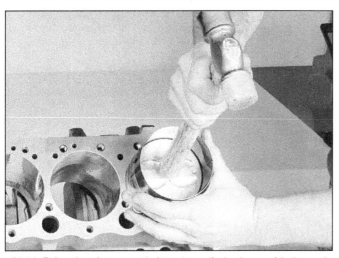

24.11 Drive the piston gently into the cylinder bore with the end of a wooden or plastic hammer handle

24.13 Lay the Plastigage strips on each rod bearing journal, parallel to the crankshaft centerline

pressure on the ring compressor. Work slowly, and if any resistance is felt as the piston enters the cylinder, stop immediately. Find out what's hanging up and fix it before proceeding. Do not, for any reason, force the piston into the cylinder - you might break a ring and/or the piston.

12 Once the piston/connecting rod assembly is installed, the connecting rod bearing oil clearance must be checked before the rod cap is permanently bolted in place.

13 Cut a piece of the appropriate size Plastigage slightly shorter than the width of the connecting rod bearing and lay it in place on the number one connecting rod journal, parallel with the journal axis **(see illustration)**.

14 Clean the connecting rod cap bearing face, remove the protective hoses from the connecting rod bolts and install the rod cap. Make sure the mating mark on the cap is on the same side as the mark on the connecting rod.

15 Install the nuts and tighten them to the torque listed in this Chapter's Specifications. Work up to it in three steps. **Note:** *Use a thin-wall socket to avoid erroneous torque readings that can result if the socket is wedged between the rod cap and nut. If the socket tends to wedge itself between the nut and the cap, lift up on it slightly until it no longer contacts the cap. Do not rotate the crankshaft at any time during this operation.*

16 Remove the nuts and detach the rod cap, being very careful not to disturb the Plastigage.

17 Compare the width of the crushed Plastigage to the scale printed on the Plastigage envelope to obtain the oil clearance **(see illustration)**. Compare it to this Chapter's Specifications to make sure the clearance is correct.

18 If the clearance is not as specified, the bearing inserts may be the wrong size (which means different ones will be required). Before deciding different inserts are needed, make sure no dirt or oil was between the bearing inserts and the connecting rod or cap when

the clearance was measured. Also, recheck the journal diameter. If the Plastigage was wider at one end than the other, the journal may be tapered (refer to Section 18).

Final connecting rod installation

19 Carefully scrape all traces of the Plastigage material off the rod journal and/or bearing face. Be very careful not to scratch the bearing - use your fingernail or the edge of a credit card.

20 Make sure the bearing faces are perfectly clean, then apply a uniform layer of clean moly-base grease or engine assembly lube to both of them. You'll have to push the piston into the cylinder to expose the face of the bearing insert in the connecting rod - be sure to slip the protective hoses over the rod bolts first.

21 Slide the connecting rod back into place on the journal, remove the protective hoses from the rod cap bolts, install the rod cap and tighten the nuts to the torque specified in this Chapter. Again, work up to the torque in three steps.

22 Repeat the entire procedure for the remaining pistons/connecting rods.

23 The important points to remember are:

a) *Keep the back sides of the bearing inserts and the insides of the connecting rods and caps perfectly clean when assembling them.*

b) *Make sure you have the correct piston/rod assembly for each cylinder.*

c) *The arrow or mark on the piston must face the front (timing chain/belt end) of the engine.*

d) *Lubricate the cylinder walls with clean oil.*

e) *Lubricate the bearing faces when installing the rod caps after the oil clearance has been checked.*

24 After all the piston/connecting rod assemblies have been properly installed, rotate the crankshaft a number of times by

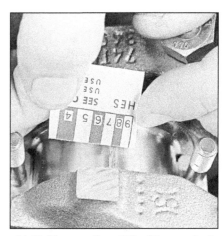

24.17 Measure the width of the crushed Plastigage to determine the rod bearing oil clearance (be sure to use the correct scale - inch and metric ones are included)

hand to check for any obvious binding.

25 As a final step, the connecting rod endplay must be checked. Refer to Section 13 for this procedure.

26 Compare the measured endplay to this Chapter's Specifications to make sure it's correct. If it was correct before disassembly and the original crankshaft and rods were reinstalled, it should still be right. If new rods or a new crankshaft were installed, the endplay may be inadequate. If so, the rods will have to be removed and taken to an automotive machine shop for resizing.

25 Initial start-up and break-in after overhaul

1 Once the engine has been installed in the vehicle, double-check the engine oil and coolant levels.

2 With the spark plugs out of the engine and the ignition system disabled (see Section 2), crank the engine until oil pressure reg-

isters on the gauge.

3   Install the spark plugs, hook up the plug wires and restore the ignition system functions (Section 2).

4   Start the engine. It may take a few moments for the gasoline to reach the carburetor or injectors, but the engine should start without a great deal of effort.

5   After the engine starts, it should be allowed to warm up to normal operating temperature. While the engine is warming up, make a thorough check for oil and coolant

leaks.

6   Shut the engine off and recheck the engine oil and coolant levels.

7   Drive the vehicle to an area with minimum traffic, accelerate at full throttle from 30 to 50 mph, then allow the vehicle to slow to 30 mph with the throttle closed. Repeat the procedure 10 or 12 times. This will load the piston rings and cause them to seat properly against the cylinder walls. Check again for oil and coolant leaks.

8   Drive the vehicle gently for the first 500

miles (no sustained high speeds) and keep a constant check on the oil level. It's not unusual for an engine to use oil during the break-in period.

9   At approximately 500 to 600 miles, change the oil and filter.

10   For the next few hundred miles, drive the vehicle normally. Don't pamper it or abuse it.

11   After 2000 miles, change the oil and filter again and consider the engine fully broken in.

Chapter 3
Cooling, heating and air conditioning systems

Contents

Specifications

General

Coolant capacity	See Chapter 1
Thermostat opening temperature	180-degrees F

Torque specifications

	Ft-lbs
Thermostat housing cover bolts	12 to 15
Water pump mounting bolts	
Four-cylinder engine	
G180Z and G200Z	13.7
4ZD1 and 4ZE1	18.0
V6 engine	
Small bolts	7
Large bolts	18

1 Cooling system - general information

Refer to illustration 1.2

The components of the cooling system are the radiator, upper and lower radiator hoses, water pump, thermostat, radiator cap (with pressure relief valve) and heater hoses.

The principle of the system is that coolant in the bottom of the radiator circulates up through the lower radiator hose to the water pump, where the pump impeller pushes it around the block and heads through the various cast-in passages to cool the cylinder bores, combustion surfaces and valve seats **(see illustration)**. When sufficient heat has been absorbed by the coolant, and the engine has reached operating temperature, the coolant moves from the cylinder head past the now open thermostat into the top radiator hose and into the radiator header tank. The coolant then travels down the radiator tubes where it is rapidly cooled by the natural flow of air as the vehicle moves down the road. A multi-blade fan, mounted on the water pump pulley, assists this cooling action. The coolant now reaches the bottom of the radiator and the cycle is repeated.

When the engine is cold, the thermostat remains closed until the coolant reaches a predetermined temperature (see the Specifications). This assists rapid warm-up.

The system is pressurized by a spring-loaded radiator filler cap, which prevents premature boiling by increasing the boiling point of the coolant. If the coolant temperature goes above this increased boiling point, the extra pressure in the system forces the radiator cap internal spring loaded valve off its seat and exposes the overflow hose, down which displaced coolant escapes into the coolant recovery reservoir.

The coolant recovery system consists of a plastic reservoir into which the overflow coolant from the radiator flows when the engine is hot. When the engine cools, coolant is drawn back into the radiator from the reservoir and maintains the system at full capacity.

Aside from cooling the engine during operation, the cooling system also provides the heat for the vehicle's interior heater and heats the intake manifold. On vehicles equipped with an automatic transmission, the transmission fluid is cooled by a cooler attached to the base of the radiator.

On vehicles equipped with an air conditioning system, a condenser is placed ahead of the radiator.

The radiator cooling fan has either four or seven fan blades. Seven-blade fans normally have a thermostatically controlled fan clutch and are installed on vehicles with air conditioning to prevent engine overheating at idle as well as to minimize loss of engine power at high speeds.

Warning: *The radiator cap should not be removed while the engine is hot. The*

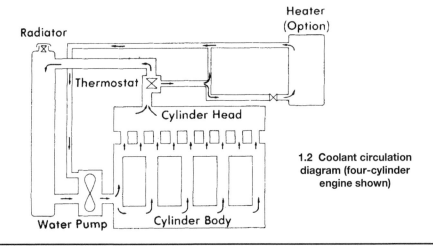

1.2 Coolant circulation diagram (four-cylinder engine shown)

proper way to remove the cap is to wrap a thick cloth around it, rotate the cap slowly counterclockwise to the detent and allow any residual pressure to escape. Do not press the cap down until all hissing has stopped, then push down and twist it off.

2 Antifreeze - general information

Warning: *Don't allow antifreeze to come in contact with your skin or painted surfaces of the vehicle. Rinse off spills immediately with plenty of water. Never leave antifreeze lying around in an open container or in a puddle on the driveway or garage floor. Children and animals are attracted by its sweet smell. Antifreeze is toxic, so use common sense when disposing of it. Some communities maintain toxic material disposal sites and/or offer regular pick-up of hazardous materials. Antifreeze is also combustible, so don't store or use it near open flames.*

The cooling system should be filled with a water/ethylene glycol based antifreeze solution, which will prevent freezing down to at least -20-degrees F, or lower if local climate requires it. It also provides protection against corrosion and increases the coolant boiling point.

The cooling system should be drained, flushed and refilled at the specified intervals (see Chapter 1). Old or contaminated antifreeze solutions are likely to cause damage and encourage the formation of rust and scale in the system. Use distilled water with the antifreeze.

Before adding antifreeze, check all hose connections, because antifreeze tends to search out and leak through very minute openings. Engines don't normally consume coolant, so if the level goes down, find the cause and correct it.

The exact mixture of antifreeze-to-water which you should use depends on the relative weather conditions. The mixture should contain at least 50 percent antifreeze, but should never contain more than 70 percent antifreeze. Consult the mixture ratio chart on

the antifreeze container before adding coolant. Hydrometers are available at most auto parts stores to test the coolant. Use antifreeze which meets the vehicle manufacturer's specifications.

3 Thermostat - removal and installation

Refer to illustrations 3.5a and 3.5b

1   The thermostat is a restriction valve which is actuated by a heat sensitive element. It's mounted inside a housing on the engine and is designed to open and close at predetermined temperatures to allow coolant to warm up or cool the engine.

Removal

2   Drain the cooling system (see Chapter 1). If the coolant is in good condition (see Chapter 1), save it and reuse it.
3   Follow the upper radiator hose to the engine to locate the thermostat housing cover. Disconnect the upper radiator hose from the cover. On carburetor-equipped models, you'll probably need to remove the air cleaner assembly first (see Chapter 4).
4   Remove the bolts that retain the thermostat housing cover.
5   Lift off the cover along with the gasket and thermostat. Note how the thermostat is positioned in the recess - it must be installed the same way **(see illustrations)**. If the cover is stuck, tap it with a soft-face hammer to jar it loose.

Installation

6   Before installation, use a gasket scraper or putty knife to carefully remove all traces of the old gasket. Stuff a rag into the thermostat housing to prevent gasket material from falling into the engine.
7   Clean the sealing surfaces with lacquer thinner or acetone.
8   Place the thermostat in position in the housing. Make sure the air bleeder or jiggle valve is on top.

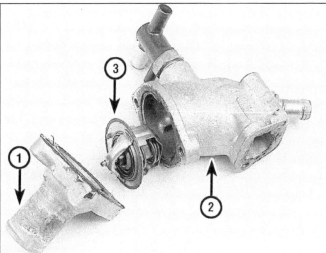

3.5b Note the installed order of the thermostat and related components (4ZE1/4ZD1 four-cylinder engines shown)

1   Thermostat housing cover
2   Thermostat housing
3   Thermostat

3.5a Correct installed position of the thermostat (4ZE1/4ZD1 four-cylinder engines shown - removed from the engine for clarity)

9   Apply a thin, uniform layer of RTV sealant to both sides of the gasket, then position it on the cover.
10   Install the cover and bolts. Tighten the bolts to the torque listed in this Chapter's Specifications.
11   Reinstall the upper radiator hose.
12   Refill the radiator with the proper mixture of antifreeze and water. Refer to Chapter 1 if necessary.
13   With the radiator cap removed, start the engine and run it until the upper radiator hose becomes hot. At this point, the thermostat will be in the open position. Check the coolant level and add as necessary.
14   Reinstall the radiator cap.

4   Radiator - removal and installation

Refer to illustrations 4.4a and 4.4b

1   With the engine cold, remove the battery. Always disconnect the negative cable first, followed by the positive cable.
2   Remove the lower splash pan.
3   Open the drain valve on the underside of

the radiator and drain the coolant into a container (see Chapter 1). If the coolant is in good condition, it can be reused.
4   Disconnect the upper and lower radiator hoses from the radiator **(see illustrations)**. If they're stuck, grasp each hose near the end with a pair of Channelock pliers and twist it to break the seal, then pull it off - be careful not to distort the radiator fittings! If the hoses are old and deteriorated, cut them off and install new ones.
5   Disconnect the reservoir hose from the radiator filler neck.
6   Remove the screws that attach the shroud to the radiator and slide the shroud toward the engine.
7   If the vehicle is equipped with an automatic transmission, disconnect the cooler hoses from the radiator. Place a drip pan under the hoses to catch any fluid that may spill.
8   Remove the bolts that attach the radiator to the body.
9   Lift out the radiator - be careful not to spill coolant on the vehicle or scratch the paint. Also be careful not to damage the radiator's cooling fins on the fan or fan shroud.
10   With the radiator removed, it can be inspected for leaks or damage. If in need of

repairs, have a professional radiator shop or dealer perform the work as special techniques are required.
11   Bugs and dirt can be cleaned from the radiator with compressed air and a soft brush. Don't bend the cooling fins as this is done.
12   Installation is the reverse of the removal procedure.
13   After installation, fill the cooling system with the proper mixture of antifreeze and water. Refer to Chapter 1 if necessary.
14   Start the engine and check for leaks. Allow the engine to reach normal operating temperature, indicated by the upper radiator hose becoming hot. Recheck the coolant level and add more if required.
15   On automatic transmission equipped models, check and add fluid as needed.

5   Engine cooling fan and clutch - check, removal and installation

Refer to illustration 5.7

Fan clutch check

1   Rock the fan back and forth to check for

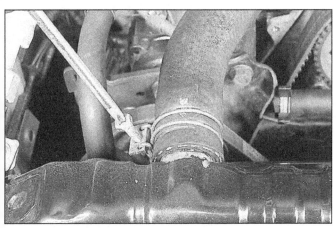

4.4a Remove the upper . . .

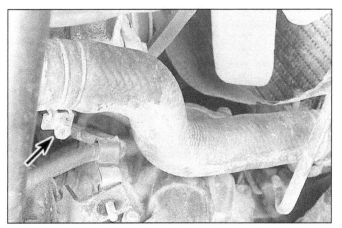

4.4b . . . and lower radiator hose clamps (arrow)

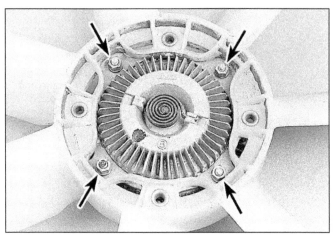

5.7 If the fan clutch or fan must be replaced, remove the four nuts (arrows) and separate the two components (four-cylinder engine shown)

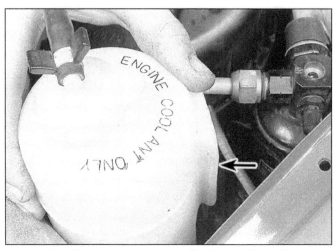

6.2 Coolant reservoir mounting bracket (arrow)

excessive bearing play.

2   With the engine cold, turn the fan blades by hand. The fan should turn freely.

3   Visually inspect for substantial fluid leakage from the clutch assembly. If problems are noted, replace the clutch assembly.

4   With the engine completely warmed up, turn the fan by hand. Some drag should be evident. If the fan turns easily, replace the fan clutch.

Removal and installation

5   Unbolt the radiator shroud (see the previous Section) and push it back over the fan.

6   On models without a fan clutch, remove the four bolts at the front of the fan and separate the fan from the water pump.

7   Models equipped with a fan clutch have four nuts securing the fan and clutch assembly to the water pump. Remove these nuts and pull the assembly off the water pump. After the assembly is removed, the fan and clutch assembly can be separated by removing the nuts **(see illustration)**.

8   Installation is the reverse of removal.

6   Coolant reservoir - removal and installation

Refer to illustration 6.2

1   Disconnect the reservoir hose at the radiator neck.

2   Lift the reservoir up. The reservoir is held by a grooved bracket **(see illustration)**.

3   Remove the reservoir from the engine compartment.

4   Drain the coolant into a container. If the reservoir is being replaced, disconnect the hose from the top.

5   Installation is the reverse of the removal procedure.

6   Refill the reservoir with the proper mixture of antifreeze and water. Refer to Chapter 1 if necessary.

8.9a Remove the five bolts (arrows) retaining the water pump to the engine block (4ZE1 engine shown)

7   Water pump - check

1   A failure in the water pump can cause serious engine damage due to overheating.

2   There are three ways to check the operation of the water pump while it is installed on the engine. If the pump is suspect, it should be replaced with a new or rebuilt unit.

3   With the engine running and warmed to normal operating temperature, squeeze the upper radiator hose. If the water pump is working properly, a pressure surge should be felt as the hose is released.

4   Water pumps are equipped with weep or vent holes. If a failure occurs in the pump seal, coolant will leak from this hole. In most cases it will be necessary to use a flashlight to find the hole on the water pump by looking through the space just below the pump to see evidence of leakage.

5   If the water pump shaft bearings fail there may be a squealing sound at the front of the engine while it is running. Shaft wear can be felt if the water pump pulley is rocked up and down. Do not mistake drivebelt slippage, which also causes a squealing sound, for water pump failure.

8   Water pump - removal and installation

Refer to illustrations 8.9a and 8.9b

1   Disconnect the negative cable at the battery.

2   Jack up the vehicle and support it securely on jackstands.

3   Drain the cooling system (Chapter 1).

4   Refer to Sections 4 and 5 to remove the fan and shroud.

5   Refer to Chapter 1 and remove the drivebelts. **Note:** *The V6 engine is equipped with a serpentine belt system.*

6   Remove the water pump pulley.

7   On timing belt engines, remove the upper and lower timing belt covers (refer to Chapter 2A). On V6 engines, remove the power steering pump and air conditioning compressor and their brackets, if necessary (see Section 16 and Chapter 10).

8   On timing belt engines, loosen the timing belt tensioner and move it out of the way of the water pump bolts. Be careful not to disturb the timing belt or sprockets or the valve timing could be altered (see Chapter 2A).

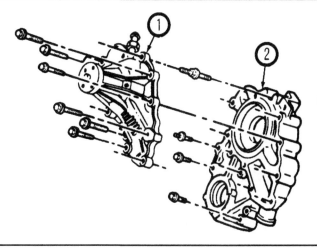

8.9b V6 engine water pump and related components - exploded view

1 *Water pump*
2 *Timing chain cover*

9.2 Coolant temperature sending unit location - 4ZE1 engine

9   Remove the water pump mounting bolts **(see illustrations)**.

10   Remove the water pump from the engine block and scrape all gasket and sealant material from the mounting surfaces.

11   Upon installation, apply gasket sealant to both sides of the new gasket and install it and the water pump on the engine block.

12   Tighten all the water pump bolts a little at a time, to the torque listed in this Chapter's specifications.

13   The remainder of installation is the reverse of the removal procedure.

14   Adjust the drivebelt to the proper tension. Refer to Chapter 1 if necessary.

15   Fill the cooling system with the proper mixture of antifreeze and water, again referring to Chapter 1 if necessary. Then start the engine and allow it to idle until it reaches normal operating temperature. This is indicated by the upper radiator hose becoming hot. Check around the water pump and radiator for any leaks.

16   Recheck the coolant level and add more if necessary.

9   Coolant temperature sending unit - check and replacement

Refer to illustration 9.2

Check

1   The coolant temperature sending unit provides a variable electrical ground for the coolant temperature gauge.

2   The coolant temperature sending unit is located on the intake manifold **(see illustration)**.

3   If the gauge does not register, check the electrical fuses. Then unplug the wiring connector from the coolant temperature sending unit and connect a jumper wire between it and a clean electrical ground. Turn the ignition on briefly and the gauge should go to full scale (Hot).

4   If the gauge went to Hot, the sending unit is defective; replace it.

5   If the gauge did not move, the gauge or connecting circuitry is probably faulty.

Replacement

6   Allow the engine to cool completely.

7   Drain the coolant (Chapter 1).

8   Unplug the wiring connector from the sending unit.

9   Using an appropriate size deep socket or wrench, unscrew the sending unit from the intake manifold.

10   Install the new unit and tighten it securely.

10   Heating system - general information

Refer to illustration 10.1

The main components of the heating system include the heater unit (which contains the heater core and cable-operated valves) the blower motor, the control assembly (mounted in the dash) and the air ducts which deliver the air to the various outlet locations **(see illustration)**.

Either outside air or interior (recirculated) air (depending on the settings) is drawn into the system through the blower unit. From there the blower motor forces the air into the heater unit.

The lever settings on the control assembly operate the valves in the heater unit, which determines the mix of heated and outside air by regulating how much air is passed through the heater core. The hotter the setting the more air is passed through the core.

The air ducts carry the heated air from the heater unit to the desired location. Again, valves within the duct system regulate where in the vehicle the air will be delivered.

The heater core is heated by engine coolant passing through it. The heater hoses carry the coolant from the engine to the heater core and then back again.

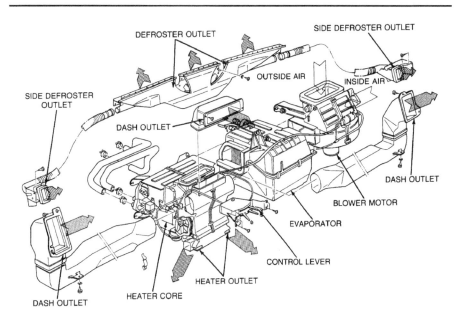

10.1 Heating and air conditioning units, control assembly and associated components

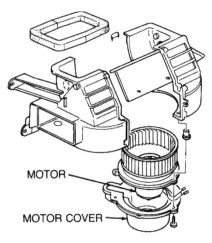

MOTOR

MOTOR COVER

11.3 Blower assembly and related components - exploded view

11.4 Remove the mounting bolts and carefully lower the blower motor and cover from the blower unit

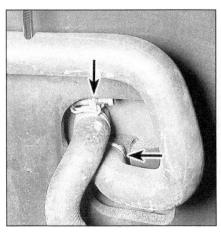

12.3 Remove the clamps that retain the heater hoses (arrows)

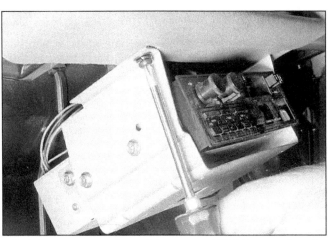

12.4 Remove the radio retaining screws (1988 Trooper shown)

12.8 Four screws like this (arrow) secure the heater unit

11 Blower unit - removal and installation

Refer to illustrations 11.3 and 11.4
1   Disconnect the negative battery cable from the battery.
2   Disconnect the electrical connector at the blower motor.
3   Remove the bolts that retain the motor cover and motor to the blower unit assembly **(see illustration)**.
4   Carefully lower the motor and motor cover from the unit **(see illustration)**.
5   Remove the outlet duct from the side of the blower unit.
6   Remove the blower unit mounting screws and lower the unit out.
7   Installation is the reverse of the removal procedure.

12 Heater core - removal and installation

Refer to illustrations 12.3, 12.4 and 12.8
1   Drain the coolant into a clean container

(see Chapter 1).
2   Disconnect the negative battery cable.
3   Disconnect the heater hoses at the fire-wall **(see illustration)**.
4   Working in the interior, remove the radio, if equipped **(see illustration)**.
5   Disconnect the control cables from the heater unit assembly.
6   Remove the air flow ducts from the sides of the heater unit assembly.
7   Remove any electrical connectors and vacuum lines.
8   Remove the four heater unit assembly mounting screws and pull the unit out **(see illustration)**.
9   Once removed, the heater core can be detached from the heater unit assembly **(see illustration 10.1)**.
10   If the heater unit assembly must be disassembled in order to reach the heater core, disconnect the rod from the swivel joint and the cold position lever control. Remove the screws that retain the heater core cover and remove the heater core. **Note:** *During reassembly, the control rods must be adjusted* (see Section 18).
11   Installation is the reverse of removal.

13 Air conditioning system - general information

Refer to illustration 13.3
Warning: *The air conditioning system is under high pressure. Do not loosen any hose fittings or remove any components until after the system has been discharged by a service station or automotive air conditioning shop.*

The air conditioning system used in these vehicles maintains proper temperature by cycling the compressor on and off according to the pressure within the system, and by maintaining a mix of cooled, outside and heated air, using the same blower, heater core and outlet duct system that the heating system uses.

A fast idle control device regulates the engine idle speed when the air conditioner is operating.

The main components of the system include a belt-driven compressor, a condenser (mounted in front of the radiator), a receiver/drier and an evaporator **(see illustration)**.

The system operates by air (outside or

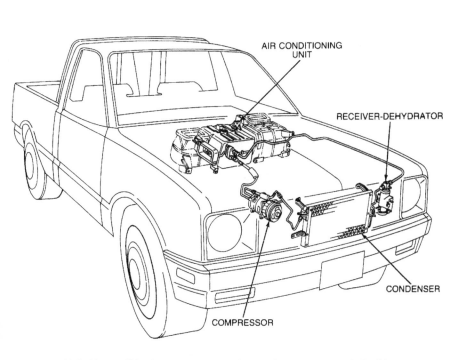

13.3 Air conditioning system on a pick-up (the evaporator is inside
the air conditioning unit)

**14.3 Top view of the receiver/drier
showing the sight glass (arrow)**

recirculated) entering the evaporator core by the action of the blower, where it receives cooling. When the air leaves the evaporator, it enters the heater/air conditioner duct assembly and by means of a manually controlled deflector, it either passes through or bypasses the heater core in the correct proportions to provide the desired vehicle temperature.

Distribution of this air into the vehicle is then regulated by a manually operated deflector, and is directed either to the floor vents, dash vents or defroster vents according to settings.

14 Air conditioning system - check and maintenance

Refer to illustration 14.3

1   The following maintenance steps should be performed on a regular basis to ensure that the air conditioner continues to operate at peak efficiency.

a) *Check the tension of the A/C compressor drivebelt and adjust it if necessary (refer to Chapter I).*
b) *Visually inspect the condition of the hoses, looking for any cracks, hardening and other deterioration. **Note:** Don't remove any hoses until the system has been discharged.*
c) *Make sure the fins of the condenser aren't covered with foreign material, such as leaves or bugs. A soft brush and compressed air can be used to remove them.*

d) *Be sure the evaporator drain is open by slipping a wire into the drain tube occasionally.*

2   The A/C compressor should be run about 10 minutes at least once every month. This is especially important to remember during the winter months because long-term non-use can cause hardening of the seals.

3   Due to the complexity of the air conditioning system and the special equipment required to effectively work on it, accurate troubleshooting is beyond the scope of the home mechanic and should be left to a professional. If the system should lose its cooling action, some causes can be diagnosed by the home mechanic. Look for other symptoms of trouble such as those in the following list. In all cases, it's a good idea to have the system serviced by a professional.

a) *If bubbles appear in the sight glass (located on top of the receiver/drier* **(see illustration)**, *this is an indication of either a small refrigerant leak or air in the refrigerant. If air is in the refrigerant, the receiver/drier is suspect and should be replaced.*
b) *If the sight glass takes on a mist-like appearance or shows many bubbles, this indicates a large refrigerant leak. In such a case, do not operate the compressor at all until the fault has been corrected.*
c) *Sweating or frosting of the expansion valve inlet indicates that the expansion valve is clogged or defective. It should be cleaned or replaced as necessary.*
d) *Sweating or frosting of the suction line*

(which runs between the suction throttle valve and the compressor) indicates that the expansion valve is stuck open or defective. It should be corrected or replaced as necessary.
e) *Frosting on the evaporator indicates a defective suction throttle valve, requiring replacement of the valve.*
f) *Frosting of the high pressure liquid line (which runs between the condenser, receiver/drier and expansion valve) indicates that either the drier or the high pressure line is restricted. The line will have to be cleared or the receiver/drier replaced.*
g) *The combination of bubbles in the sight glass, a very hot suction line and, possibly overheating of the engine is an indication that either the condenser is not operating properly or the refrigerant is overcharged. Check the tension of the drivebelt and adjust if necessary (Chapter 1). Check for foreign matter covering the fins of the condenser and clean if necessary. Also check for proper operation of the cooling system. If no fault can be found in these checks, the condenser may have to be replaced.*

15 Air conditioning receiver/drier - removal and installation

Refer to illustration 15.4
Warning: *The air conditioning system is under high pressure. Do not loosen any hose fittings or remove any components until after the system has been discharged by a service station or automotive air conditioning shop.*

1   Disconnect the negative battery cable from the battery.

2   If equipped, unplug the wires from the pressure switch.

3   Disconnect the refrigerant lines from the receiver/drier. Use a backup wrench to keep from twisting the line.

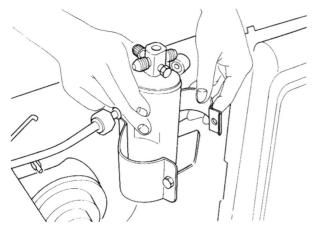

15.4 Open the mounting bracket slightly before removing the receiver/drier

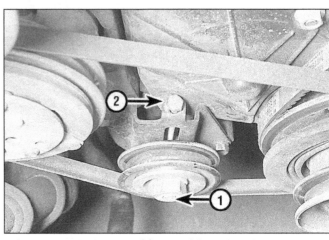

16.5 Loosen the large nut (1) on the idler pulley, then loosen the adjusting bolt (2) to relieve the tension on the belt

4   Unbolt the receiver/drier bracket and lift the receiver/drier out **(see illustration)**

5   Installation is the reverse of the removal procedure. When reconnecting the hoses be sure that new O-rings are used and that they are installed properly.

6   Take the vehicle to a dealer service department or automotive air conditioning shop to have the system evacuated and recharged.

16   Air conditioning compressor - removal and installation

Refer to illustrations 16.5, 16.7 and 16.9
Warning: *The air conditioning system is under high pressure. Do not loosen any hose fittings or remove any components until after the system has been discharged by a service station or automotive air conditioning shop.*

1   Operate the compressor, if possible, for about ten minutes at idle speed.

2   The air conditioning controls should be set at maximum cooling and on the high blower speed, with all windows open. This will return the oil within the system to the compressor.

3   Have the system discharged by a dealer service department or automotive air conditioning shop.

4   Disconnect the negative battery cable from the battery.

5   Loosen the bolt at the center of the compressor idler pulley and turn the adjusting bolt to loosen the A/C compressor drivebelt **(see illustration)**.

6   Remove the drivebelt from the compressor.

7   Disconnect the refrigerant lines at the compressor **(see illustration)**.

8   Disconnect all wires attached to the compressor.

9   Remove the bolts that retain the compressor to the bracket and detach the compressor **(see illustration)**.

10   Remove the compressor from the engine compartment. **Note:** *The compressor should not be left on its side or upside-down for more than ten minutes at a time, as compressor oil could enter the low pressure chambers and cause internal damage. If this*

should happen, the oil can be expelled from the chambers by positioning the compressor right-side up and hand cranking it several times.

11   Installation is the reverse of the removal procedure. When reconnecting the lines to the compressor, be sure to use new O-rings at each connection.

12   Once the compressor and all A/C lines have been securely connected, the vehicle must once again be taken to a dealer service department or an air conditioning shop to have the system evacuated and charged.

17   Air conditioning condenser - removal and installation

Refer to illustration 17.4
Warning: *The air conditioning system is under high pressure. Do not loosen any hose fittings or remove any components until after the system has been discharged by a service station or an automotive air conditioning shop.*

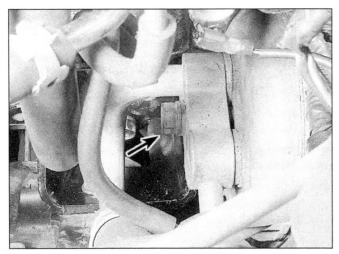

16.7 Remove the center bolt to release the lines from the air conditioning compressor (arrow)

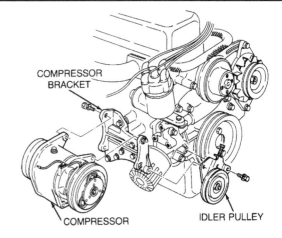

16.9 Typical air conditioning compressor mounting details (G180Z and G200Z four-cylinder engines shown)

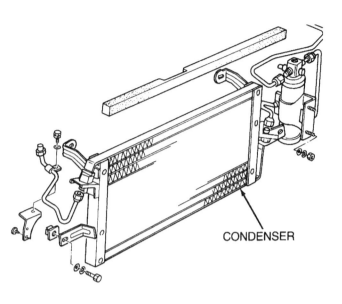

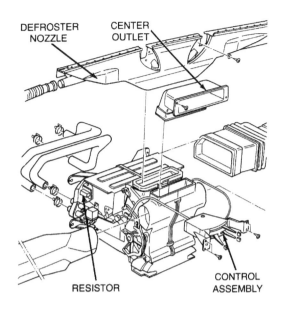

17.4 Typical condenser mounting details

18.2 Exploded view showing the heater control assembly

1   Remove the grille as described in Chapter 11 and the optional grille guard if so equipped.
2   Remove the radiator as described in Section 4.
3   Remove the left headlight assembly.
4   Disconnect all tubes and rubber A/C hoses leading from the condenser **(see illustration)**.
5   Remove the bolts that hold the condenser to the body and lift out the condenser.
6   If the condenser fins or air passages are clogged with foreign material, such as dirt, insects or leaves, use compressed air and a soft brush to clean the condenser. If the condenser is in need of other repairs, have a dealer service department or professional radiator shop perform the work.
7   Installation is the reverse of the removal

procedure. When reconnecting the hoses be sure that new O-rings are used and that they are installed correctly.
8   Once all A/C lines have been securely connected, the vehicle must once again be taken to a dealer service department or an air conditioning shop to have the system evacuated and recharged.

18   Heater/air conditioner control assembly - removal, installation and adjustment

Removal and installation
Refer to illustrations 18.2, 18.4 and 18.5
1   Disconnect the negative battery cable.

2   Disconnect the control cables from the heater unit and the blower unit **(see illustration)**.
3   Unscrew the cigarette lighter socket from the control assembly and disconnect the electrical connector behind the assembly.
4   Detach the knobs from the control levers **(see illustration)** and remove the control unit panel.
5   Remove the heater/A/C control unit mounting screws **(see illustration)**. Lower the unit from the dash and unplug any wiring harnesses leading to the unit.
6   The fan switch can be tested by performing a continuity test, referring to Chapter 12, if necessary. For replacement, it can be detached from the rest of the control unit.
7   Installation is the reverse of the removal procedure. **Note:** *Prior to reconnecting the*

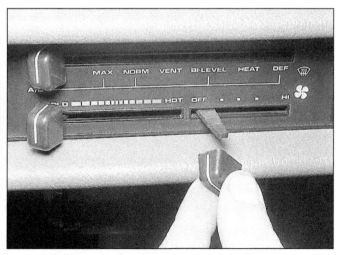

18.4 Remove the knobs from the control levers by pulling them off

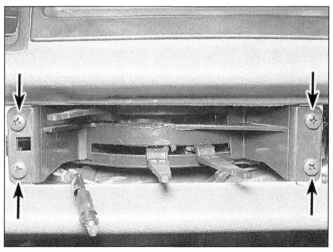

18.5 Remove the control unit mounting screws (arrows)

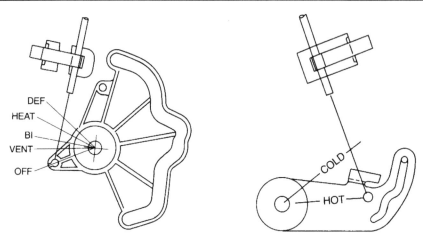

18.8 Mode selector lever positions

18.9 Temperature lever positions

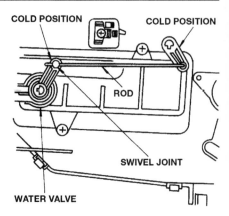

18.10 Air mix door and water valve lever adjustment details

control cables to the heater and blower units, they must be adjusted as follows:

Adjustment

Refer to illustrations 18.8, 18.9 and 18.10

8   Connect the mode selector cable to the mode selector lever with the mode lever on the control assembly set in the Off position. Secure the outer tube with a cable clip **(see illustration)**.

9   Connect the temperature control cable to the temperature lever with the temperature lever on the control assembly set in the Hot position. Secure the outer tube with a cable clip **(see illustration)**.

10   Set the air mix door lever in the Cold position by turning it counterclockwise. Set the water valve lever to the Cold position by turning it clockwise. Secure the link rod to the swivel joint with a screw **(see illustration)**.

Chapter 4
Fuel and exhaust systems

Contents

Specifications

Fuel Pressure
I-TEC fuel injection
    With vacuum hose removed from pressure regulator ... 42 psi
    With vacuum hose attached to pressure regulator ... 35 PSI
TBI injection ... 9 to 13 psi

Torque specifications — Ft-lbs
Fuel pump mounting nuts (mechanical pump) ... 20
Carburetor mounting nuts ... 18
Throttle body mounting nuts ... 18
Common chamber mounting nuts ... 18
Front exhaust pipe-to-exhaust manifold bolts ... 16

1 General information

Fuel system

Carburetor equipped vehicles

Carburetor equipped versions of this vehicle use a fuel system of conventional design. Fuel is pumped from the fuel tank, through the fuel lines and fuel filter into the carburetor, where it's mixed with air for combustion. The fuel system, as well as the exhaust system, is interrelated with various emissions control systems covered in Chapter 6. Thus, some elements that relate directly to the fuel system and carburetor functions are covered in that Chapter.

All carburetors are down-draft, two-barrel types. Specific information on the carburetor can be found in Section 9.

Two types of fuel pumps are used. A conventional diaphragm-type mechanical pump is attached to the side of the cylinder head on earlier models. The operating arm of the pump extends into the cylinder head where it is actuated by an eccentric cam mounted on the front of the camshaft. This eccentric cam, when rotating, moves the operating arm up and down, providing the pump action.

An electric pump is used on later models. It is mounted inside the fuel tank on most vehicles.

Besides the fuel feed line leading to the fuel filter and pump, the tank also has an emission vent hose leading to the charcoal canister, an air ventilation line that connects back into the fuel filler hose and a fuel return hose to route excess fuel back to the tank.

1.8 Multi-port (I-TEC) fuel injection system component locations

1	*Crank angle sensor (built into distributor)*	*4*	*Oxygen sensor (on exhaust manifold)*	*8*	*Canister purge valve*
2	*Ignition coil*	*5*	*Throttle body*	*9*	*Throttle valve switch*
3	*Air flow sensor*	*6*	*Common chamber*	*10*	*Power switch*
		7	*Fuel pressure regulator*		

Throttle Body Injection (TBI) equipped vehicles

The throttle body system utilizes one injector, centrally mounted in a carburetor-like housing. The injector is an electric solenoid which controls fuel flow into the throttle body airstream. Fuel is delivered to the injector at a constant pressure level. To maintain the fuel pressure at a constant level, excess fuel is returned to the fuel tank.

A signal from the ECM opens the solenoid, allowing fuel to spray through the injector into the throttle body airstream. The amount of time the injector is held open by the ECM determines the fuel/air mixture ratio.

Multi-port fuel injection (I-TEC) equipped vehicles

Refer to illustration 1.8

The multi-port fuel injection (I-TEC) system utilizes one injector per intake port **(see**

illustration). The throttle body serves only to control the amount of air passing into the system. Because each cylinder is equipped with an injector mounted immediately adjacent to the intake valve, much better control of the fuel/air mixture ratio is possible.

Fuel is circulated from the fuel tank to the fuel injection system, and back to the fuel tank, through a pair of metal lines running along the underside of the vehicle. An electric fuel pump is attached to the fuel sending unit inside the fuel tank. To reduce the likelihood of vapor lock, a vapor return system routes all vapors and hot fuel back to the fuel tank through a separate return line.

Exhaust system

The exhaust system, which is similar for both four-cylinder and V6 powered vehicles, includes an exhaust manifold fitted with an exhaust oxygen sensor, a catalytic converter, an exhaust pipe, and a muffler.

2   Fuel pressure relief procedure (fuel injected vehicles)

1   Before servicing any component on a fuel injected vehicle, it is necessary to relieve the fuel pressure to minimize the risk of fire or personal injury. Remove the fuel filter cap to relieve the vapor pressure in the fuel tank.

Multi-port fuel injection (ITEC) equipped vehicles

Refer to illustration 2.3
2   Locate the fuse and relay panel under the hood of the vehicle (see Chapter 12).
3   Locate the fuel pump relay on the panel. Most vehicles are equipped with a panel cover that is marked on top with the designated relay location **(see illustration)**.
4   With the ignition turned to Off, remove the fuel pump relay.

2.3 The fuel pump relay location is printed on the fuse panel cover

3.8 Disconnect the outlet hose from the fuel pressure regulator (arrow)

5   Start the engine. It will run until the fuel supply remaining in the fuel lines is depleted.
6   After the engine stops, engage the starter again for another three seconds to assure that any remaining pressure is dissipated.
7   Turn the ignition off. The fuel pressure is now relieved.

Throttle Body Injection (TBI) equipped vehicles

8   These vehicles have a constant-bleed feature which relieves pressure when the ignition key is turned to OFF. No special relief procedure is required. Disconnect the negative battery cable and re-connect it after repairs are complete.

---

3   Fuel pressure testing

Warning: *Gasoline is extremely flammable, so take extra precautions when you work on any part of the fuel system. Don't smoke or allow open flames or bare light bulbs near the work area, and don't work in a garage where a natural gas-type appliance (such as a water heater or clothes dryer) with a pilot light is present. If you spill any fuel on your skin, rinse it off immediately with soap and water. When you perform any kind of work on the fuel system, wear safety glasses and have a Class B type fire extinguisher on hand.*

Carbureted vehicles (mechanical pump)

1   Check that there is adequate fuel in the fuel tank.
2   With the engine running, examine all fuel lines between the fuel tank and fuel pump for leaks, loose connections, kinks or flattening in the rubber hoses. Do this quickly, before the engine gets hot. Air leaks upstream of the fuel pump can seriously affect the pump's output.
3   Check the pump diaphragm flange for leaks.

4   Disconnect the fuel line at the carburetor. Disconnect the ignition coil wire from the coil and ground it on the engine block (use a jumper wire to prevent sparks) so the engine can be cranked without it firing. Place a clean container such as a coffee can at the end of the detached fuel line and crank the engine for several seconds. There should be a strong spurt of gasoline from the line on every second revolution.
5   If little or no gasoline emerges from the line during engine cranking, then either the line is clogged or the fuel pump is not working properly. Disconnect the fuel line from the pump and blow air through it to be sure the line is clear. If the line is clear then the pump is suspect and needs to be replaced with a new one.
6   If the fuel pump flow capacity is adequate, check the fuel pressure by attaching a fuel gauge to the same fuel line and cranking the engine. The fuel pressure should be approximately 3.5 psi.

I-TEC equipped vehicles (electric pump)

Refer to illustrations 3.8 and 3.9
7   Relieve the fuel pressure (see Section 2).
8   Disconnect the fuel outlet hose from the fuel pressure regulator **(see illustration)**.
9   Connect a fuel pressure gauge to the outlet hose **(see illustration)**.
10   Start the engine. Detach the vacuum hose from the pressure regulator. With the engine idling, measure the fuel pressure. It should as listed in this Chapter's Specifications.
11   If the fuel pressure is not within specifications, check the following:

a)   If the pressure is higher than specified, check for a faulty regulator or a pinched or clogged fuel return hose or pipe.
b)   If the pressure is lower than specified:
   1)   Inspect the fuel filter - make sure it's not clogged (see Chapter 1).
   2)   Look for a pinched or clogged fuel hose between the fuel tank and fuel pump.

3.9 Connect a fuel pressure gauge to the outlet hose - make sure all connections are tight to avoid fuel leaks

   3)   Look for leaks in the fuel line.
   4)   If there are no problems with the above components, the fuel pump is not delivering sufficient pressure or the pressure regulator is malfunctioning.

12   Reconnect the vacuum hose to the pressure regulator. Start the engine and measure the fuel pressure at idle. The pressure should drop to the value listed in this Chapter's Specifications. If the pressure does not drop, check that the vacuum hose is not pinched, deteriorated, collapsed or otherwise damaged. If the vacuum hose is okay, the pressure regulator is probably faulty.

---

4   Fuel lines and fittings - inspection and replacement

Warning: *On fuel injected vehicles, the fuel system pressure must be relieved before disconnecting fuel lines and fittings (see Section 2). Gasoline is extremely flammable, so take extra precautions when you work on any part of the fuel system. Don't smoke or allow open*

5.4 Disconnect the fuel tank electrical connector

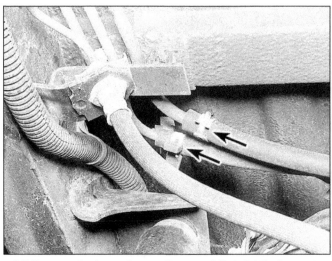

5.5 Remove the hose clamps, then carefully twist the hoses to free them from the fittings (arrows)

flames or bare light bulbs near the work area, and don't work in a garage where a natural gas-type appliance (such as a water heater or clothes dryer) with a pilot light is present. If you spill any fuel on your skin, rinse it off immediately with soap and water. When you perform any kind of work on the fuel system, wear safety glasses and have a Class B type fire extinguisher on hand.

Inspection

1   Once in a while, you will have to raise the vehicle to service or replace some component (an exhaust pipe hanger, for example). Whenever you work under the vehicle, always inspect the fuel lines and fittings for possible damage or deterioration.

2   Check all hoses and pipes for cracks, kinks, deformation or obstructions.

3   Make sure all hose and pipe clips attach their associated hoses or pipes securely to the underside of the vehicle.

4   Verify all hose clamps attaching rubber hoses to metal fuel lines or pipes are snug enough to assure a tight fit between the hoses and pipes.

Replacement

5   If you must replace any damaged sections, use original equipment replacement hoses or pipes constructed from exactly the same material as the section you are replacing. Do not install substitutes constructed from inferior or inappropriate material or you could cause a fuel leak or a fire.

6   Always, before detaching or disassembling any part of the fuel line system, note the routing of all hoses and pipes and the orientation of all clamps and clips to assure that replacement sections are installed in exactly the same manner.

7   Before detaching any part of the fuel system, be sure to relieve the fuel tank pressure by removing the fuel filler cap. On fuel injected vehicles, also relieve the fuel system pressure (see Section 2).

5.7 On most Trooper models, two mounting bolts (arrows) are located on each side of the fuel tank

8   While you're under the vehicle, it's a good idea to check the following related components:

   a) *Check the condition of the fuel filter - make sure that it's not clogged or damaged (see Chapter 1).*

   b) *Inspect the evaporative emission control system. Verify that all hoses are attached and in good condition (see Chapter 6).*

5   Fuel tank - removal and installation

Refer to illustrations 5.4, 5.5 and 5.7

Warning: *Gasoline is extremely flammable, so take extra precautions when you work on any part of the fuel system. Don't smoke or allow open flames or bare light bulbs near the work area, and don't work in a garage where a natural gas-type appliance (such as a water heater or clothes dryer) with a pilot light is present. If you spill any fuel on your skin, rinse it off immediately with soap and water. When you perform any kind of work on the fuel tank, wear safety glasses and have a Class B type fire extinguisher on hand.*

Note: *The following procedure is much easier to perform if the fuel tank is empty. Some tanks have a drain plug for this purpose. If the tank does not have a drain plug, simply run the engine until the tank is empty.*

1   Relieve the fuel pressure on fuel injected vehicles (see Section 2).

2   Detach the cable from the negative terminal of the battery.

3   Raise the vehicle and place it securely on jackstands.

4   Locate the electrical connector for the electric fuel pump and/or fuel gauge sending unit **(see illustration)** on the right frame member just in front of the tank, and unplug it.

5   Disconnect the fuel feed and return lines and the vapor return line **(see illustration)**. Label them so they can be returned to their original locations.

6   Support the fuel tank with a floor jack or some other suitable means of support.

7   Disconnect the fuel tank mounting bolts. On most Pick-Up models, there are two front bolts and three rear bolts. On most Trooper models, there are two bolts on each side of the tank **(see illustration)**.

8   Lower the tank enough to disconnect

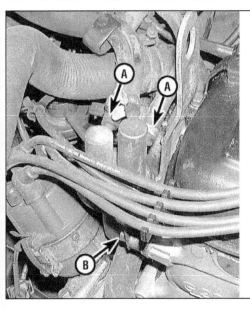

7.4 Fuel pump mounting details

A *Fuel hoses*
B *Mounting bolt (other bolt not visible in this photo)*

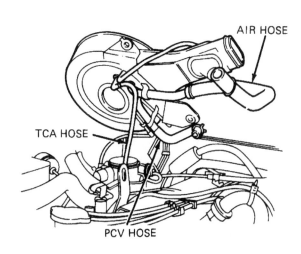

8.3 Air cleaner assembly on carburetor-equipped vehicles

the electrical wires and ground strap from the fuel pump/fuel gauge sending unit, if you have not already done so.
9   Remove the tank from the vehicle.
10 Installation is the reverse of removal.

6   Fuel tank cleaning and repair - general information

1   All repairs to the fuel tank or filler neck should be carried out by a professional who has experience in this critical and potentially dangerous work. Even after cleaning and flushing of the fuel system, explosive fumes can remain and ignite during repair of the tank.
2   If the fuel tank is removed from the vehicle, it should not be placed in an area where sparks or open flames could ignite the fumes coming out of the tank. Be especially careful inside garages where a natural gas-type appliance is located, because the pilot light could cause an explosion.

7   Fuel pump - removal and installation

Warning: *Gasoline is extremely flammable, so take extra precautions when you work on any part of the fuel system. Don't smoke or allow open flames or bare light bulbs near the work area, and don't work in a garage where a natural gas-type appliance (such as a water heater or clothes dryer) with a pilot light is present. If you spill any fuel on your skin, rinse it off immediately with soap and water. When you perform any kind of work on the fuel system, wear safety glasses and have a Class B type fire extinguisher on hand.*
Note: *On carbureted 4ZD1 engines, it may be necessary to remove the intake manifold (see Chapter 2A) to gain access to the fuel pump.*

Mechanical pump (carbureted models)

Refer to illustration 7.4
1   Detach the cable from the negative battery terminal.
2   Remove the fuel tank filler cap to relieve the fuel tank pressure, then remove the spark plug wires and distributor (see Chapters 1 and 5).
3   Wrap shop towels around the fuel lines at the fuel pump to absorb any spilled fuel.
4   Detach the fuel lines from the fuel pump **(see illustration)**.
5   Unscrew the two fuel pump mounting bolts and remove the fuel pump and gasket.
6   Carefully scrape away all traces of old gasket material from the fuel pump and cylinder head sealing surfaces.
7   Installation is the reverse of removal. Be sure to use a new gasket and tighten the mounting bolts securely.

Electric pump (fuel injected models)

8   Relieve the fuel pressure (Section 2).
9   Remove the cable from the negative battery terminal.
10 Remove the fuel tank (Section 5).
11 The fuel pump/sending unit assembly is located inside the fuel tank. It's held in place by screws around the perimeter of a circular steel mounting plate on top of the tank.
12 To remove the fuel pump/sending unit assembly, remove the screws and pull the assembly out of the tank. The fuel level float and sending unit are delicate. Do not bump them on the fuel tank during removal or the accuracy of the sending unit may be affected.
13 Inspect the condition of the rubber gasket around the mouth of the fuel pump opening in the fuel tank. If it is dried, cracked or deteriorated, replace it.
14 Inspect the strainer on the lower end of the fuel pump. If it is dirty, remove it, clean it

with a suitable solvent and blow it out with compressed air. If it is too dirty to be cleaned, replace it.
15 If it is necessary to separate the fuel pump and sending unit, remove the pump from the sending unit by pulling the fuel pump assembly into the rubber connector and sliding the pump away from the bottom support. Care should be taken to prevent damage to the rubber insulator and fuel strainer during removal. After the pump assembly is clear of the bottom support, pull the pump assembly out of the rubber connector.
16 Insert the fuel pump/sending unit assembly into the fuel tank.
17 Install the screws and tighten them securely.
18 Install the fuel tank (Section 5).

8   Air cleaner housing assembly - removal and installation

1   Detach the cable from the negative terminal of the battery.

Carbureted engines

Refer to illustration 8.3
2   Remove the air cleaner top plate and remove the air filter (see Chapter 1).
3   Remove the PCV hose from the air cleaner body. Disconnect the AIR hose from the air pump **(see illustration)**.
4   Remove the bolts which attach the air cleaner housing to the engine.
5   Detach the TCA hose and remove the air cleaner housing.
6   Installation is the reverse of removal.

Fuel injected engines

Refer to illustrations 8.8 and 8.11
7   Detach the cable from the negative terminal of the battery.

8.8 Unplug the electrical connector from the air flow sensor

8.11 Remove the three bolts at the bottom of the air cleaner housing (arrows)

8   Unplug the electrical connector from the air flow sensor **(see illustration)**.

9   Loosen the hose clamps at either end of the intake duct and separate it from the air flow sensor.

10   Remove the top portion of the air cleaner together with the air flow sensor and remove the air filter.

11   Remove the three bolts at the bottom of the air cleaner housing **(see illustration)**.

12   Loosen the lower air cleaner housing bolt and remove the air cleaner housing.

13   Installation is the reverse of removal.

9   Carburetor - general information

Refer to illustration 9.2

The carburetor is a downdraft, two-bar-rel type. The primary throttle valve is mechanically operated while the secondary one is vacuum operated by a diaphragm unit which is actuated by the vacuum in the carburetor venturi.

An electrically assisted, bimetal-type automatic choke is incorporated **(see illustration)**. This operates a choke valve which closes one of the venturi tubes and is synchronized with the primary throttle plate.

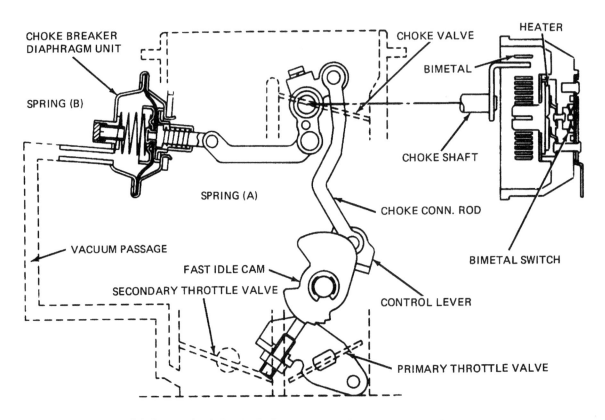

9.2 Automatic choke details (carburetor-equipped engines only)

When the engine is cold, the choke valve closes and the throttle plate opens sufficiently to provide a rich mixture and an increased idle speed for easy starting.

For idling and low-speed operation, fuel passes through the slow jet, the primary slow air bleed and the secondary slow air bleed. The fuel is finally ejected from the bypass and the idle holes. An antidieseling (run-on) solenoid valve is incorporated to ensure that the fuel supply is cut-off when the ignition is switched Off, thus preventing the engine from running-on.

The accelerator pump is synchronized with the throttle valve. During acceleration, the pump, which is of simple piston and valve construction, provides an additional metered quantity of fuel to enrich the normal mixture.

The secondary system provides a mixture for normal driving conditions by means of a main jet and air bleed. On some carburetors, an additional high-speed circuit is incorporated. It consists of an additional jet, air bleed and nozzle. It allows additional fuel to be drawn into the secondary bore as the air velocity through that bore increases.

Carburetors used on certain models use an altitude compensator to correct an otherwise too rich mixture which can occur at high altitude.

The float chamber is fed with fuel pumped by the mechanically operated pump on the cylinder head. The level in the chamber is critical and must at all times be maintained as specified.

10 Carburetor - removal and installation

Warning: *Gasoline is extremely flammable, so take extra precautions when you work on any part of the fuel system. Don't smoke or allow open flames or bare light bulbs near the work area, and don't work in a garage where a natural gas-type appliance (such as a water heater or clothes dryer) with a pilot light is present. If you spill any fuel on your skin, rinse it off immediately with soap and water. When you perform any kind of work on the fuel system, wear safety glasses and have a Class B type fire extinguisher on hand.*

Removal

1   Detach the negative battery cable, then remove the air cleaner assembly (see Section 8).
2   Detach the accelerator linkage.
3   If equipped with an automatic transmission, detach the transmission throttle valve (TV) cable (see Chapter 7 Part B).
4   Detach the cruise control cable, if equipped.
5   Clearly label, then detach, all vacuum lines.
6   Disconnect the fuel line at the carburetor inlet.
7   Clearly label, then unplug all electrical connectors.
8   Remove the four carburetor mounting bolts.
9   Remove the carburetor from the intake manifold. Remove the carburetor mounting gasket. Stuff a shop rag into the intake manifold openings to prevent the entry of foreign material.

Installation

10   Use a gasket scraper to remove all traces of gasket material and sealant from the intake manifold (and the carburetor, if it's being reinstalled), then remove the shop rag from the manifold openings. Clean the mating surfaces with lacquer thinner or acetone.
11   Place a new gasket on the intake manifold.
12   Position the carburetor on the gasket and install the mounting bolts.
13   To prevent carburetor distortion or damage, tighten the bolts in a criss-cross pattern, 1/4-turn at a time to the torque listed in this Chapter's Specifications.
14   The remaining installation steps are the reverse of removal.
15   Check and, if necessary, adjust the idle speed (see Chapter 1).
16   If the vehicle is equipped with an automatic transmission, refer to Chapter 7, Part B for the TV cable adjustment procedure.
17   Attach the negative battery cable.
18   Start the engine and check carefully for fuel leaks.

11 Carburetor - diagnosis, overhaul and adjustments

Warning: *Gasoline is extremely flammable, so take extra precautions when you work on any part of the fuel system. Don't smoke or allow open flames or bare light bulbs near the work area, and don't work in a garage where a natural gas-type appliance (such as a water heater or clothes dryer) with a pilot light is present. If you spill any fuel on your skin, rinse it off immediately with soap and water. When you perform any kind of work on the fuel system, wear safety glasses and have a Class B type fire extinguisher on hand.*

Diagnosis

1   A thorough road test and check of carburetor adjustments should be done before any major carburetor service work. Specifications for some adjustments are listed on the Vehicle Emissions Control Information (VECI) label found in the engine compartment.
2   Carburetor problems usually show up as flooding, hard starting, stalling, severe backfiring and poor acceleration. A carburetor that's leaking fuel and/or covered with wet looking deposits definitely needs attention.
3   Some performance complaints directed at the carburetor are actually a result of loose, out-of-adjustment or malfunctioning engine or electrical components. Others develop when vacuum hoses leak, are disconnected or are incorrectly routed. The proper approach to analyzing carburetor problems should include the following items:

 a) *Inspect all vacuum hoses and actuators for leaks and correct installation (see Chapters 1 and 6).*
 b) *Tighten the intake manifold and carburetor mounting bolts evenly and securely.*
 c) *Perform a cylinder compression test (see Chapter 2).*
 d) *Clean or replace the spark plugs as necessary (see Chapter 1).*
 e) *Check the spark plug wires (see Chapter 1).*
 f) *Inspect the ignition primary wires.*
 g) *Check the ignition timing (see Chapter 1).*
 h) *Check the fuel pressure (see Section 3).*
 i) *Check the thermo-sensor assembly in the air cleaner for proper operation (see Chapter 6).*
 j) *Check/replace the air filter element (see Chapter 1).*
 k) *Check the PCV system (see Chapters 1 and 6).*
 l) *Check/replace the fuel filter (see Chapter 1). Also, the strainer in the tank could be restricted.*
 m) *Check for a plugged exhaust system.*
 n) *Check EGR valve operation (see Chapter 6).*
 o) *Check the choke - it should be completely open at normal engine operating temperature (see Chapter 1).*
 p) *Check for fuel leaks and kinked or dented fuel lines (see Chapters 1 and 4).*
 q) *Check accelerator pump operation with the engine off (remove the air cleaner cover and operate the throttle as you look into the carburetor throat - you should see a stream of gasoline enter the carburetor).*
 r) *Check for incorrect fuel or bad gasoline.*
 s) *Check the valve clearances (if applicable) and camshaft lobe lift (see Chapters 1 and 2).*
 t) *Have a dealer service department or repair shop check the electronic engine and carburetor controls.*

4   Diagnosing carburetor problems may require that the engine be started and run with the air cleaner off. While running the engine without the air cleaner, backfires are possible. This situation is likely to occur if the carburetor is malfunctioning, but just the removal of the air cleaner can lean the fuel/air mixture enough to produce an engine backfire. **Warning:** *Don't position any part of your body, especially your face, directly over the carburetor during inspection and servicing procedures. Wear eye protection!*

Overhaul

Refer to illustration 11.5

5   If you are going to overhaul the carburetor yourself, first obtain a good quality carburetor rebuild kit (which will include all necessary gaskets, internal parts, instructions and

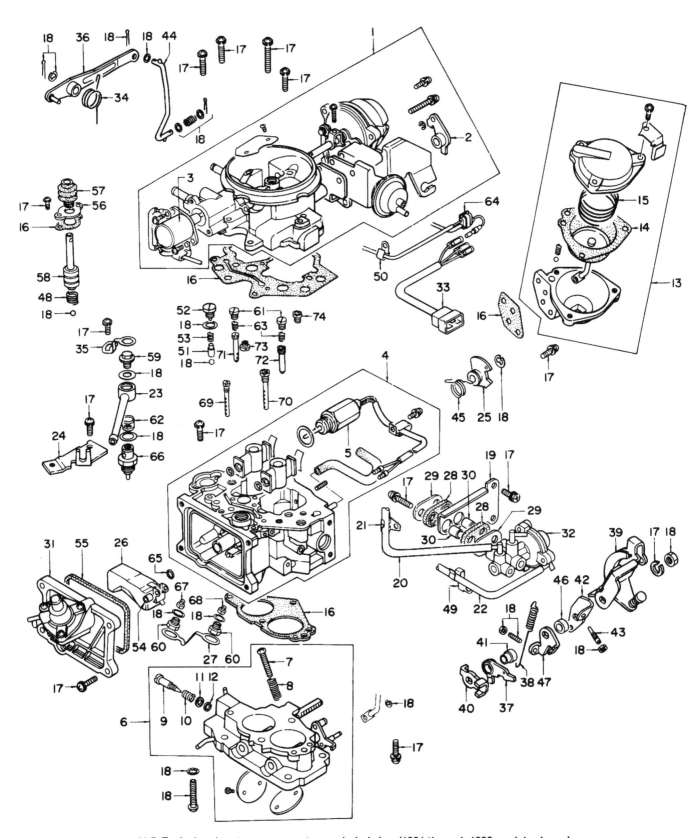

11.5 Typical carburetor components - exploded view (1981 through 1989 models shown)

1   Chamber assembly
2   Counter lever
3   Vent switching valve
4   Float chamber assembly
5   Slow cut valve
6   Throttle chamber assembly
7   Throttle adjust screw
8   Throttle adjust spring
9   Idle adjust screw
10 Spring
11 Washer
12 Rubber seal
13 Diaphragm chamber assembly
14 Diaphragm
15 Diaphragm spring
16 Overhaul gasket kit
17 Screw and washer kit (A)
18 Screw and washer kit (B)
19 Actuator bracket
20 Rubber hose
21 Hose clamp
22 Rubber hose
23 Fuel fitting
24 Stopping plate
25 Fast idle cam
26 Float
27 Drain plug lock plate
28 Mounting rubber
29 Plate
30 Collar
31 Main actuator
32 Slow actuator
33 Connector harness assembly
34 Pump lever spring
35 Lock lever
36 Pump lever
37 Throttle return plate
38 Secondary throttle spring
39 Throttle lever
40 Throttle adjust lever
41 Throttle sleeve (A)
42 Fast adjust lever
43 Fast idle screw
44 Pump rod
45 Fast idle cam spring
46 Throttle sleeve (B)
47 Throttle kick lever
48 Piston return spring
49 Hose clamp (B)
50 Wire holder
51 Injector weight
52 Pump set screw
53 Injector spring
54 Float set collar (C)
55 Level gauge rubber seal
56 Cylinder plate
57 Dust cover
58 Pump piston
59 Fitting set screw
60 Drain plug
61 Slow jet taper plug
62 Needle valve filter
63 Slow jet spring
64 Wire connector
65 O-ring
66 Needle valve
67 Primary main jet
68 Secondary main jet
69 Primary main air bleed
70 Secondary main air bleed
71 Primary slow jet
72 Secondary slow jet
73 Primary slow air bleed
74 Secondary slow air bleed

a parts list) **(see illustration)**. You will also need some solvent and a means of blowing out the internal passages of the carburetor with air.

6   Because carburetor designs are constantly modified by the manufacturer in order to meet emissions regulations, it isn't feasible for us to do a step-by-step overhaul of each type. You'll receive a detailed set of instructions with any quality carburetor overhaul kit. They will apply in a more specific manner to the carburetor on your vehicle.

7   An alternative is to obtain a new or rebuilt carburetor. They are readily available from dealers and auto parts stores. Make sure the exchange carburetor is identical to the original. A tag is usually attached to the top of the carburetor. It will aid in determining the exact type of carburetor you have. When obtaining a rebuilt carburetor or a rebuild kit, take time to make sure that the kit or carburetor matches your application exactly. Seemingly insignificant differences can make a large difference in the performance of your engine.

8   If you choose to overhaul your own carburetor, allow enough time to disassemble the carburetor carefully, soak the necessary parts in the cleaning solvent (usually for at least one-half day or according to the instructions listed on the carburetor cleaner) and reassemble it, which will usually take much longer than disassembly. When disassembling the carburetor, match each part with the illustration in the carburetor kit and lay the parts out in order on a clean work surface.

Adjustments

9   Because there are a number of different configurations for Federal and California carburetors and because a considerable number of special tools and tuning equipment is necessary to adjust these carburetors, it is impossible to include a detailed step-by-step procedure outlining every adjustment. Aside from idle speed adjustment (see Chapter 1) and other adjustments shown in the carburetor overhaul kit, do not attempt to adjust the carburetor on your vehicle. If adjustments are needed other than those listed above, take the vehicle to a professional mechanic.

12 Fuel injection system - general information

Electronic fuel injection provides optimum air/fuel mixture ratios at all stages of combustion and offers better throttle response characteristics than carburetion. It also enables the engine to run at the leanest possible air/fuel mixture ratio, greatly reducing exhaust gas emissions.

One of two types of fuel injection systems is used on non-carbureted models. Some four-cylinder engines use a multi-port fuel injection system also called the I-TEC system. V6 engines use a Throttle Body Injection (TBI) system.

Both systems are electronic and employ an Electronic Control Module (ECM) and information sensors which monitor various engine functions and send data back to the ECM.

These electronic systems are analogous to the central nervous system in the human body. The sensors (nerve endings) constantly relay information to the ECM (brain), which processes the data and, if necessary, sends out a command to change the operating parameters of the engine (body).

Here's a specific example of how one portion of an electronic fuel injection system operates: An oxygen sensor, located in the exhaust manifold, constantly monitors the oxygen content of the exhaust gas. If the percentage of oxygen in the exhaust gas is incorrect, an electrical signal is sent to the ECM. The ECM takes this information, processes it and then sends a command to the fuel injection system, telling it to change the fuel/air mixture. This happens in a fraction of a second and it goes on continuously when the engine is running. The end result is a fuel/air mixture ratio which is constantly maintained at a predetermined ratio, regardless of driving conditions.

One might think that a system which uses an on-board computer and electrical sensors would be difficult to diagnose. This is not necessarily the case. The systems on most models have a built-in diagnostic feature which indicates a problem by flashing a Check Engine light on the instrument panel (see Section 13). When this light comes on during normal vehicle operation, a fault in one of the information sensor circuits or the ECM itself has been detected. More importantly, the source of the malfunction is stored in the ECM's memory.

13 Self-diagnostic codes - retrieving

Refer to illustration 13.1

Note: *Not all models are equipped with the self-diagnosis feature. If the vehicle has a Check Engine light on the dashboard, it's equipped with self-diagnosis. If there's no Check Engine light, it's not equipped with self-diagnosis.*

1   To retrieve self-diagnosis information from the ECM memory, you must turn the ignition switch to ON and ground the diagnostic terminals by connecting them together. There are two different diagnostic terminals in the passenger compartment and only one is used, depending on which engine is installed in the vehicle:

a) *DIAG warning light harness - This is used on four-cylinder engines* **(see illustration on following page)**.

b) *ALDL connector - This is used on V6 engines.*

2   The DIAG warning light harness is located in different areas of the passenger compartment, depending on the year and

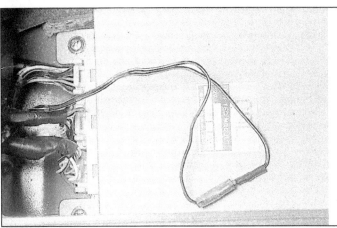

13.1 Typical DIAG hook-up - make sure the ignition switch is on before connecting the terminals

model:

a) *On 1981 through 1983 pick-ups, the DIAG harness connectors are located next to the ECM.*

b) *On 1984 through 1987 pick-ups, the DIAG harness connectors are located directly under the dash on the driver's side, next to the hood release handle. The male and female connectors hang loosely from the wire harness.*

c) *On 1988 and later pick-ups, the DIAG harness connectors are located in the corner of the floor on the driver's side under the ECM.*

d) *On 1985 Troopers, the DIAG harness connectors are located under the dash-board console near the radio on the passenger's side. The male and female connectors are tucked up onto the wire harness and attached by a piece of tape.*

e) *On 1986 through 1987 Troopers, the DIAG harness connectors are located*

under the dashboard near the cigarette lighter on the driver's side. The male and female connectors are tucked up onto the wire harness and held in with a piece of tape.

f) *On 1988 and later Troopers, the DIAG harness connectors are located under the center console, near the ECM.*

3   On vehicles equipped with a V6 engine, the wiring connector known as the Assembly Line Diagnostics Link (ALDL) is normally located in the center console. It is activated by using a jumper wire. Remove the plastic cover by sliding it toward you. With the connector exposed, push one end of the jumper wire into the diagnostic terminal (terminal B - second terminal from the right of the ALDL connector's top row) and the other end into the ground terminal (terminal A - the terminal directly to the right of terminal B on the top row of the ALDL connector.

4   When the diagnostic terminal is

grounded with the ignition on and the engine stopped, the system will enter the Diagnostic Mode. In this mode the ECM will display a "Code 12" by flashing the Check Engine light, indicating the system is operating. A code 12 is simply one flash, followed by a brief pause, then two flashes in quick succession. This code will be flashed three times. If no other codes are stored, Code 12 will continue to flash until the diagnostic terminal ground is removed.

5   After flashing Code 12 three times, the ECM will display any stored trouble codes. Each code will be flashed three times, then Code 12 will be flashed again, indicating the display of stored trouble codes has been completed.

6   When the ECM sets a trouble code, the Check Engine light will come on and a trouble code will be stored in the memory. If the problem is intermittent, the light will go out after 10-seconds, when the fault goes away. However, the trouble code will stay in the ECM memory until the battery voltage to the ECM is interrupted. Removing battery voltage for 30-seconds will clear all stored trouble codes. Trouble codes should always be cleared after repairs have been completed. **Caution:** *To prevent damage to the ECM, the ignition switch must be off when disconnecting power to the ECM.*

7   The following is a list of the typical trouble codes which may be encountered while diagnosing the computerized system. Also included are simplified troubleshooting procedures. If the problem persists after these checks have been made, more detailed service procedures will have to be done by a dealer service department.

Trouble codes for four-cylinder engines

Note: *These codes do not apply to 1981 through 1983 pick-up models.*

Trouble code	Circuit or system	Probable cause
Code 12 (1 flash, pause, 2 flashes)	Diagnostic mode	This code will flash whenever the diagnostic terminal is grounded with the ignition turned on and the engine not running. If additional trouble codes are stored in the ECM, they will appear after this code has flashed three times.
Code 13 (1 flash, pause, 3 flashes)	Oxygen sensor circuit	The wiring is broken or the oxygen sensor is deteriorated. Check the wiring, then replace the sensor.*
Code 14 (1 flash, pause, 4 flashes)	Coolant temperature sensor	Insufficient signal from the sensor or a shorted circuit. Check the wiring, then replace the sensor.*
Code 15 (1 flash, pause, 5 flashes)	Open coolant temperature	Excessive signal from the sensor. Check for an open circuit in the sensor circuit coolant temperature sensor circuit wiring.
Code 21 (2 flashes, pause, 1 flash)	Throttle valve switch	Replace the throttle valve switch.*
Code 22 (2 flashes, pause, 2 flashes)	Fuel cut solenoid circuit	The fuel cut solenoid circuit is open or shorted (except 1988 and 1989 Troopers).

Trouble code	Circuit or system	Probable cause
Code 22 (2flashes, pause, 2 flashes)	Starter signal system	The starter signal circuit is open or grounded (1988 and 1989 Troopers).
Code 23 (2 flashes, pause, 3 flashes)	Mixture control solenoid circuit	The mixture control solenoid circuit is open or shorted (except 1988 and 1989 Troopers).
Code 23 (2 flashes, pause, 3 flashes)	Power transistor	The power transistor for the ignition system is shorted at the output terminal (1988 and 1989 Troopers). Check the circuit for a short.
Code 25 (2 flashes, pause, 5 flashes)	Air vacuum switching valve	The air vacuum switching valve circuit is open or grounded. Check the circuit.
Code 26 (2 flashes, pause, 6 flashes)	Air vacuum switching valve	The air vacuum switching valve circuit for the canister purge system is open or grounded. Check the circuit.
Code 27 (2 flashes, pause, 7 flashes)	Driver transistor	The driver transistor in the ECM is open. Have the ECM checked by a dealer service department.
Code 31 (3 flashes, pause, 1 flash)	Ignition signals	Ignition circuit shorted. Have the vehicle checked by a dealer service department.
Code 32 (3 flashes, pause, 2 flashes)	EGR sensor and circuit	The EGR sensor is faulty or the circuit is shorted.
Code 33 (3 flashes, pause, 3 flashes)	Fuel injector circuit	The fuel injector circuit is open or grounded. Check the circuit.
Code 34 (3 flashes, pause, 4 flashes)	EGR temperature sensor circuit	The EGR temperature sensor circuit is shorted. Check the circuit for shorts.
Code 35 (3 flashes, pause, 5 flashes)	Power transistor circuit	The power transistor circuit in the ignition system is open. Check the wiring.
Code 41 (4 flashes, pause, 1 flash)	Crank angle sensor	Check the crank angle sensor wiring. Replace the crank angle sensor.*
Code 43 (4 flashes, pause, 3 flashes)	Throttle valve switch	The throttle valve switch continuously makes contact at idle. Replace the switch.*
Code 44 (4 flashes, pause, 4 flashes)	Fuel metering system	The signal is at low voltage (lean). Check the oxygen sensor.
Code 45 (4 flashes, pause, 5 flashes)	Fuel metering system	The signal is at high voltage (rich). Check the oxygen sensor.
Code 51 (5 flashes, pause, 1 flash)	ECM	The fuel cut solenoid circuit is shorted or the ECM is faulty. Have the vehicle checked by a dealer service department.
Code 52 (5 flashes, pause, 2 flashes)	ECM	Faulty ECM. Have the vehicle checked by a dealer service department.
Code 53 (5 flashes, pause, 3 flashes)	ECM or vacuum switching valve	Faulty ECM or vacuum switching valve. Have the vehicle checked by a dealer service department.
Code 54 (5 flashes, pause, 4 flashes)	ECM	Shorted vacuum control solenoid or defective transistor in the ECM. Have the vehicle checked by a dealer service department
Code 55 (5 flashes, pause, 5 flashes)	ECM	Faulty A/D converter in ECM. Have the vehicle checked by a dealer service department

Trouble codes for four-cylinder engines (continued)

Trouble code	Circuit or system	Probable cause
Code 61 (6 flashes, pause, 1 flash)	Air flow sensor circuit	Check the air flow sensor circuit for a short or a broken hot wire.
Code 62 (6 flashes, pause, 2 flashes)	Air flow sensor circuit	Check the air flow sensor circuit for a broken cold wire.
Code 63 (6 flashes, pause, 3 flashes)	Vehicle speed sensor circuit	The vehicle speed sensor is broken or the circuit is shorted.
Code 64 (6 flashes, pause, 4 flashes)	Driver transistor	The driver transistor in the ECM is faulty. Have the vehicle checked by a dealer service department.
Code 65 (6 flashes, pause, 5 flashes)	Throttle valve switch	The throttle valve switch continuously makes contact. Replace the switch.*

* Component replacement may not cure problem in all cases. For this reason, you may want to seek professional advice before purchasing replacement parts.

Trouble codes for V6 engines

Trouble code	Circuit or system	Probable cause
Code 12 (1 flash, pause, 2 flashes)	Diagnostic	This code will flash whenever the diagnostic terminal is grounded with the ignition turned On and the engine not running. If additional trouble codes are stored in the ECM they will appear after this code has flashed three times.
Code 13 (1 flash, pause, 3 flashes)	Oxygen sensor circuit (open circuit)	Check the wiring and connectors from the oxygen sensor. Replace the oxygen sensor.*
Code 14 (1 flash, pause, 4 flashes)	Coolant sensor circuit. (high temperature)	If the engine is experiencing overheating problems the problem must be rectified before continuing. Check all wiring and connectors associated with the coolant temperature sensor. Replace the coolant temperature sensor.*
Code 15 (1 flash, pause, 5 flashes)	Coolant sensor circuit (low temperature)	See above, then check the wiring connections at the ECM.
Code 21 (2 flashes, pause, 1 flash)	Throttle position sensor (signal voltage high)	Check for a sticking or misadjusted TPS plunger. Check all wiring and connections between the TPS and the ECM. Adjust or replace the TPS.*
Code 22 (2 flashes, pause, 2 flashes)	Throttle position sensor (signal voltage low)	Check the TPS adjustment. Check the ECM connector. Replace the TPS.*
Code 24 (2 flashes, pause, 4 flashes)	Vehicle speed sensor	A fault in this circuit should be indicated only when the vehicle is in motion. Disregard Code 24 if it is set when the drive wheels are not turning. Check the connections at the ECM.
Code 32 (3 flashes, pause, 2 flashes)	EGR system	Check vacuum hoses and connections for leaks an restrictions. Replace the EGR solenoid or valve.*
Code 33 (3 flashes, pause, 3 flashes)	MAP sensor (signal voltage high)	Replace the MAP sensor.*
Code 34 (3 flashes, pause, 4 flashes)	MAP (signal voltage low)	Check for a loose or damaged air duct, misadjusted minimum idle speed and vacuum leaks. Inspect the MAP sensor and The electrical connections.

Trouble code	Circuit or system	Probable cause
Code 42 (4 flashes, pause, 2 flashes)	Electronic Spark Timing	Faulty connections or ignition module.
Code 43 (4 flashes, pause, 3 flashes)	Electronic Spark Control	Faulty knock sensor or MEM CAL.*
Code 44 (4 flashes, pause, 4 flashes)	Lean exhaust	Check the ECM wiring connections from the oxygen sensor. Check for vacuum leaks at the hoses and intake manifold gasket.*
Code 45 (4 flashes, pause, 5 flashes)	Rich exhaust	Check the evaporative charcoal canister and its components for the presence of fuel.
Code 51 (5 flashes, pause, 1 flash)	Prom or MEM-CAL	Make sure the MEM-CAL or PROM is properly installed in the ECM. Replace the MEM CAL or PROM.*
Code 52 (5 flashes, pause, 2 flashes)	CAL-PAK	Check the CAL-PAK to insure proper installation. Replace the CAL-PAK.*
Code 54 (5 flashes, pause, 4 flashes)	Fuel pump circuit (low voltage)	Check the fuel pump relay. Replace if defective*.
Code 55 (5 flashes, pause, 5 flashes)	ECM	Check for faulty connections or replace the ECM.*

Component replacement may not cure the problem in all cases. For this reason, you may want to seek professional advice before purchasing replacement parts.

14 Fuel injectors (I-TEC fuel injection) - replacement

Warning: *Gasoline is extremely flammable, so take extra precautions when you work on any part of the fuel system. Don't smoke or allow open flames or bare light bulbs near the work area, and don't work in a garage where a natural gas-type appliance (such as a water heater or clothes dryer) with a pilot light is present. If you spill any fuel on your skin, rinse it off immediately with soap and water. When you perform any kind of work on the fuel system, wear safety glasses and have a Class B type fire extinguisher on hand.* **Note:** *This procedure applies to the I-TEC fuel injection system used on some four-cylinder models.*

1   Relieve the fuel pressure (see Section 2).
2   Remove the throttle chamber and the common chamber (Section 15).
3   Remove the fuel line that attaches to the fuel rail.
4   Remove the fuel rail mounting bolts, then remove the fuel rail by pulling up carefully.
5   Carefully mark each injector connector and disconnect it from the injector.
6   Remove the injectors.
7   Installation is the reverse of removal. Be sure to use new rubber O-rings on the injectors when installing. Coat the O-rings with a little fuel prior to installation.

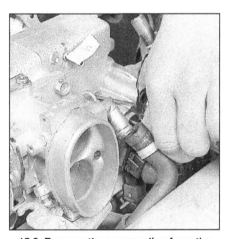

15.2 Remove the vacuum line from the throttle body

15 Throttle body and common chamber (I-TEC fuel injection) - removal and installation

Refer to illustrations 15.2, 15.4, 15.8 and 15.9
Note: *This procedure applies to the I-TEC fuel injection system used on some four-cylinder engines.*

Throttle body

1   Remove the air intake ducts that straddle the valve cover.

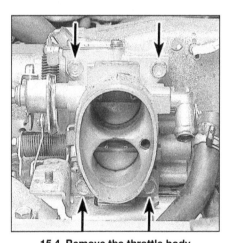

15.4 Remove the throttle body bolts (arrows)

2   Carefully mark the vacuum lines and electrical connectors attached to the throttle body, then remove them **(see illustration)**.
3   Disconnect the accelerator cable (see Section 17).
4   Remove the throttle body bolts **(see illustration)**, then remove the throttle body.
5   Stuff shop towels into the holes in the common chamber. Using a gasket scraper or putty knife, remove all traces of old gasket material from the gasket mating surfaces on the throttle body and common chamber. Be careful not to damage the delicate aluminum surfaces.

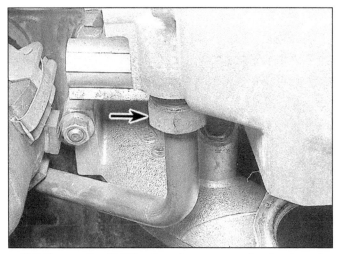

15.8 Remove the EGR line from the common chamber (arrow)

**15.9 Remove the common chamber retaining nuts (arrows) -
some are located underneath the common chamber**

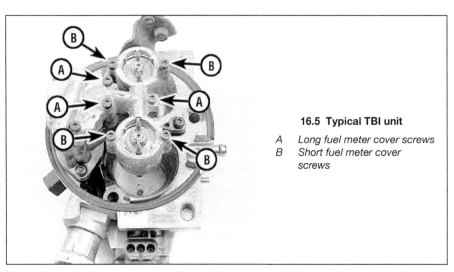

16.5 Typical TBI unit

A   Long fuel meter cover screws
B   Short fuel meter cover
    screws

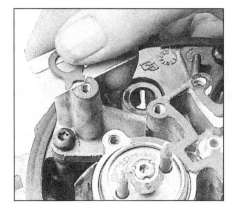

**16.6 Carefully peel away the old fuel
meter outlet passage and cover gaskets
with a razor blade**

6   Installation is the reverse of removal.
Install a new gasket and tighten the mounting
bolts to the torque listed in this Chapter's
Specifications.

Common chamber

7   Remove the throttle body.
8   Disconnect the EGR line located on the
lower section of the common chamber **(see
illustration)**.
9   Remove the common chamber retaining
nuts **(see illustration)**. Some of the nuts are
located underneath the common chamber.
10   Lift the common chamber off the intake
manifold.
11   Stuff shop towels into the holes in the
intake manifold. Using a gasket scraper or
putty knife, remove all traces of old gasket
material from the gasket mating surfaces on
the intake manifold and common chamber.
Be careful not to damage the delicate aluminum surfaces.
12   Installation is the reverse of removal.
Install a new gasket and tighten the mounting
nuts to the torque listed in this Chapter's
Specifications.

16   Throttle Body Injection (TBI) unit
- removal, overhaul and
installation

*Refer to illustrations 16.5, 16.6, 16.7, 16.14,
16.16, 16.21, 16.22, 16.23, 16.24, 16.25,
16.37, 16.41a, 16.41b, 16.41c, 16.50, 16.52,
16.65 and 16.66*
Note: *Because of its relative simplicity, the
throttle body assembly does not need to be
removed from the intake manifold or disassembled for component replacement. However, for the sake of clarity, the following procedures are shown with the TBI assembly
removed from the vehicle.*
1   Relieve system fuel pressure (see Section 2).
2   Detach the cable from the negative terminal of the battery.
3   Remove the air cleaner duct and
adapter.

Fuel meter cover/fuel pressure
regulator assembly

Note: *The fuel pressure regulator is housed in*

*the fuel meter cover. Whether you are replac-
ing the meter cover or the regulator itself, the
entire assembly must be replaced. The regulator must not be removed from the cover.*
4   Unplug the electrical connectors to the
fuel injectors.
5   Remove the long and short fuel meter
cover screws **(see illustration)** and remove
the fuel meter cover.
6   Remove the fuel meter outlet passage
gasket, cover gasket and pressure regulator
seal. Carefully remove any old gasket material that is stuck with a razor blade **(see illustration)**. **Caution:** *Do not attempt to re-use
either of these gaskets.*
7   Inspect the cover for dirt, foreign material and casting warpage. If it is dirty, clean it
with a clean shop rag soaked in solvent. Do
not immerse the fuel meter cover in cleaning
solvent - it could damage the pressure regulator diaphragm and gasket. **Warning:** *Do not
remove the four screws* **(see illustration)**
*securing the pressure regulator to the fuel
meter cover. The regulator contains a large
spring under compression which, if accidentally released, could cause injury. Disassembly might also result in a fuel leak between the
diaphragm and the regulator housing. The*

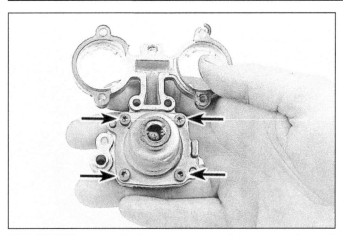

16.7 DO NOT remove the four pressure regulator screws (arrows) from the fuel meter cover

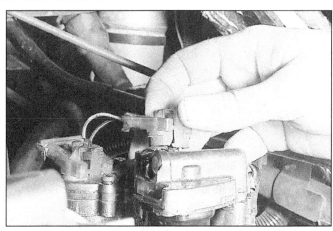

16.14 To remove either injector electrical connector, depress the tabs at the front and rear and pull straight up

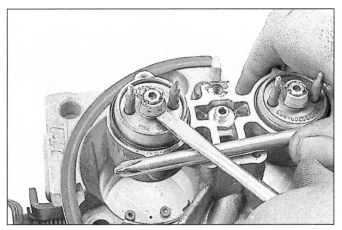

16.16 To remove an injector, slip the tip of a screwdriver under the lip on top of the injector and, using another screwdriver as a fulcrum, carefully pry the injector up and out

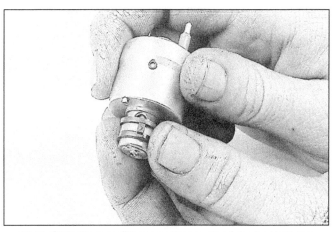

16.21 Slide the new filter onto the nozzle of the fuel injector

new fuel meter cover assembly will include a new pressure regulator.

8   Install the new pressure regulator seal, fuel meter outlet passage gasket and cover gasket.

9   Install the fuel meter cover assembly using Loctite 262 or equivalent on the screws. **Note:** The short screws go next to the injectors.

10   Attach the electrical connectors to both injectors.

11   Attach the cable to the negative terminal of the battery.

12   With the engine off and the ignition on, check for leaks around the gasket and fuel line couplings.

13   Install the air cleaner duct and adapter.

Fuel injector assembly

14   To unplug the electrical connectors from the fuel injectors, squeeze the plastic tabs and pull straight up **(see illustration)**.

15   Remove the fuel meter cover/pressure regulator assembly. *Note: Do not remove the fuel meter cover assembly gasket - leave it in place to protect the casting from damage during injector removal.*

16   Use a screwdriver and fulcrum **(see**

16.22 Lubricate the lower O-ring with transmission fluid, then place it on the shoulder in the bottom of the injector cavity

illustration) to pry out the injector.

17   Remove the upper (larger) and lower (smaller) O-rings and filter from the injector.

18   Remove the steel back-up washer from the top of the injector cavity.

19   Inspect the fuel injection filter for evidence of dirt and contamination. If present, check for the presence of dirt in the fuel lines and fuel tank.

20   Be sure to replace the fuel injector with

an identical part. Injectors from other models can fit in the Model 220 TBI assembly but are calibrated for different flow rates.

21   Slide the new filter into place on the nozzle of the injector **(see illustration)**.

22   Lubricate the new lower (smaller) O-ring with automatic transmission fluid and place it on the small shoulder at the bottom of the fuel injector cavity in the fuel meter body **(see illustration)**.

16.23 Place the steel back-up washer on the shoulder near the top of the injector cavity

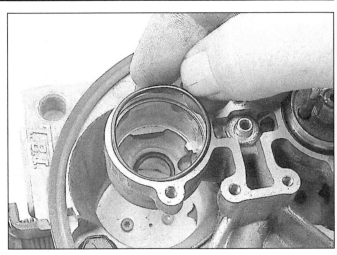

16.24 Lubricate the upper O-ring with transmission fluid, then install it on top of the steel washer

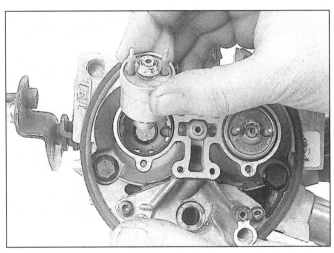

16.25 Make sure the lug is aligned with the groove in the bottom of the fuel injector cavity

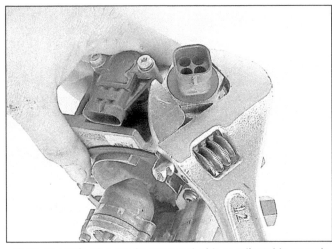

16.37 The IAC valve can be removed with an adjustable wrench (shown) or a 1-1/4 inch wrench

23   Install the steel back-up washer in the injector cavity **(see illustration)**.

24   Lubricate the new upper (larger) O-ring with automatic transmission fluid and install it on top of the steel back-up washer **(see illustration)**. **Note:** *The back-up washer and the large O-ring must be installed before the injector. If they aren't, improper seating of the large O-ring could cause fuel leakage.*

25   To install the injector, align the raised lug on the injector base with the notch in the fuel meter body cavity **(see illustration)**. Push down on the injector until it is fully seated in the fuel meter body. **Note:** *The electrical terminals should be parallel with the throttle shaft.*

26   Install the fuel meter cover assembly and gasket.

27   Attach the electrical connectors to the fuel injector(s).

28   Attach the cable to the negative terminal of the battery.

29   With the engine off and the ignition on, check for fuel leaks.

30   Install the air cleaner duct and adapter.

Throttle Position Sensor (TPS)

31   Remove the two TPS attaching screws and retainers and remove the TPS from the throttle body.

32   If you intend to re-use the same TPS, do not attempt to clean it by soaking it in any liquid cleaner or solvent. The TPS is a delicate electrical component and can be damaged by solvents.

33   Install the TPS on the throttle body while lining up the TPS lever with the TPS drive lever.

34   Install the two TPS attaching screws and retainers.

35   Install the air cleaner duct and adapter.

36   Attach the cable to the negative terminal of the battery. Have the TPS output adjusted by a dealer service department or other qualified shop.

Idle Air Control (IAC) valve

37   Unplug the electrical connector from the IAC valve and remove the IAC valve **(see illustration)**.

38   Remove and discard the old IAC valve gasket. Clean any old gasket material from the surface of the throttle body assembly to insure proper sealing of the new gasket.

39   All pintles in IAC valves on model 220 TBI units have the same dual taper. However, the pintles on some units have a 12mm diameter and the pintles on others have a 10mm diameter. A replacement IAC valve must have the appropriate pintle taper and diameter for proper seating of the valve in the throttle body.

40   Measure the distance between the tip of the pintle and the housing mounting surface with the pintle fully extended. If dimension "A" is greater than the specified dimension, it must be reduced to prevent damage to the valve.

41   If the pintle must be adjusted, determine whether your valve is a Type I (collar around the electrical terminal) or a Type II (no collar around the electrical terminal).

a) *To adjust the pintle of an IAC valve with a collar, grasp the valve and exert firm pressure on the pintle with the thumb.*

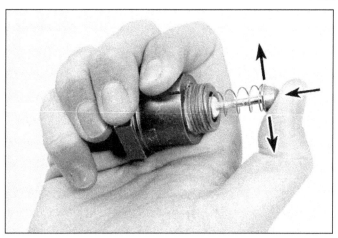

16.41a To adjust an IAC valve with a collar, retract the pintle by exerting firm pressure while moving it from side to side

16.41b To adjust an IAC valve without a collar, compress the retaining spring while turning the valve clockwise . . .

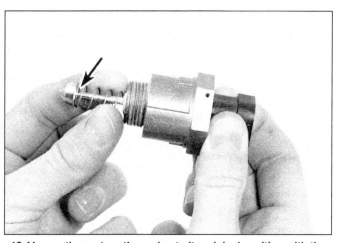

16.41c . . . then return the spring to its original position with the straight portion aligned in the slot under the flat surface of the valve

16.50 Remove the fuel inlet and outlet nuts from the fuel meter body

Use a slight side-to-side movement on the pintle as you press it in with your thumb **(see illustration)**.
b) *To adjust the pintle of an IAC valve without a collar, compress the retaining spring while turning the pintle clockwise* **(see illustration)**. *Return the spring end to its original position with the straight portion aligned in the slot under the flat surface of the valve* **(see illustration)**.

42 Install the IAC valve and tighten it securely. Attach the electrical connector.
43 Install the air cleaner duct and adapter.
44 Attach the cable to the negative terminal of the battery.
45 Start the engine and allow it to reach operating temperature, then turn it off. No adjustment of the IAC valve is required after installation. The IAC valve is reset by the ECM when the engine is turned off.

Fuel meter body assembly

46 Unplug the electrical connectors from the fuel injectors.
47 Remove the fuel meter cover/pressure regulator assembly, fuel meter cover gasket,

16.52 Once the fuel inlet and outlet nuts are off, pull the fuel meter body straight up to separate it from the throttle body

fuel meter outlet gasket and pressure regulator seal.
48 Remove the fuel injectors.
49 Unscrew the fuel inlet and return line threaded fittings, detach the lines and remove the O-rings.
50 Remove the fuel inlet and outlet nuts and gaskets from the fuel meter body assem-

bly **(see illustration)**. Note the locations of the nuts to ensure proper reassembly. The inlet nut has a larger passage than the outlet nut.
51 Remove the gasket from the inner end of each fuel nut.
52 Remove the fuel meter body-to-throttle body attaching screws and remove the fuel

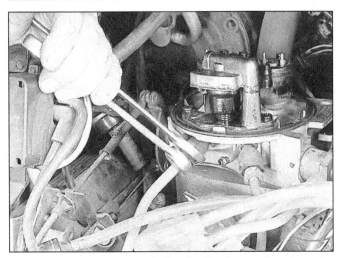

16.65 When disconnecting the fuel feed and return lines from the fuel inlet and outlet nuts, be sure to use a back-up wrench to prevent damage to the lines

16.66 To Detach the throttle body from the intake manifold, remove the three bolts (arrows)

meter body from the throttle body **(see illustration)**.

53   Install the new throttle body-to-fuel meter body gasket. Match the cut-out portions in the gasket with the openings in the throttle body.

54   Install the fuel meter body on the throttle body. Coat the fuel meter body-to-throttle body attaching screws with thread locking compound before installing them.

55   Install the fuel inlet and outlet nuts, with new gaskets, in the fuel meter body and tighten the nuts to the specified torque. Install the fuel inlet and return line threaded fittings with new O-rings. Use a back-up wrench to prevent the nuts from turning.

56   Install the fuel injectors.

57   Install the fuel meter cover/pressure regulator assembly.

58   Attach the electrical connectors to the fuel injectors.

59   Attach the cable to the negative terminal of the battery.

60   With the engine off and the ignition on, check for leaks around the fuel meter body, the gasket and around the fuel line nuts and threaded fittings.

61   Install the air cleaner housing assembly, adapters and gaskets.

Throttle body assembly

62   Unplug all electrical connectors - the IAC valve, TPS and fuel injectors. Detach the grommet with the wires from the throttle body.

63   Detach the throttle linkage, return spring(s), transmission control cable (automatics) and, if equipped, cruise control.

64   Clearly label, then detach, all vacuum hoses.

65   Using a back-up wrench, detach the inlet and outlet fuel line nuts **(see illustration)**. Remove the fuel line O-rings from the nuts and discard them.

66   Remove the TBI mounting bolts **(see illustration)** and lift the TBI unit from the intake manifold. Remove and discard the TBI manifold gasket.

67   Place the TBI unit on a holding fixture, available at most auto part stores, or equivalent). **Note:** *If you don't have a holding fixture, and decide to place the TBI directly on a work bench surface, be extremely careful when servicing it. The throttle valve can be easily damaged.*

68   Remove the fuel meter body-to-throttle body attaching screws and separate the fuel meter body from the throttle body.

69   Remove the throttle body-to-fuel meter body gasket and discard it.

70   Remove the TPS.

71   Invert the throttle body on a flat surface for greater stability and remove the IAC valve.

72   Clean the throttle body assembly in a cold immersion cleaner. Clean the metal parts thoroughly and blow dry with compressed air. Be sure that all fuel and air passages are free of dirt or burrs. **Caution:** *Do not place the TPS, IAC valve, pressure regulator diaphragm, fuel injectors or other components containing rubber in the solvent or cleaning bath. If the throttle body requires cleaning, soaking time in the cleaner should be kept to a minimum. Some models have throttle shaft dust seals that could lose their effectiveness by extended soaking.*

73   Inspect the mating surfaces for damage that could affect gasket sealing. Inspect the throttle lever and valve for dirt, binds, nicks and other damage.

74   Invert the throttle body on a flat surface for stability and install the IAC valve and the TPS.

75   Install a new throttle body-to-fuel meter body gasket and place the fuel meter body assembly on the throttle body assembly. Coat the fuel meter body-to-throttle body attaching screws with thread locking compound and tighten them securely.

76   Install the TBI unit and tighten the mounting bolts to the specified torque. Use a new TBI-to-manifold gasket.

77   Install new O-rings on the fuel line nuts. Install the fuel line and outlet nuts by hand to prevent stripping the threads. Using a back-up wrench, tighten the nuts securely once they have been correctly threaded into the TBI unit.

78   Attach the vacuum hoses, throttle linkage, return spring(s), transmission control cable (automatics) and, if equipped, cruise control cable. Attach the grommet, with wire harness, to the throttle body.

79   Plug in all electrical connectors, making sure that the connectors are fully seated and latched.

80   Check to see if the accelerator pedal is free by depressing the pedal to the floor and releasing it with the engine off.

81   Connect the negative battery cable, and, with the engine off and the ignition on, check for leaks around the fuel line nuts.

82   Adjust the minimum idle speed and have the TPS output checked by a dealer service department.

83   Install the air cleaner housing assembly, adapter and gaskets.

Minimum idle speed adjustment

Note: *This adjustment must be performed by a dealer service department or other qualified shop because of the expensive "SCAN" tool that must be used.*

17   Accelerator cable - removal and installation

Refer to illustrations 17.2, 17.3 and 17.4

1   Remove the air cleaner on carbureted engines.

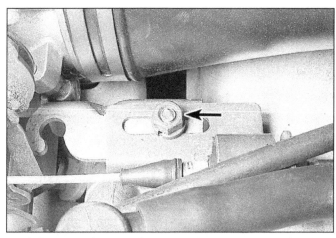

17.2 Remove this nut to disconnect the accelerator cable guide from the rocker arm cover (4ZE1 engine shown)

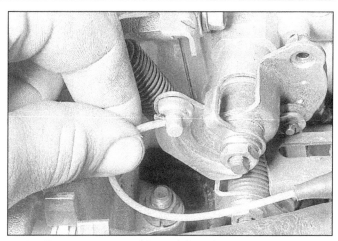

17.3 Lift the accelerator cable up and slide it through the slot on the side of the throttle valve lever

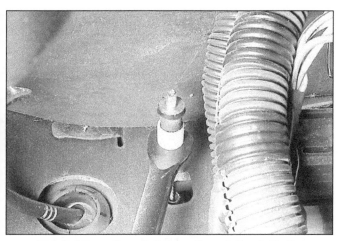

17.4 Pull the collar out, then up through the slot in the accelerator pedal arm

18.3 Remove the thermal valve from the thermostat housing

2   Remove the nut that attaches the accelerator cable guide to the rocker arm cover **(see illustration)**.
3   Disconnect the accelerator cable from the carburetor or throttle body **(see illustration)**.
4   Working under the dash, disengage the accelerator cable nylon collar from the upper end of the accelerator pedal by pulling it out and up **(see illustration)**.
5   Installation is the reverse of the removal procedure. During installation of the cable on the pedal, apply a coat of multi-purpose grease to the nylon collar.
6   Make sure that the throttle valve opens fully when the accelerator pedal is fully depressed and returns to idle when released.
7   Make sure the throttle cable does not contact any components in close proximity to it.

18   Fast idle system (I-TEC fuel injection) – check

Refer to illustrations 18.3, 18.4 and 18.8
1   The fast idle system improves cooling

when the engine is hot by allowing additional air into the common chamber. The system consists of a thermal valve that senses coolant temperature and a fast idle solenoid that controls flow to the common chamber.

Thermal valve
2   Drain the coolant from the radiator (see Chapter 3).
3   Remove the thermal valve **(see illustration)** and place it in a pan of heated water.
4   Heat the water to 200-degrees F and check to make sure air passes from A through C **(see illustration)**.
5   Allow the valve to cool and check that air passes from A through B **(see illustration 18.4)**.
6   If the thermal valve does not pass the tests, replace the unit with a new one. If the valve passes the tests, check the fast idle solenoid.

Fast idle solenoid
7   Disconnect the air hoses and the electrical connector on the fast idle solenoid.
8   Connect a jumper wire between the battery positive terminal and terminal 0.5G

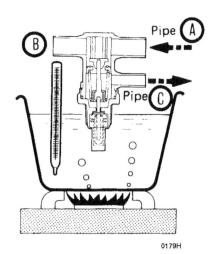

0179H

18.4 With the coolant temperature above 200-degrees F, air should flow from pipe A to C - with the valve cool, air should flow from pipe A to pipe B

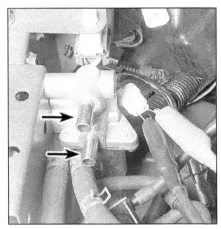

18.8 Use jumper wires to connect the fast idle solenoid to the battery (see text) - air should flow through the hose fittings on the valve (arrows)

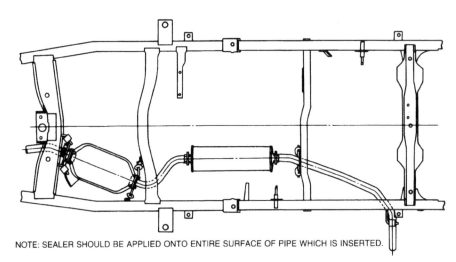

NOTE: SEALER SHOULD BE APPLIED ONTO ENTIRE SURFACE OF PIPE WHICH IS INSERTED.

19.1 Exhaust system on early pick-ups

(green wire) at the back side of the connector. Ground the other terminal (black wire) **(see illustration)**.

9   Blow air to make sure it passes through the two hose fittings.

10   Replace the unit if it does not pass the test.

19   Exhaust system servicing - general information

Refer to illustrations 19.1 and 19.4

Warning: *Inspection and repair of exhaust system components should be done only after enough time has elapsed after driving the vehicle to allow the system components to cool completely. Also, when working under the vehicle, make sure it is securely supported on jackstands.*

1   The exhaust system **(see illustration)** consists of the exhaust manifold(s), the catalytic converter, the muffler, the tailpipe and all connecting pipes, brackets, hangers and clamps. The exhaust system is attached to the body with mounting brackets and rubber hangers. If any of the parts are improperly installed, excessive noise and vibration will be transmitted to the body.

2   Conduct regular inspections of the exhaust system to keep it safe and quiet. Look for any damaged or bent parts, open seams, holes, loose connections, excessive corrosion or other defects which could allow exhaust fumes to enter the vehicle. Deteriorated exhaust system components should not be repaired; they should be replaced with new parts.

3   If the exhaust system components are extremely corroded or rusted together, welding equipment will probably be required to remove them. The convenient way to accomplish this is to have a muffler repair shop remove the corroded sections with a cutting torch. If, however, you want to save money by doing it yourself (and you don't have a welding outfit with a cutting torch), simply cut off the old components with a hacksaw. If you have compressed air, special pneumatic cutting chisels can also be used. If you do decide to tackle the job at home, be sure to wear safety goggles to protect your eyes from metal chips and work gloves to protect your hands.

4   Here are some simple guidelines to follow when repairing the exhaust system:

a) *Work from the back to the front when removing exhaust system components.*

b) *Apply penetrating oil to the exhaust system component fasteners to make them easier to remove* **(see illustration)**.

c) *Use new gaskets, hangers and clamps when installing exhaust systems components.*

d) *Apply anti-seize compound to the threads of all exhaust system fasteners during reassembly.*

e) *Be sure to allow sufficient clearance between newly installed parts and all points on the underbody to avoid overheating the floor pan and possibly damaging the interior carpet and insulation. Pay particularly close attention to the catalytic converter and heat shield.*

19.4 Apply penetrating oil to the exhaust pipe flange nuts before attempting to remove them

Chapter 5
Engine electrical systems

Contents

Specifications

Ignition system

Primary voltage	12-volts
Distributor	
Pick-up coil air gap	
4ZD1	0.012 to 0.020 in (0.3 to 0.5 mm)
G180Z and G200Z	0.008 to 0.016 in (0.2 to 0.4 mm)
Pick-up coil resistance	
G180Z and G200Z	140 to 180 ohms
V6 engine	500 to 1500 ohms
Signal generator air gap (V6 engines)	0.008 to 0.0016 in (0.2 to 0.4 mm)
Ignition coil	
Primary resistance at 68-degrees F (20-degrees C)	
G180Z, G200Z and 4ZD1	1.13 to 1.53 ohms
4ZD1 (1988 and 1989 pick-ups)	1.2 to 1.4 ohms
4ZE1	0.9 ohms
Secondary resistance at 68-degrees F (20-degrees C)	
G180Z, G200Z and 4ZD1	10,200 to 13,800 ohms
4ZD1	8,600 to 13,000 ohms
4ZE1	9,400 ohms
Insulation resistance (to ground)(all models)	Greater than 10,000 ohms
Spark plug wire resistance	9,600 to 22,600 ohms
Spark plug type and gap	Refer to Chapter 1
Firing order and distributor rotation	Refer to Chapter 1
Ignition timing	Refer to the vehicle Emission Control Information Label in the engine compartment

Starter minimum brush length

G180Z and G200Z	0.472 to 0.629 in (12 to 16 mm)
4ZD1 and 4ZE1	0.374 to 0.571 in (9 to 14 mm)
V6 engine	Not available

Alternator minimum brush length

4ZD1	0.551 to 0.787 in (14 to 20 mm)
4ZE1	0.217 to 0.709 in (6 to 18 mm)
V6 engine	Not available

Torque specifications   Ft-lbs

Starter mounting bolts	30
Alternator pulley nut	40

1   Ignition system - general information

In order for the engine to run correctly, it is necessary for an electrical spark to ignite the fuel/air mixture in the combustion chamber at exactly the right moment in relation to engine speed and load. The ignition system is based on feeding low tension (LT) voltage from the battery to the coil, where it is converted to high tension (HT) voltage. The high tension voltage is powerful enough to jump the spark plug gap in the cylinders many times a second under high compression pressures, providing that the system is in good condition and that all adjustments are correct.

The ignition system is divided into two circuits; the low tension circuit and the high tension circuit.

The low tension (sometimes known as the primary) circuit consists of the ignition switch, ignition and accessory relay, the primary windings of the ignition coil(s), the transistorized IC ignition unit, the pickup assembly in the distributor and all connecting wires.

The high tension circuit consists of the high tension or secondary windings of the ignition coil(s), the heavy ignition lead from the center of the coil to the center of the distributor cap, the rotor, the spark plug wires and spark plugs.

The system functions in the following manner. Low tension voltage fed to the coils is changed into high tension voltage by the magnetic pick-up in the distributor. When the ignition switch is turned on, current flows through the primary circuit. A reluctor on the distributor shaft is aligned with the magnetic stator element of the pick-up assembly within the distributor housing, and as it turns it induces a low voltage in the pick-up coil. When the teeth on the reluctor and the magnets in the stator line up, a signal passes to the IC ignition unit which opens the coil's primary circuit. When the primary circuit is opened by the transistor unit, the magnetic field built up in the primary winding collapses and induces a very high voltage in the secondary winding. The high voltage current then flows from the coil, along the heavy ignition lead, to the carbon brush in the distributor cap. From the carbon brush, current flows

to the distributor rotor which distributes the current to one of the terminals in the distributor cap. The spark occurs while the high tension voltage jumps across the spark plug gap. This process is repeated for each power stroke of the engine.

The system features a long spark duration and the dwell period automatically increases with engine speed. This is desirable for firing the leaner mixtures provided by the emissions control systems.

The distributor used on all models is equipped with both mechanical and vacuum advance mechanisms. The mechanical governor mechanism compresses two weights which move out from the distributor shaft due to centrifugal force as the engine speed increases. The weights are held in position by two light springs, and it is the tension of the springs which is largely responsible for the correct spark advancement. The vacuum control consists of a diaphragm, one side of which is connected via a small-bore tube to a vacuum source, and the other side to the magnetic pick-up assembly. Vacuum in the intake manifold, which varies with engine speed and throttle opening, causes the diaphragm to move, which, in turn, moves the magnetic pick-up assembly, advancing or retarding the spark.

Because 1988 and later models are equipped with ECCS (see Chapter 4), the distributor used on these models differs in design and function from earlier distributors. A *crank angle sensor* inside the distributor on all four-cylinder engines monitors engine speed and piston position, then sends a signal to the ECCS control unit (ECU). The ECU uses this signal to determine ignition timing, fuel injector duration and other functions. The crank angle sensor assembly consists of a rotor plate, a "wave forming" circuit, a light emitting diode (LED) and a photo diode.

The rotor plate, which is attached to the distributor shaft, is in the base of the distributor housing. There are 360 slits machined into the outer edge of the rotor plate. These slits correspond to each degree of crankshaft rotation.

The wave forming circuit is positioned underneath the rotor plate. A small housing attached to one side of the wave forming circuit encloses the upper and lower outer

edges of the rotor plate. A light emitting diode (LED) is located in the upper half and a photo diode is located in the lower half of the small housing. When the engine is running, the LED emits a continuous beam of light directly at the photo diode. As the outer edge of the rotor plate passes through the housing, the slits allow the light beam to pass through to the photo diode, but the solid spaces between the slits block the light beam. This constant interruption generates pulses which are converted into on-off signals by the wave forming circuit and sent to the ECU. The ECU uses the signal from the outer row of slits to determine engine speed and crankshaft position. It uses the signal generated by the inner, larger slits to determine when to fire each cylinder. This information is then relayed to the coil which builds secondary voltage and sends it to the distributor cap in the conventional manner, where it is distributed by the rotor to the appropriate cylinder. **Warning:** *Because of the higher voltage generated by the electronic ignition system, extreme caution should be taken whenever an operation is performed involving ignition components. This not only includes the distributor, coil, control module and ignition wires, but related items which are connected to the system as well, such as the plug connections, tachometer and any testing equipment. Consequently, before any work is performed such as replacing ignition components or even connecting test equipment, the ignition should be turned off or the battery ground cable disconnected. Never disconnect any of the ignition HT leads when the engine is running or the transistor ignition unit will be permanently damaged.*

2   Ignition system - testing

Refer to illustrations 2.7a, 2.7b, 2.8a, 2.8b, 2.9, 2.10, 2.11a, 2.11b and 2.12

Warning: *Never touch the spark plug or coil wires with your bare hands while the engine is running or being cranked. Even insulated parts can cause a shock if they are moist. Wear dry, insulated gloves or wrap the part in dry cloth before handling.*

1   If the engine will turn over but will not start, the first check of the ignition system should be to visually inspect the condition of

2.7a Plug wire resistance is checked by connecting one ohmmeter lead to the spark plug end of the wire and the other lead to the corresponding distributor cap inner terminal

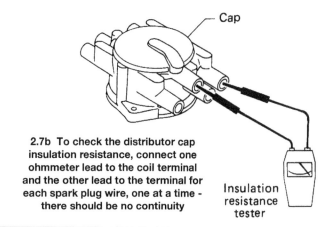

2.7b To check the distributor cap insulation resistance, connect one ohmmeter lead to the coil terminal and the other lead to the terminal for each spark plug wire, one at a time - there should be no continuity

the spark plugs, spark plug wires, distributor cap and rotor as described in Chapter 1. Also check the distributor air gap as described in Section 3.

2   If these are all in good condition, and the plug wires are secure in their connections, the next check should be to see if current is flowing through the high tension circuit and causing spark at the plugs.

a) *Turn the ignition switch to the Off position.*

b) *Relieve the fuel pressure (see Chapter 4).*

c) *Disconnect the ignition coil wire from the distributor cap and, using an insulated tool, hold it approximately 3/16 to 1/4-inch (4 to 5 mm) from a clean metal area of the engine. Have an assistant crank the engine over and check if a spark occurs between the coil wire and the engine.*

d) *If a spark occurs, the ignition system is okay, and the problem lies in another system. If no spark occurs, or occurs intermittently, proceed with further ignition system tests.*

3   In order to accurately diagnose problems in the ignition system, a voltmeter which measures in the 0 to 20 volts DC and 0 to 10 volts AC ranges, and an ohmmeter which measures in the 0 to 1000 ohms and 0 to 5000 ohms ranges are needed.

4   If possible, start the engine and let it run about 5 to 15 minutes with the hood closed to bring all components to normal operating temperature. Turn off the engine.

5   Checking the battery voltage with no load:

a) *With the ignition key in the Off position, connect a voltmeter so the positive lead is on the positive battery terminal and the negative lead is on the negative battery terminal.*

b) *Note the reading on the voltmeter. If the reading is between 11.5 volts and 12.5 volts, the battery is okay and you should proceed to Step 7.*

c) *If the reading is below 11.5 volts, the battery is insufficiently charged. It should be brought to a full charge either*

by running the engine or by using a battery charger. If the vehicle has been used on a regular basis and there is no obvious cause for the battery to be discharged (such as leaving the lights on), then the condition of the battery, charging system and starting system should be checked as described in this Chapter or Chapter 1.

6   Checking the battery voltage while the engine is cranking:

a) *Leave the voltmeter connected to the battery as in the previous test.*

b) *Disconnect the ignition coil wire(s) from the distributor cap and ground it to the engine.*

c) *Have an assistant crank the engine over for about 15 seconds and note the reading on the voltmeter.*

d) *If the voltage is more than 9.6 volts, the battery is okay and you should proceed to Step 8. If the voltage is below 9.6 volts, the battery is insufficiently charged. Refer to Step 6.*

7   Checking the distributor cap and secondary (spark plug) wires:

a) *Disconnect the spark plug wires from the plugs.*

b) *Disconnect the ignition coil wire(s) from the coil(s).*

c) *Remove the distributor cap, with spark plug and coil wires still attached.*

d) *Connect an ohmmeter so that one lead is inserted in the spark plug end of one of the plug wires and the other lead is contacting the inner distributor cap terminal that the wire is connected to (see illustration).*

e) *If the reading on the ohmmeter is 9,600 to 22,600 ohms, the cap terminal and wire are okay.*

f) *If the reading is more than 22,600 ohms, the resistance is too high. Check the cap and wire individually and replace the appropriate part.*

g) *Repeat this test on each of the spark plug and coil wires.*

h) *Check the cap and rotor for dust, carbon deposits and cracks. Measure the insulation resistance between the terminals for the coil and spark plug wires (see illustration). If continuity exists, replace the cap.*

8   Checking the ignition coil secondary circuit:

a) *With the ignition key in the Off position, connect one ohmmeter lead to the center high tension terminal and the other lead to the negative primary terminal (see illustrations).*

b) *If the reading is within the specified range, the secondary coil windings are okay.*

c) *If the ohmmeter reading is not within specification, replace the ignition coil.*

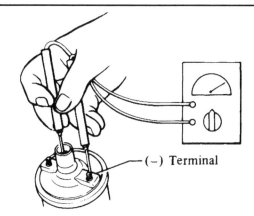

(−) Terminal

2.8a Place one ohmmeter lead inside the coil high tension terminal and touch the other to the (-) terminal to check the secondary resistance of the coil on G180Z, G200Z and 4ZD1 engines

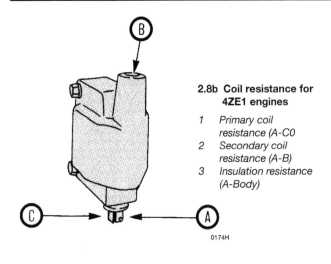

2.8b Coil resistance for 4ZE1 engines

1. Primary coil resistance (A-C0
2. Secondary coil resistance (A-B)
3. Insulation resistance (A-Body)

0174H

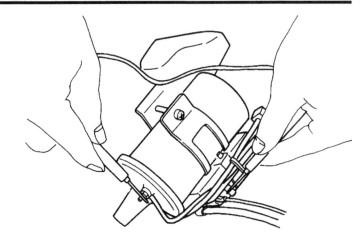

2.9 Checking power feed voltage on the ignition coil

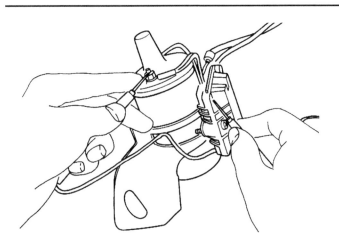

2.10 To check the power transistor, connect the positive lead of the voltmeter to the negative terminal of the coil and the negative voltmeter lead to the outside of the transistor body (ground)

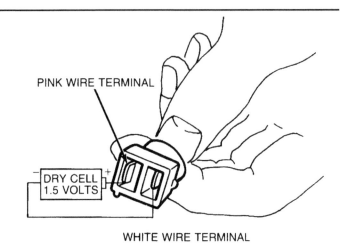

PINK WIRE TERMINAL

DRY CELL 1.5 VOLTS

WHITE WIRE TERMINAL

2.11a Activate the power transistor with a 1.5-volt dry cell battery as shown

9 Checking the power supply circuit at the coil (1981 through 1987 models):

a) Connect the positive lead of the voltmeter to the positive terminal on the ignition coil (see illustration). Note: When performing this or any of the following tests, it is not necessary to disconnect the wiring connector when hooking up the voltmeter, providing they have probes that can be inserted into the rear of the connector.
b) Turn the ignition switch to On.
c) The voltmeter should indicate that power source line voltage is approximately 12 volts.

10 Checking the power transistor in the ignitor (1981 through 1987 models):

a) Using a voltmeter, connect the positive probe to the negative terminal on the ignition coil and the negative probe to the body ground (see illustration).
b) Turn the ignition switch to On.
c) The voltmeter should indicate approximately 12 volts.

d) If the reading is incorrect, replace the transistor.

11 Checking the power transistor using a dry cell battery (1981 through 1987 models):

a) Disconnect the wiring connector from the distributor. Using a dry cell battery (1.5 volts), connect the positive pole of the battery to the pink wire terminal and the negative pole of the battery to the white wire terminal (see illustration). Note: Do not apply the voltage more than 5 seconds otherwise the power transistor in the ignitor may be damaged.
b) Turn the ignition switch to On.
c) Using a voltmeter, connect a positive probe to the ignition coil negative terminal and the negative probe to the body ground. The voltmeter should indicate approximately 5 volts (see illustration).
d) If the readings are incorrect, replace the transistor.

12 Checking the ignition coil primary circuit:

a) With the ignition key in the Off position and the ignition coil wire and primary wires removed from the coil, connect an ohmmeter (set in the 1X range) between the positive and negative primary termi-

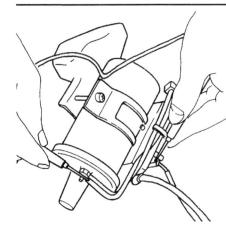

2.11b Checking transistor voltage

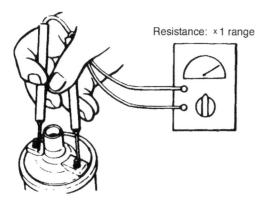

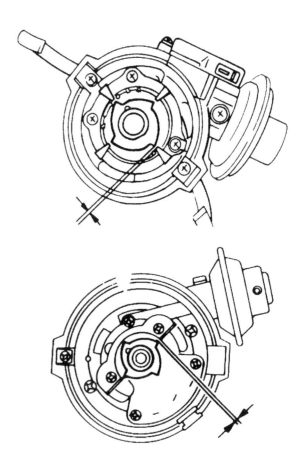

Resistance: × 1 range

2.12 To check ignition coil primary circuit resistance on 1981 through 1987 models, connect the ohmmeter leads to the positive and negative primary terminals of the coil as shown (be sure to set the ohmmeter to the 1X range)

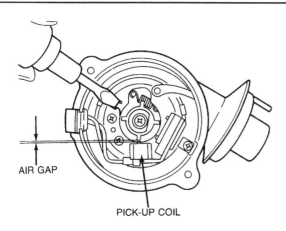

AIR GAP

PICK-UP COIL

3.4b To adjust the air gap, loosen the pick-up coil assembly mounting screws, insert a screwdriver into the notch and move the pick-up coil closer to or farther away from the reluctor tooth (G180Z or G200Z engine shown, others similar)

3.4a To check the distributor air gap, position one of the raised segments of the reluctor directly opposite the stator magnet (pole piece) protruding from the pick-up coil. and, using a feeler gauge, measure the gap between them (4ZDI engine)

---

nals on the ignition coil (see illustration). **Note:** *Refer to illustration 2.8b for the 4ZE1 engine coil.*
b) *If the ohmmeter is within the specified range, the coil is okay.*
c) *If the ohmmeter reading is not within these specification, replace the coil.*

3 Distributor air gap (G180Z, G200Z and 4ZD1 engines) – check and adjustment

Refer to illustrations 3.4a and 3.4b
Note: *To ensure that the ignition system functions correctly, the air gap (distance between the pick-up coil and reluctor) must be maintained as specified. To do this, proceed as follows:*
1   Disengage the two spring retaining clips and remove the cap from the distributor.
2   Remove the rotor from the end of the distributor shaft.
3   Position one of the raised segments of

the reluctor directly opposite the pole piece protruding from the pick-up coil. This is best carried out by removing the spark plugs (to relieve compression) and rotating the crankshaft by using a ratchet and socket on the crankshaft bolt at the front of the engine.
4   Using feeler gauges, measure the gap between the pole and the reluctor segment **(see illustrations).** Check with the specifications listed in this Chapter. If the air-gap requires adjustment, loosen the pick-up coil retaining screws and move the coil in the required direction.
5   When the correct air-gap has been obtained, tighten the pick-up coil retaining screws and install the rotor and the distributor cap.

4 Distributor rotor housing (4ZE1 engine) - check and replacement

1   The rotor and cap can be changed as a standard service procedure. The distributor

housing is an integral design and cannot be disassembled. If ignition related problems are occurring, take the vehicle to a dealer service department to have the distributor diagnosed. If the distributor rotor housing has been diagnosed as faulty, then the entire distributor must be replaced, as internal parts are not available separately.
2   See Section 7 and 8 for distributor removal and installation.

5 Stator and magnet (pick-up assembly) - check and replacement (G180Z, G200Z, 4ZD1 and V6 engines)

Check

Refer to illustrations 5.1a and 5.1b
Note: *On 4ZD1 engines, have the distributor diagnosed by a dealer service department before replacing any internal parts.*
1   Disconnect the distributor electrical

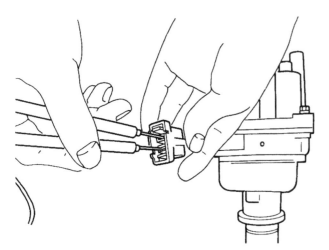

5.1a Probe the terminals of the pick-up coil with an ohmmeter and compare your reading to the value listed in the Specifications (G180Z and G200Z engine shown)

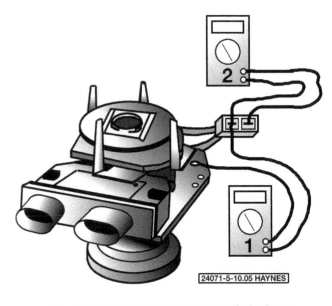

24071-5-10.05 HAYNES

5.1b Testing the pick-up coil on the V6 distributor

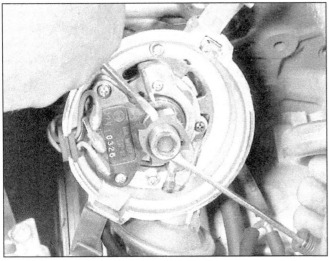

5.6 To remove the reluctor from the distributor on G180Z and G200Z engines, pry it off with a pair of screwdrivers - be careful not to damage the teeth on the reluctor

5.13 Before removing the pickup coil, unplug the lead from the ignition module

connector and check the resistance of the pick-up assembly using an ohmmeter **(see illustrations)**. On V6 models, skip to Step 13 and disconnect the pick-up coil electrical connectors from the module. Values should be as listed in this Chapter's Specifications.

2   If the readings are incorrect, replace the pick-up assembly.

Replacement

G180Z and G200Z engines

Refer to illustration 5.6

3   Disconnect the ignition coil wire from the top of the distributor cap.

4   Remove the distributor cap from the distributor and position it out of the way.

5   Pull the rotor off the distributor shaft.

6   Using two screwdrivers (one on either side of the reluctor), carefully pry the reluctor

from the distributor shaft. Be careful not to damage the teeth on the reluctor **(see illustration)**.

7   Remove the screws retaining the stator and magnet (pick-up coil assembly) to the distributor breaker plate and lift them off.

8   Disconnect the distributor wiring harness and lift out the pick-up coil assembly.

9   Installation is the reverse of the removal procedure. When installing the reluctor, be sure the pin is lined up with the flat side of the distributor shaft. Also, before tightening the pick-up coil assembly, check and adjust the air gap as described in Section 3.

V6 engine

Refer to illustrations 5.13, 5.17a, 5.17b, 5.17c, 5.18 and 5.19

10   Detach the cable from the negative terminal of the battery.

11   Remove the distributor cap and rotor

(see Chapter 1).

12   Remove the distributor from the engine (see Section 6).

13   Detach the pick-up coil leads from the module **(see illustration)**.

14   Remove the spring from the distributor shaft.

15   Mark the distributor drive gear and shaft so that they can be reassembled in the same position.

16   If the distributor is equipped with a Hall Effect switch, remove it.

17   Carefully mount the distributor in a soft-jawed vise and, using a hammer and punch, remove the roll pin from the distributor shaft and gear **(see illustrations)**. Remove the distributor shaft **(see illustration)**.

18   To remove the pick-up coil, remove the thin "C" washer or the retaining clip **(see illustration)**.

19   Lift the pick-up coil assembly straight

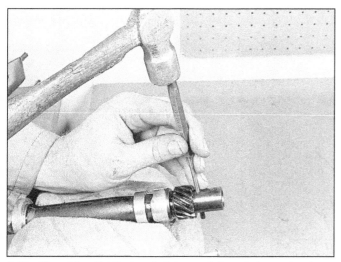

5.17a Mount the distributor shaft in a soft-jawed vise and, using a drift punch and hammer, knock out the roll pin

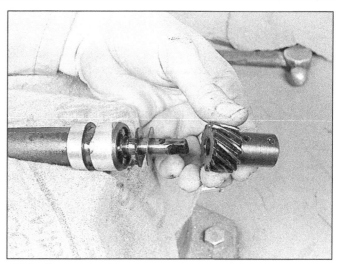

5.17b Remove the driven gear and spacer washers from the end of the shaft, making sure to note the order in which you remove any spacers

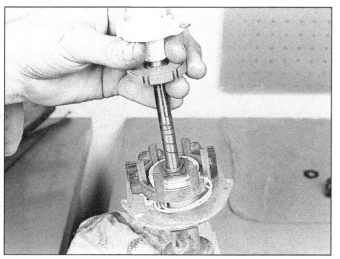

5.17c Remove the shaft from the distributor

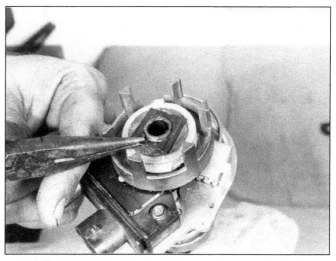

5.18 To remove the pick-up coil from a V6 engine distributor, remove the retaining clip

up and remove it from the distributor. Note the order in which you remove the pieces **(see illustration)**.

20   Reassembly is the reverse of disassembly.

---

6   Ignition module (V6 engine) - check and replacement

Refer to illustrations 6.4 and 6.15
Note: *It is not necessary to remove the distributor to check or replace the module.*

Check

1   Disconnect the tachometer lead (if so equipped) at the distributor.
2   Check for a spark at the coil and spark plug wires (Section 2).
3   If there is no spark, remove the distributor cap. Remove the ignition module from the distributor but leave the connector plugged in.

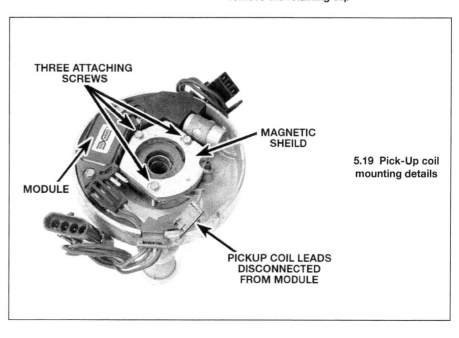

THREE ATTACHING SCREWS

MAGNETIC SHEILD

MODULE

PICKUP COIL LEADS DISCONNECTED FROM MODULE

5.19 Pick-Up coil mounting details

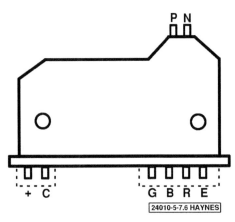

6.4 Ignition module terminal identification

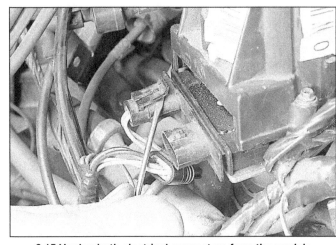

6.15 Unplug both electrical connectors from the module

4   With the ignition switch turned On, check for voltage at the module positive terminal **(see illustration)**.

5   If the reading is less than ten volts, there is a fault in the wire between the module positive (+) terminal and the ignition coil positive connector or the ignition coil and primary circuit-to-ignition switch.

6   If the reading is ten volts or more, check the "C" terminal on the module **(see illustration 6.4)**.

7   If the reading is less than one volt, there is an open or grounded lead in the distributor-to-coil "C" terminal connection or ignition coil or an open primary circuit in the coil itself.

8   If the reading is one to ten volts, replace the module with a new one and check for a spark (see Section 2). If there is a spark the module was faulty and the system is now operating properly. If there is no spark, there is a fault in the ignition coil.

9   If the reading in Step 4 is 10 volts or more, unplug the pick-up coil connector from the module. Check the "C" terminal voltage with the ignition switch On and watch the voltage reading as a test light is momentarily

(five seconds or less) connected between the battery positive (+) terminal and the module "P" terminal **(see illustration 6.4)**.

10   If there is no drop in voltage, check the module ground and, if it is good, replace the module with a new one.

11   If the voltage drops, check for spark at the coil wire as the test light is removed from the module terminal. If there is no spark, the module is faulty and should be replaced with a new one. If there is a spark, the pick-up coil or connections are faulty or not grounded.

Replacement

12   Detach the cable from the negative terminal of the battery.

13   Remove the distributor cap and rotor (see Chapter 1).

14   Remove both module attaching screws and lift the module up and away from the distributor.

15   Disconnect both electrical connectors from the module **(see illustration)**. Note that the connectors cannot be interchanged.

16   Do not wipe the grease from the module or the distributor base if the same module is to be reinstalled. If a new module is to be

installed, a package of silicone grease will be included with it. Wipe the distributor base and the new module clean, then apply the silicone grease on the face of the module and on the distributor base where the module seats. This grease is necessary for heat dissipation.

17   Install the module and attach both electrical connectors.

18   Install the distributor rotor and cap (see Chapter 1).

19   Attach the cable to the negative terminal of the battery.

---

7   Distributor - removal

Refer to illustrations 7.3a and 7.3b

1   Remove the ignition coil wire that leads to the distributor from the coil. On V6 models, remove the air cleaner housing assembly.

2   Remove the distributor cap from the distributor and position it out of the way.

3   Mark the rotor in relationship to the distributor housing and the distributor housing in relation to the block or cylinder head **(see illustrations)**.

7.3a Before removing the distributor, be sure to mark the relationship of the rotor to the distributor body to ensure that the two components are correctly aligned when reinstalled

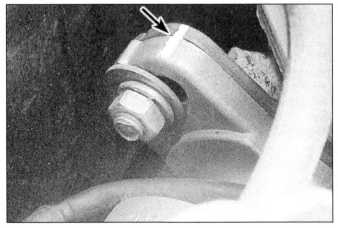

7.3b Mark the position of the distributor in relation to the block or the cylinder head (arrow) before loosening the distributor hold down clamp

9.5 Distributor assembly on G180Z and G200Z engines

1   Cap
2   Rotor
3   Centrifugal advance assembly
4   Spring A
5   Spring B
6   Weight
7   Reluctor assembly
8   Shaft assembly
9   Plate assembly
10 Pick-up assembly
11 Vacuum advance
12 Housing assembly
13 Clamp
14 Gasket
15 Collar
16 Screw
17 Screw
18 Screw
19 Washer
20 Washer
21 Washer
22 Washer
23 Pin
24 Bolt
25 Screw
26 Washer
27 O-Ring

4   Disconnect the wires or wiring connector leading to the distributor.
5   Remove the vacuum line from the vacuum advance canister.
6   Mark the position of the distributor adjusting bolt, remove the distributor adjusting bolt and lift out the distributor.

8   Distributor - installation

If the crankshaft was not turned after removal

1   Position the rotor exactly as it was when the distributor was removed.
2   Lower the distributor down into the engine. To mesh the groove in the bottom of the distributor shaft with the drive spindle (which turns the distributor shaft), it may be necessary to turn the rotor slightly.
3   With the base of the distributor all the way down against the engine block and the mounting screw holes lined up, the rotor should be pointing to the mark made on the distributor housing during removal. If the rotor is not in alignment with the mark, repeat

the previous Steps.
4   Align the mark on the distributor base with the mark on the block or cylinder head. Install the distributor hold-down bolt(s) and tighten them securely.
5   Connect the vacuum line to the vacuum advance canister (if equipped).
6   Reconnect the wires to the distributor.
7   Reinstall the distributor cap and connect the ignition coil wire(s).
8   Check the ignition timing as described in Chapter 1.

If the crankshaft was turned after removal

9   Position the number one piston to TDC, referring to Chapter 2, if necessary.
10 Temporarily install the distributor cap on the distributor. Note where the number one spark plug terminal inside the cap is located in relation to the distributor body, and make a mark on the outside of the body at this point.
11 Remove the distributor cap again. Turn the rotor until it is aligned with the mark. In this position it should be firing the number one spark plug.
12 Proceed with installation as detailed in

Steps 2 through 8. Disregard the rotor mark references in Step 3.

9   Distributor - disassembly and reassembly

Refer to illustrations 9.5, 9.10, 9.16a and 9.16b

1   Remove the distributor from the vehicle as described in Section 7.
2   Remove the distributor cap.
3   Pull off the rotor.
4   Disconnect the wiring connector(s) leading to the pick-up assembly.
5   Remove the screws **(see illustration)** and lift out the stator and magnet assembly.
6   Remove the screws retaining the vacuum controller and lift it off.
7   On 4ZD1 engines, using two screwdrivers pry the reluctor off the distributor shaft **(see illustration 5.6)**. Be careful not to damage the reluctor teeth.
8   Remove the pick-up coil assembly.
9   Remove the breaker plate set screws and lift out the breaker plate assembly.
10 Drill off the staked end of the roll pin

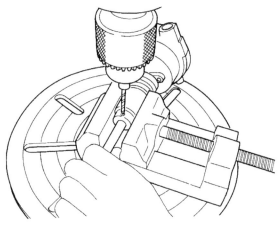

9.10 Drill off the staked end of the roll pin

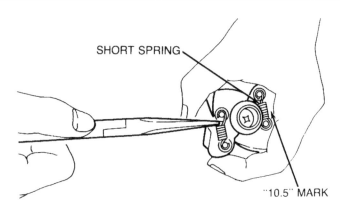

9.16a Install the weight onto the governor shaft and align the "10.5" mark with the stopper (G180Z and G200Z engines)

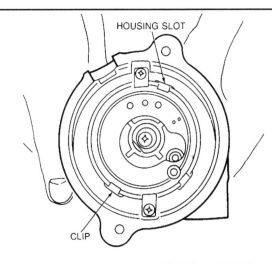

9.16b Assembling the breaker plate (G180Z and G200Z engines)

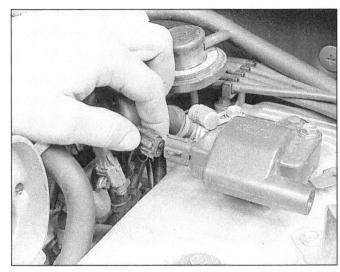

10.3 Remove the electrical connector from the coil (4ZE1 engine)

(see illustration) and using a punch, drive the knock pin from the collar and pull off the collar.

11   Pull the rotor shaft assembly and drive shaft out through the top of the distributor.

12   Mark the relative position of the rotor shaft and driveshaft. Remove the packing from the top of the rotor shaft, remove the rotor shaft set screw and separate the two shafts.

13   Mark the relationship of one of the governor springs to its bracket. Also mark the relationship of one of the governor weights to its pivot pin.

14   Carefully unhook and remove the governor springs.

15   Remove the governor weights. A small amount of grease should be applied to the weights after removal.

16   Reassembly is the reverse of the disassembly procedure with the following notes:

a) *Be sure to correctly align all positioning marks made during disassembly so that all parts are assembled in their original position* **(see illustration).**

b) *When assembling the governor shaft to the distributor housing, install the new pin and collar. Align the hole and install the pin with a hammer. Always use a new roll pin. Peen both ends of the pin in a vise.*

c) *When installing the breaker plate on G180Z and G200Z engines, fit the four clips on the breaker plate into the housing slots* **(see illustration).**

d) *While assembling, make sure the distributor is clean and free from dirt or moisture. This is very important.*

e) *Before tightening the stator plate, adjust the air gap as described in Section 3.*

---

10   Ignition coil - removal and installation

Refer to illustration 10.3

1   Remove the coil wire leading from the distributor cap.

2   Remove the dust boot.

3   Remove the nuts and detach the primary wires from the coil **(see illustration).**

4   Remove the screws that hold the coil bracket to the body and detach the coil.

5   If the coil must be removed from the bracket, loosen the bracket screws until the coil can be pulled out.

6   Installation is the reverse of the removal procedure.

---

11   Starting system - general information

The function of the starting system is to crank the engine. This system is composed of a starting motor, solenoid and battery. The battery supplies the electrical energy to the solenoid, which then completes the circuit to the starting motor which does the actual work of cranking the engine.

Early models use a non-reduction gear type of starter motor. On this starter, the solenoid is mounted on top of the starter motor. When the ignition switch is operated, the solenoid's gearshift lever moves the starter drive pinion gear into engagement with the teeth on the flywheel.

Later models are equipped with a reduction gear type of starter. This starter operates in a similar manner to the type described above except that in order to offset extremely low temperatures, increased rotating torque is provided through a reduction gear located between the armature and pinion gear.

An overrunning clutch is installed on both types of starter motors to transmit the driving torque and to prevent the armature from overrunning when the engine fires and starts.

The solenoid and starting motor are mounted together at the rear right side of the engine. No periodic lubrication or maintenance is required for the starter system components.

The electrical circuitry of the vehicle is arranged so that the starter motor can only be operated on automatic transmission models when the lever is at P or N. **Caution:** *Never operate the starter motor for more than 15 seconds at a time without pausing to allow it to cool for at least two minutes. Excessive cranking can cause overheating, which can seriously damage the starter.*

12 Starting system - check

1   If the starter motor does not rotate when the switch is operated on an automatic transmission equipped model, make sure that the shifter is in the N or P position. On standard transmission models, make sure the clutch pedal is depressed.
2   Check that the battery is well charged and all cables, both at the battery and starter solenoid terminals, are clean, free from corrosion and secure.
3   If the starter motor can be heard spinning but the engine is not being cranked, then the overrunning clutch in the starter motor is slipping and the assembly must be removed from the engine and replaced.
4   Often when the starter fails to operate, a click can be heard coming from the starter solenoid when the ignition switch is turned to the Start position. If this is heard, proceed to Step 12.
5   Disconnect the ignition wire from the solenoid terminal. Connect a test light between this lead and the negative battery terminal.
6   Have an assistant turn the key, and note whether the test light comes on or not. If the light comes on proceed to Step 12.
7   If the test light does not come on, connect a voltmeter (set on the low scale) between the positive battery terminal and the bayonet connector at the solenoid. The ignition wire should be connected.
8   Disconnect the ignition coil wire(s) from the distributor cap and ground it to the engine, so the engine will not start.
9   Have an assistant attempt to crank the engine over and note the reading on the voltmeter. If less than 1.5 volts is shown, there is

an open circuit in the starter and it should be replaced with a new or rebuilt unit.
10   If the voltmeter shows more than 1.5 volts, connect a jumper wire between the positive battery terminal and the S terminal on the solenoid.
11   Turn the ignition switch to the Start position and listen for a click from the solenoid. If there still is no click and the starter does not turn, both the solenoid and starter are defective and should be replaced. If a click is heard and the starter turns, then there is an open circuit in the ignition switch, inhibitor switch or relay (on automatic transmission equipped models) or in the wires or connectors.
12   If a click is heard while the ignition switch is turned to the Start position in Step 4, or if the starter turns at all, then the starter current should be tested. Disconnect the positive battery cable at the starter.
13   Connect an ammeter (set on the 500A range) between this cable and its terminal on the starter.
14   Remove the ignition coil wire(s) from the distributor cap to keep the engine from starting.
15   Have an assistant attempt to crank the engine and note the reading on the ammeter and the speed at which the starter turns.
16   If the ammeter reading is less than specified and the starter speed is normal, the starter is okay and the problem lies elsewhere.
17   If the reading is less than specified but the starter motor speed is slow, proceed to Step 22.
18   If the ammeter reading is more than specified, reinstall the ignition coil wire in the distributor cap.
19   Have an assistant start the engine but hold the key in the Start position so that the starter motor doesn't stop operating. Note the ammeter reading. (Do not allow the starter motor to turn for more than 30 seconds at a time).
20   If the ammeter reading is less than specified and the starter turns fast, then the problem is a mechanical one such as a tight engine. Make sure that the engine oil is not too thick, and check for other causes of resistance within the engine.
21   If the ammeter reading exceeds the specified no load current the starter is shorted and must be replaced with a new one.
22   If the ammeter reading in Step 15 is less than specified, but the starter turns slowly, test for a voltage drop in the starter positive circuit.
23   Connect a voltmeter (set on the low scale) so the positive lead is on the positive battery post (or cable) and the negative lead is connected to the M solenoid terminal (this is the terminal with the braided copper strap leading to the starter motor).
24   Remove the ignition coil wire(s) from the distributor cap and ground them on the engine.
25   Have an assistant attempt to crank the engine over and note the reading on the volt-

meter. If less than one volt is shown, proceed to Step 28.
26   If more than one volt is shown, connect the negative voltmeter lead to the B solenoid terminal (this is the terminal which connects to the battery).
27   Have an assistant crank the engine and note the reading on the voltmeter. If more than one volt is shown, then the problem is a bad connection between the battery and the starter solenoid. Check the positive battery cable for looseness or corrosion. If the reading indicated less than one volt, then the solenoid is defective and should be replaced.
28   If the voltmeter test in Step 25 shows less than one volt test for a voltage drop in the starter ground circuit.
29   Connect the negative lead of the voltmeter (set on the low scale) to the negative battery terminal and hold the positive voltmeter lead to the starter housing. Be sure to make a good connection.
30   Have an assistant crank the engine and note the voltmeter reading. If the reading shows more than 0.5 volts, then there is a bad ground connection. Check the negative battery cable for looseness or corrosion. Also check the starter motor ground connections and the tightness of the starter motor mounting bolts.
31   If the voltmeter reading shows less than 0.5 volts, return to Step 7 and follow the procedure outlined there.

13 Starter motor - removal and installation

Refer to illustrations 13.2 and 13.3
1   Disconnect the negative battery cable from the battery. Raise the vehicle and support it on jackstands.
2   Remove the wires from the starter solenoid **(see illustration)**. Label the wires and terminals to prevent confusion during installation.

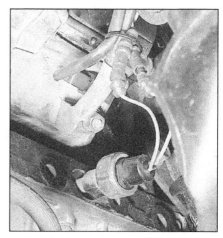

13.2 The starter solenoid is only accessible from under the engine - be sure to mark the locations of the wires to the starter before disconnecting them

13.3 To remove the starter from the V6 engine, detach the electrical wires from the solenoid terminals, then remove both mounting bolts (arrows)

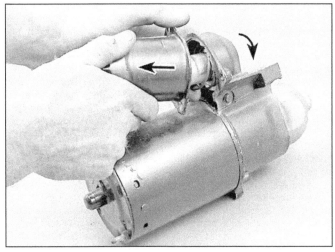

14.3 To remove the starter solenoid from the starter motor on V6 engines, remove the solenoid to starter frame mounting screws, then turn the solenoid in a clockwise direction and pull

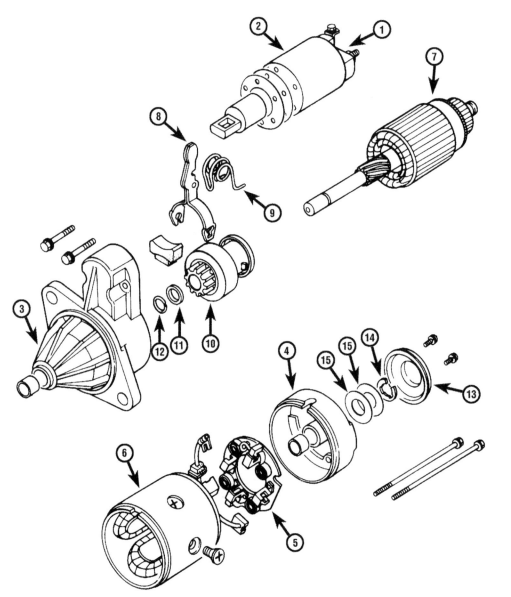

15.3 Starter motor details - G180Z and G200Z engines

1   Solenoid terminal
2   Solenoid
3   Gear case assembly
4   Rear cover assembly
5   Brush holder assembly
6   Field coil assembly
7   Armature
8   Shift lever
9   Torsion spring
10   Pinion
11   Pinion stop
12   Pinion stop clip
13   Dust cover
14   Snap-ring
15   Thrust washers

3   Remove the starter motor mounting bolts **(see illustration)**.
4   Lift out the starter motor and solenoid.
5   Installation is the reverse of the removal procedure.

14   Starter solenoid - removal and installation

Refer to illustration 14.3

1   Remove the starter motor as described in Section 12.
2   Remove the nut that holds the braided copper strap to the solenoid.
3   Remove the two mounting screws and detach the solenoid from the starter motor **(see illustration)**.
4   Installation is the reverse of the removal procedure. **Note:** *During installation be sure to engage the solenoid plunger with the starter motor shift lever.*

15   Starter motor - brush replacement

Non-reduction gear starter

Refer to illustrations 15.3 and 15.8

1   Remove the starter motor (see Section 12).
2   Remove the starter solenoid (see Section 13).
3   Remove the rear dust cover from the starter motor **(see illustration)**.
4   Remove the snap-ring and thrust washer.
5   Remove the two brush holder set screws from the rear plate.
6   Remove the two through bolts from the rear of the starter.
7   Remove the rear cover.
8   Lift the brush holder plate and disconnect the springs **(see illustration)**. Remove each of the brushes from the holder plate by disengaging them from their springs. Measure the length of the brushes. Compare them to the lengths listed in this Chapter's Specifications. If they are shorter, they

should be replaced with new ones. On some models it is necessary to use a soldering gun to remove and install the brushes.
9   Check the brush springs for rusting, distortion or breakage.
10   Assembly is the reverse of the above procedure.

Reduction gear starter

11   Remove the starter as described in Section 12.
12   Remove the solenoid as described in Section 13.
13   Remove the through bolts by unscrewing them and drawing them out through the rear.
14   Remove the rear cover from the starter motor. Be careful not to damage the overrunning clutch.
15   Remove the yoke, armature and brush holder as an assembly from the center housing. Be careful not to knock the brushes, commutator or coil against any adjacent part.
16   Lift the brush springs and remove the brushes from the brush holder.
17   Measure the length of the brushes. If they are shorter than the specified length they should be replaced with new ones. On some models it is necessary to use a soldering gun to remove and install the brushes.
18   Reassembly of the starter motor is the reverse of the disassembly procedure.

16   Charging system - general information

The charging system consists of the alternator, the voltage regulator and the battery. These components work together to supply electrical power for the engine ignition, lights, radio, etc.
The alternator is turned by a drivebelt at the front of the engine. When the engine is operating, voltage is generated by the alternator to be sent to the battery for storage.
The alternator uses a solid-state regulator mounted inside the alternator housing. The purpose of the voltage regulator is to limit alternator voltage to a pre-set value, pre-

venting power surges, circuit overloads, etc., during peak voltage output. The regulator voltage setting cannot be adjusted.
The charging system does not ordinarily require periodic maintenance. The drivebelts, electrical wiring and connections should, however, be inspected during normal tune-ups.

17   Alternator - maintenance and special precautions

1   Alternator maintenance consists of occasionally wiping away any dirt or oil which may have collected on it.
2   Check the tension of the drivebelt (refer to Chapter 1).
3   No lubrication is required, as alternator bearings are sealed for the life of the unit.
4   Take extreme care when making circuit connections to a vehicle equipped with an alternator and observe the following precautions: When making connections to the alternator from a battery, always match correct polarity. Before using electric-arc welding equipment to repair any part of the vehicle, disconnect the wiring from the alternator and disconnect the positive battery terminal. Never start the engine with a battery charger connected.

18   Charging system - check

1   If a malfunction occurs in the charging circuit, don't automatically assume that the alternator is causing the problem. First check the following items:

a) *Check the drivebelt tension and condition (Chapter 1). Replace it if it's worn or deteriorated.*
b) *Make sure the alternator mounting and adjustment bolts are tight.*
c) *Inspect the alternator wiring harness and the connectors at the alternator and voltage regulator. They must be in good condition and tight.*
d) *Check the fusible link (if equipped) located between the starter solenoid and the alternator. If it's burned, determine the cause, repair the circuit and replace the link (the vehicle won't start and/or the accessories won't work if the fusible link blows). Sometimes a fusible link may look good, but still be bad. If in doubt, remove it and check for continuity.*
e) *Start the engine and check the alternator for abnormal noises (a shrieking or squealing sound indicates a bad bearing).*
f) *Check the specific gravity of the battery electrolyte. If it's low, charge the battery (doesn't apply to maintenance free batteries).*
g) *Make sure the battery is fully charged (one bad cell in a battery can cause overcharging by the alternator).*

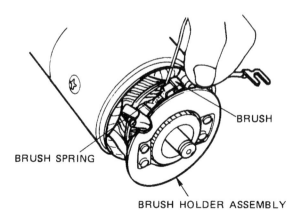

15.8 Removing the brush and holder assembly

BRUSH

BRUSH SPRING

BRUSH HOLDER ASSEMBLY

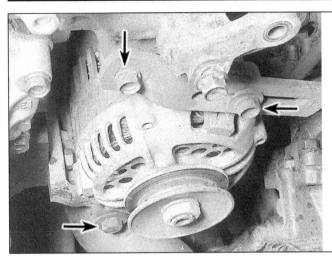

19.6 Remove the alternator mounting bolts (arrows) and bracket

h) *Disconnect the battery cables (negative first, then positive). Inspect the battery posts and the cable clamps for corrosion. Clean them thoroughly if necessary (see Chapter 1). Reconnect the cable to the positive terminal.*

i) *With the key off, connect a test light between the negative battery post and the disconnected negative cable clamp.*
   1) *If the test light does not come on, reattach the clamp and proceed to the next step.*
   2) *If the test light comes on, there is a short (drain) in the electrical system of the vehicle. The short must be repaired before the charging system can be checked.*
   3) *Disconnect the alternator wiring harness.*

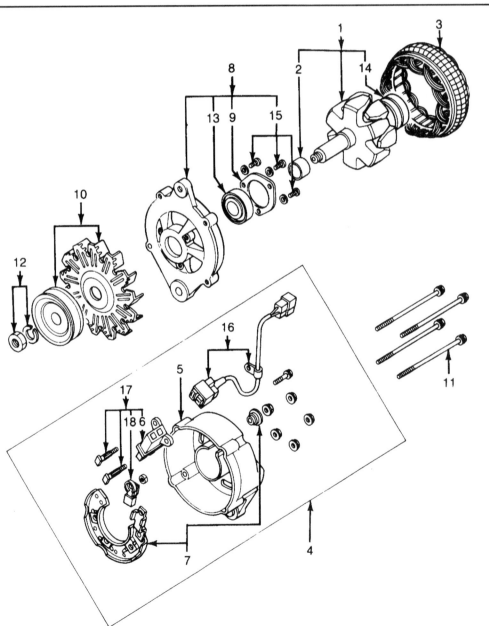

20.2 Alternator - exploded view

1	*Rotor assembly*
2	*Spacer*
3	*Stator assembly*
4	*Cover assembly*
5	*Rear cover*
6	*Holder*
7	*Diode assembly*
8	*Cover assembly*
9	*Retainer*
10	*Pulley assembly*
11	*Through bolt*
12	*Nut assembly*
13	*Bearing*
14	*Bearing*
15	*Screw kit*
16	*Wire assembly*
17	*Brush and condenser assembly*
18	*Condenser assembly*

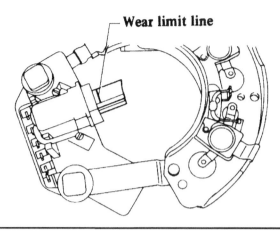

Wear limit line

20.10 The quickest way to check for brush wear is to note how close the brush has worn to the limit line

(a) If the light goes out, the alternator is bad.
(b) If the light stays on, pull each fuse until the light goes out (this will tell you which component or circuit is shorted).

2   Using a voltmeter, check the battery voltage with the engine off. If should be approximately 12-volts.

3   Start the engine and check the battery voltage again. It should now be approximately 14-to-15 volts.

4   Turn on the headlights. The voltage should drop, and then come back up, if the charging system is working properly.

5   If the voltage reading is more than the specified charging voltage, replace the voltage regulator (refer to Section 20). If the voltage is less, the alternator diode(s), stator or rectifier may be bad or the voltage regulator may be malfunctioning.

19   Alternator - removal and installation

Refer to illustration 19.6

1   Disconnect the negative battery cable from the battery.

2   On V6 engines, remove the serpentine belt (see Chapter 1).

3   On G180Z, G200Z, 4ZD1 and 4ZE1 engines, remove the air pump.

4   Clearly label then detach all wires from the rear of the alternator.

5   Refer to Chapter 1 and remove the alternator drivebelt.

6   Remove the adjusting and mounting bolts **(see illustration)**.

7   Lift the alternator from its mounting bracket.

8   Installation is the reverse of the removal procedure.

20   Alternator voltage regulator and brushes - replacement

Refer to illustrations 20.2 and 20.10
Note: *On vehicles equipped with a V6 engine, the alternator (model CS-130) cannot be repaired. If the unit is faulty, replace it with a new alternator. Also, G180Z and G200Z engines are equipped with alternators and separate voltage regulators. The voltage regulator is mounted on the firewall. On 4ZD1 and 4ZE1 engines, the alternator is equipped with an internal voltage regulator.*

1   Remove the alternator from the vehicle as described in Section 19.

2   Remove the four through-bolts from the rear cover **(see illustration)**.

3   Separate the front cover/rotor assembly from the rear cover/stator assembly by lightly tapping the front bracket with a soft-faced hammer.

4   Disconnect the wire connecting the diode set plate to the brush at the brush terminal by heating it with a soldering iron. **Note:** *When melting the solder, hold the lead wire with long nose pliers to prevent heat transferred to the diodes.*

5   Disconnect the diode set plate from the side face of the rear cover.

6   Remove the nut securing the battery terminal bolt.

7   Partially lift the diode set plate together with the stator coil from the rear cover. Remove the screw connecting the diode set plate to the brush.

8   Separate the rear cover, together with the stator coil and diode and remove the brush and IC regulator (4ZE1 and 4ZD1 engines).

9   Check for free movement of the brush and make sure the holder is clean and undamaged.

10   Check for brush wear either by noting the brush wear limit line **(see illustration)** or by measuring the length of the brush and comparing it to the wear limit in the Specifications at the beginning of this Chapter. If the brush is worn beyond either limit, it must be replaced with a new one.

11   If the voltage regulator is defective, it and the brush assembly should be replaced as one unit.

12   Reassembly of the alternator is the reverse of the disassembly procedure, with the following notes:

a) *Soldering of the stator coil lead wires to the diode assembly should be done as quickly as possible to prevent excessive heat from building up around the diode assembly.*

b) *When soldering, use a non-acidic soldering paste and keep the soldering iron temperature between 300 and 350*

Notes

Chapter 6
Emissions control systems

Contents

1 General information

Refer to illustration 1.5

As smog standards have become more stringent, the emissions control systems developed to meet these requirements have not only become increasingly more diverse and complex, but are now designed as integral parts of the operation of the engine. Where once the anti-pollution devices used were installed as peripheral "add-on" components, the present systems work closely with such other systems as the fuel, ignition and exhaust systems. All vital engine operations are controlled by the emissions control system.

Because of this close integration of systems, disconnecting or not maintaining the emissions control systems can adversely affect engine performance and life, as well as fuel economy.

This is not to say that the emissions systems are particularly difficult for the home mechanic to maintain and service. You can perform general operational checks, and do most (if not all) of the regular maintenance easily and quickly at home with common tune-up and hand tools. **Note:** *The most frequent cause of emissions problems is simply a loose or broken vacuum hose or wire, so*

always check hoses, wires and connectors before performing major repairs.

While the end result from the various emissions systems is to reduce the output of pollutants into the air, namely hydrocarbons (HC), carbon monoxide (CO), and oxides of nitrogen (NOx), the various systems function independently toward this goal. This is the

way in which this Chapter is divided.

Note: *Always refer to the Vehicle Emission Control Information (VECI) label* **(see illustration)** *for specific information regarding emissions components on your vehicle. Similarly, always use the vacuum hose routing diagram as the final word regarding hose routing for your particular vehicle.*

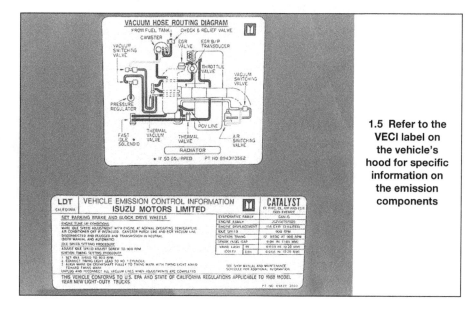

1.5 Refer to the VECI label on the vehicle's hood for specific information on the emission components

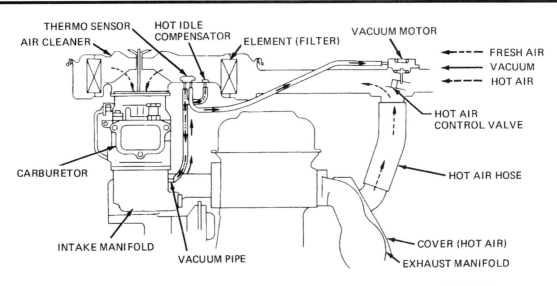

2.2 Thermostatically controlled air cleaner system (four-cylinder engine shown, V6 similar)

2   Thermostatically Controlled Air Cleaner (TCAC) system

Refer to illustration 2.2

General information

1   The Thermostatically Controlled Air Cleaner (TCAC) system improves engine efficiency and reduces hydrocarbon emissions during the initial warmup period of the engine by maintaining a controlled air temperature into the carburetor. Temperature control of the incoming air allows leaner fuel mixtures and helps prevent carburetor icing in cold weather.

2   The system **(see illustration)** uses a hot air control valve located in the snorkel of the air cleaner housing to control the ratio of cold and warm air into the carburetor or TBI unit. This valve is controlled by a vacuum motor which is, in turn, modulated by a thermo-sensor or temperature sensor in the air cleaner. This air bleed valve opens when the intake air temperature is cold, thus allowing warm air from the heat stove to reach the carburetor. When the air is hot, the valve closes, thus closing off the warm air.

3   It is during the first few miles of driving (depending on outside temperature) that this system has its greatest effect on engine performance and emissions output. When the engine is cold, the air control valve blocks off the air cleaner inlet snorkel, allowing only warm air from the exhaust manifold to enter the carburetor. Gradually, as the engine warms up, the valve opens the snorkel passage, increasing the amount of cold air allowed in. Once the engine reaches normal operating temperature, the valve completely opens, allowing only cold, fresh air to enter.

4   Because of this cold-engine-only function, it is important to periodically check this system to prevent poor engine performance when cold, or overheating of the fuel mixture

once the engine has reached operating temperatures. If the air cleaner valve sticks in the "no heat" position, the engine will run poorly, stall and waste gas until it has warmed up on its own. A valve sticking in the "heat" position causes the engine to run as if it is out of tune due to the constant flow of hot air to the carburetor.

Checking

Thermo-sensor

Refer to illustration 2.8

5   With the engine off, note the position of the hot air control valve inside the air cleaner snorkel. If the vehicle is equipped with an air duct on the end of the snorkel, it will have to be removed prior to this check. If visual access to the valve is difficult, use a mirror. The valve should be down, meaning that all air would flow through the snorkel (fresh air) and none through the exhaust manifold hot-air duct at the underside of the air cleaner housing (hot air).

6   Now have an assistant start the engine and continue to watch the valve inside the snorkel. With the engine cold and at idle, the

valve should close off all air from the snorkel (fresh air), allowing heated air from the exhaust manifold to enter the air cleaner intake. As the engine warms to operating temperature the valve should move, allowing outside air through the snorkel to be included in the mixture. Eventually, the valve should move down to the point where most of the incoming air is through the snorkel (fresh air) and not the exhaust manifold duct (hot air).

7   If the valve did not close off snorkel air (fresh air) when the cold engine was first started, disconnect the vacuum hose at the snorkel vacuum motor and place your thumb over the hose end, checking for vacuum. If there is vacuum going to the motor, check that the valve and link are not seized or binding in the air cleaner snorkel. Replace the vacuum motor if the hose routing is correct and the valve moves freely.

8   If there was no vacuum going to the motor in the above test, check the hoses to make sure they are not cracked, crimped or disconnected. If the hoses are clear and in good condition, replace the thermo-sensor inside the air cleaner housing **(see illustration)**.

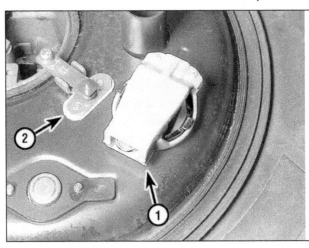

2.8 Location of thermo-sensor and idle compensator on carbureted vehicles

1   *Thermo sensor*
2   *Idle compensator*

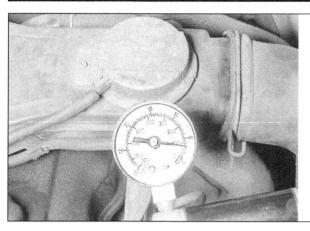

2.9 The vacuum motor should activate the hot air control valve when vacuum is applied

Vacuum motor

Refer to illustration 2.9

9   Detach the vacuum hose from the vacuum motor and connect a hand-held vacuum pump in its place **(see illustration)**.
10   Apply vacuum to the motor. The hot air control valve should move, closing off fresh air from the snorkel inlet.
11   If the vacuum motor does not perform as described it should be replaced.

Component replacement

Thermo-sensor

12   Using pliers, flatten the clip that retains the vacuum hose to the sensor and remove the clips.
13   Disconnect the hoses from the sensor.
14   Carefully note the position of the sensor. The new sensor must be installed in exactly the same position.
15   Lift the sensor from the air cleaner.
Note: *The gasket between the sensor and air cleaner is bonded to the air cleaner and should not be removed.*
16   Install the new sensor with new gaskets in the same position as the old one.
17   Press the retaining clip on the sensor. Do not damage the control mechanism in the center of the sensor.
18   Connect the vacuum hoses and install the air cleaner on the engine.

Vacuum motor

19   Remove the screws or drill out the rivets that attach the vacuum motor retainer to the air cleaner.
20   Turn the vacuum motor to disengage it from the air control valve and lift it off.
21   Installation is the reverse of the removal procedure.

3   High altitude emission control system

General information

1   1982 and 1983 vehicles can be equipped with a high altitude emission control system to compensate for the density of air at higher elevations. This system will allow the engine to perform better and the emission system to control the release of excess emissions at higher altitudes. **Note:** *If the vehicle has been in use in high altitudes and is equipped with this system, and is returning to low altitudes (permanently), reverse the procedure described below.*

Installation

Manual transmission models

2   Obtain the correct high altitude emission control system kit from an Isuzu dealer parts department.
3   Detach the hose from the air switching valve at the left front of the engine.
4   Connect the delay valve to the disconnected valve and hose, with the "A" on the valve facing the carburetor side of the hose.
5   Adjust the ignition timing to the Specification on the label that came with the kit (4-degrees more advance than the original timing specification). Check the idle speed and adjust it if necessary.
6   Attach the label in a prominent location under the hood.

Automatic transmission models

7   Follow Steps 3 and 4 and install the delay valve.
8   Remove the ignition coil and install the altitude switch and the solenoid valve assembly using a long bolt and nut.
9   Remove the three rubber caps on the altitude compensation air passages on the carburetor and install the solenoid valve.
10   Remove the rubber cap from the altitude compensation port on the air cleaner, and connect the large rubber hose from the solenoid valve to the port.
11   Connect the electrical connector on the altitude switch and the solenoid valve assembly to the body harness connector.
12   Adjust the engine speed to the specification.
13   Check the ignition timing and adjust it to the specification on the label that came with the kit (4-degrees more advance than the original timing specification).
14   Attach the label in a prominent location under the hood.

4   Positive Crankcase Ventilation (PCV) system

Refer to illustration 4.2

General information

1   The Positive Crankcase Ventilation, or PCV system, as it is more commonly called, reduces hydrocarbon emissions by circulating fresh air through the crankcase. This air combines with blow-by gases, or gases blown past the piston rings during compression, and the combination is then sucked into the intake manifold to be reburned by the engine.
2   This process is achieved by using one air pipe running from the air cleaner to the

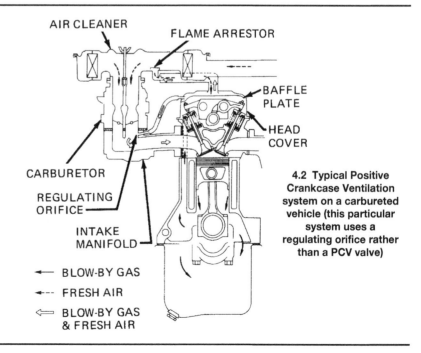

4.2 Typical Positive Crankcase Ventilation system on a carbureted vehicle (this particular system uses a regulating orifice rather than a PCV valve)

AIR CLEANER
FLAME ARRESTOR
BAFFLE PLATE
HEAD COVER
CARBURETOR
REGULATING ORIFICE
INTAKE MANIFOLD

← BLOW-BY GAS
←-- FRESH AIR
⇐ BLOW-BY GAS & FRESH AIR

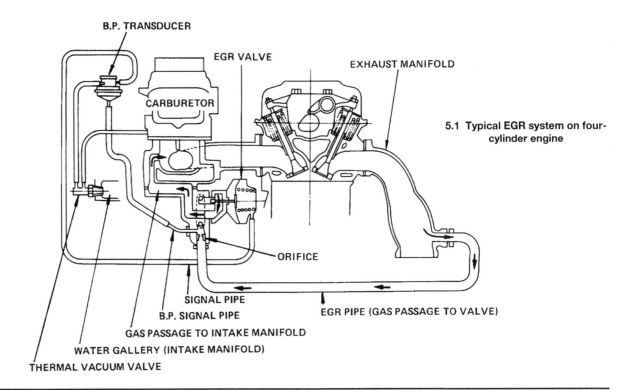

B.P. TRANSDUCER

EGR VALVE

EXHAUST MANIFOLD

CARBURETOR

5.1 Typical EGR system on four-cylinder engine

ORIFICE

SIGNAL PIPE

B.P. SIGNAL PIPE

EGR PIPE (GAS PASSAGE TO VALVE)

GAS PASSAGE TO INTAKE MANIFOLD

WATER GALLERY (INTAKE MANIFOLD)

THERMAL VACUUM VALVE

rocker arm cover, a one-way PCV valve located on the intake manifold and a second air pipe running from the crankcase to the PCV valve **(see illustration)**. **Note:** *Some models use a regulating orifice instead of a PCV valve.*

3   During the partial throttle operation of the engine, the vacuum created in the intake manifold is great enough to suck the gases from the crankcase, through the PCV valve and into the manifold. The PCV valve allows the gases to enter the manifold, but will not allow them to pass in the other direction.

4   The ventilating air is drawn into the rocker arm cover from the air cleaner, and then into the crankcase.

5   Under full-throttle operation, the vacuum in the intake manifold is not great enough to suck the gases in. Under this condition the blowby gases flow backwards into the rocker arm cover, through the air tube and into the air cleaner, where it is carried into the intake manifold in the normal air intake flow.

Checking

6   The PCV system can be checked for proper operation quickly and easily. This system should be checked regularly as carbon and sludge deposited by the blow-by gases may eventually clog the PCV valve and/or system hoses. When the flow of the PCV system is reduced or stopped, common symptoms are rough idling or reduced engine speed at idle.

7   To check for proper vacuum in the system, disconnect the rubber air hose where it exits the air cleaner.

8   With the engine idling, place your thumb lightly over the end of the hose. You should feel a slight pull of vacuum. The suction may be heard as your thumb is released. This will indicate that air is being drawn all the way through the system. If a vacuum is felt, the system is functioning properly.

9   If there is very little vacuum or none at all, at the end of the hose, the system is clogged and must be inspected further.

10   With the engine still idling, disconnect the rubber hose from the PCV valve (if the system isn't equipped with a PCV valve, disconnect the hose from the rocker arm cover - **see illustration 4.2**). Now place your finger over the end of the valve and feel for a suction. You should feel a relatively strong vacuum at this point. This indicates that the valve is good.

11   If no vacuum is felt at the PCV valve, remove the valve from the intake manifold. Shake it and listen for a clicking sound. That is the rattle of the valve's check needle. If the valve does not click freely, replace it.

12   If a strong vacuum is felt at the PCV valve, yet there is still no vacuum during the test described in Step 8, then one of the system's vent tubes is probably clogged. Both should be removed and blown out with compressed air.

13   If, after cleaning the vent tubes, there is still no suction at the air pipe, there is a blockage in an internal passage possibly at the baffle plate and steel net inside the crankcase. This requires disassembly of the engine to correct.

14   When purchasing a new PCV valve, make sure it is the proper one for your vehi-

cle. An incorrect PCV valve may pull too much or too little vacuum, possibly causing damage to the engine.

5   Exhaust Gas Recirculation (EGR) system

Refer to illustrations 5.1, 5.6a, 5.6b and 5.7

General information

1   This system is used to reduce oxides of nitrogen (NOx) emitted from the exhaust. Formation of these pollutants take place at very high temperatures; consequently, it occurs during the peak temperature period of the combustion process. To reduce peak temperatures, and thus the formation of NOx, a small amount of exhaust gas is taken from the exhaust system and recirculated in the combustion cycle **(see illustration)**.

2   In addition, within each system various EGR valves are used according to the transmission used and the altitude the vehicle is expected to be operated at. For replacement, the EGR valve can be identified by the part number stamped on the recessed portion at the top of the valve.

3   Because a malfunctioning EGR system can severely affect the performance of the engine, it is important to understand how each system works in order to accurately troubleshoot it. Common engine problems associated with the EGR system are: rough idling or stalling when at an idle, rough engine performance upon light throttle application and stalling on deceleration.

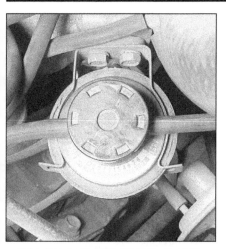

5.6a The backpressure transducer is located on the intake manifold near the distributor

EGR valve

4   The EGR valve is mounted under the intake manifold (four cylinder models) or on top of the manifold (V6 models) and controls the recycled gas flow drawn into the intake manifold from the exhaust manifold.
5   As the throttle valve is opened, vacuum is applied to the vacuum diaphragm in the EGR valve. When vacuum reaches the specified value, the diaphragm overcomes spring force and opens the exhaust gas metering valve thus allowing exhaust gas to be pulled into the engine intake.

Backpressure transducer valve (four-cylinder models only)

6   The backpressure transducer is responsive to exhaust pressure **(see illustrations)**.

Under normal operating conditions, ported vacuum leaks into the atmosphere and is applied to the EGR valve under high pressure conditions thereby modulating the volume of recirculating exhaust gas.

Thermal vacuum valve (four-cylinder models only)

7   The thermal vacuum valve is mounted on the intake manifold and is connected in series between the vacuum port in the carburetor and the EGR valve **(see illustration)**. The valve operates similarly to the coolant thermostat. As the wax melts and expands, it opens and closes the vacuum valve.
8   When the coolant temperature is below 115 to 120-degrees F, the valve is closed and the EGR system does not operate. When the coolant temperature is above 115 to 120-degrees F, the valve is open and the EGR system operates.

Checking

9   Place your finger under the EGR valve and push upward on the diaphragm plate (see Chapter 1). The diaphragm should move freely from the closed to the open position. If it doesn't, replace the EGR valve.
10   Now start the engine and run at idle speed. While the engine is still just beginning to warm up, again push upward on the EGR diaphragm with your finger. If the valve or adjacent accessories are hot, wear gloves to prevent burning your fingers. When the diaphragm is pressed (valve open to recirculate exhaust), the engine should lose speed, stumble or even stall. If the engine did not change speed, the exhaust gas pipe or passage leading to the EGR valve should be checked for blockage.

Component replacement

EGR valve

11   Disconnect the EGR exhaust tube by loosening the retaining nut (four cylinder models only).
12   Disconnect all hoses from the EGR valve, noting how they are installed. On V6 models, remove the air cleaner.
13   Remove the EGR valve mounting bolts and lift it off.
14   Clean the mating surfaces of the EGR valve and the intake manifold, removing all traces of old gasket material.
15   Installation is the reverse of the removal procedure. **Note:** *Be sure to use a new | gasket.*

Backpressure transducer valve (four-cylinder models only)

16   Disconnect the vacuum signal hoses from the top of the valve.
17   Lift the valve away from the bracket.
18   Disconnect the exhaust backpressure hose from the bottom of the valve, and remove the valve from the engine compartment.
19   Installation is the reverse of the removal procedure. **Note:** *When replacing the valve, be sure the new valve has the same type number stamped on it as the old one.*

Thermal vacuum valve (four-cylinder models only)

20   Drain about one quart of coolant from the radiator.
21   Disconnect the vacuum hoses from the switch, noting their positions for reassembly.
22   Using a wrench, remove the switch.
23   When installing the switch, apply thread sealer to the threads, being careful not to allow the sealant to touch the bottom sensor.
24   Install the switch and tighten it securely.

NORMAL OPERATING CONDITION

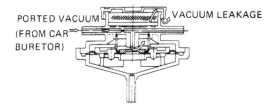

HIGH EXHAUST PRESSURE CONDITION

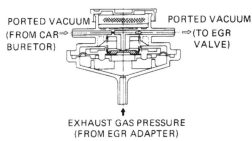

5.6b Backpressure transducer operating modes

BELOW SPECIFIED TEMPERATURE

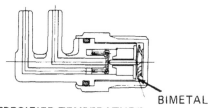

ABOVE SPECIFIED TEMPERATURE

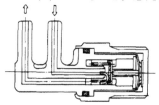

5.7 Thermal vacuum valve operation

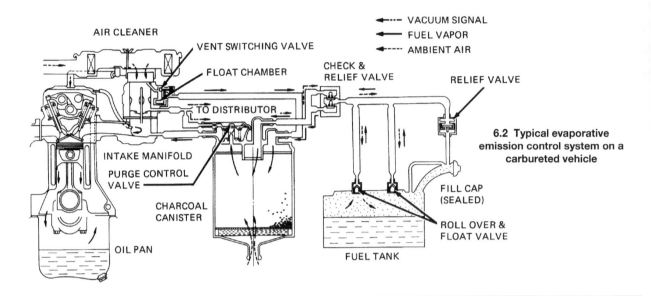

6.2 Typical evaporative emission control system on a carbureted vehicle

EGR control solenoid (V6 models only)

25 Remove the air cleaner assembly. Trace the vacuum hose from the EGR valve to the EGR control solenoid.
26 Remove the EGR control solenoid mounting bolts.
27 Disconnect the electrical connector and the vacuum hose from the solenoid.
28 Installation is the reverse of the removal procedure.

6 Fuel evaporative emission control system

Warning: *Gasoline is extremely flammable, so take extra precautions when working on any part of the fuel system. Don't smoke or allow open flames or bare light bulbs in or near the work area. Also, don't work in a garage in which a natural gas appliance such as a water heater or clothes dryer is present.*

General information

Refer to illustrations 6.2, 6.3 and 6.8
1   The evaporative emission control system is designed to prevent the escape of fuel vapors from the fuel system into the atmosphere.
2   The evaporative emission control system on four-cylinder models consists of a canister, a roll over and float valve, a purge control valve, a vent switching valve, a check and relief valve, a fuel check valve, the associated vapor lines and a specially designed fuel filler cap **(see illustration)**. On V6 models it is composed of the canister, the thermostatic vacuum switch and the associated vapor lines. The purpose of each of these components, and how to check them, is described briefly in this Section.

Canister

3   When the engine is inoperative, fuel vapors generated inside the fuel tank and the carburetor float chamber are absorbed and stored in the canister. When the engine is running, the fuel vapors absorbed in the canister are drawn into the intake manifold through the check and relief valve and an orifice **(see illustration)**.

Vent switching valve

4   The bowl vent valve controls the carburetor bowl vapors. When the engine is not running, the vent switching valve is opened allowing fuel vapor to enter the canister.
5   During engine operation, the vent switching valve is closed to prevent ambient air from flowing into the carburetor through the canister.

Purge control valve

6   The purge control valve is closed during idle to prevent vaporized fuel from entering the intake manifold. At higher engine speeds, the purge control valve opens.

Fuel filler cap

7   The fuel filler cap is equipped with a relief valve to prevent the escape of fuel vapor into the atmosphere.

Roll over and float valve

8   The fuel vapor taken up into the roll over and float valve is fed into the check and relief valve. When the pressure of the fuel vapor becomes high, the check valve opens, allowing the vapor into the canister **(see illustration)**.
9   While the check valve is held open, the valve at the air cleaner side remains closed preventing the flow of vapor into the air cleaner.
10   When the engine is running, vacuum is

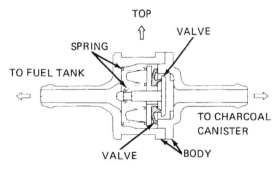

6.3 Relief valve details

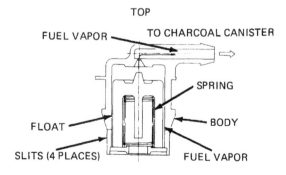

6.8 Roll over and float valve details

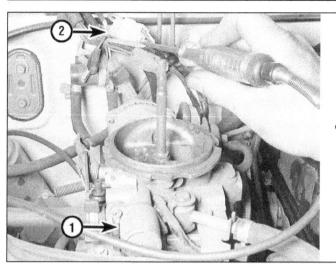

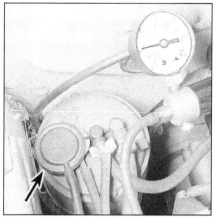

6.12 With the ignition switch On, there should be battery voltage present at the vent switching valve electrical connector

1 Vent switching valve
2 Electrical connector

6.15 Apply vacuum to the purge control valve - it should take more than 1.4 in-Hg to open the valve

developed in the tank and the difference in pressure between the relief side and the fuel tank decreases, the relief valve opens allowing ambient air from the air cleaner into the fuel tank. This ambient air replaces the vapor within the fuel tank bringing the fuel tank back into atmospheric condition.

Checking

Canister

11 See Chapter 1 for canister check and replacement procedures.

Vent switching valve

Refer to illustration 6.12

12 Remove the air cleaner (see Chapter 4). Turn the ignition switch to On and attach a test light to the electrical connector for the vent switching valve **(see illustration)**.

13 The valve should be energized with battery voltage activating the test light. If there is not any voltage present, check the circuit for an open or shorted condition.

Purge control valve

Refer to illustration 6.15

14 Remove the purge control valve hose from the valve on the canister.

15 Attach a hand vacuum pump to the vacuum fitting on the valve **(see illustration)**.

16 Apply vacuum - the valve should not open until at least 1.4 in-Hg of vacuum is applied.

Fuel filler cap

17 Inspect the gasket on the underside of the cap for deformation and deterioration. If damage is evident, replace the cap.

7 Air injection reactor system

General information

Refer to illustrations 7.1a and 7.1b

1 This is a method of injecting air (generated in an external compressor) into the exhaust manifold in order to reduce hydro-

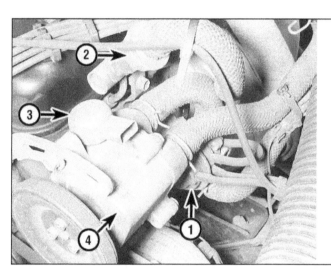

7.1a Location of the air injection reaction components (four cylinder models)

1 Air switching valve
2 Check valve
3 Relief valve
4 Air pump

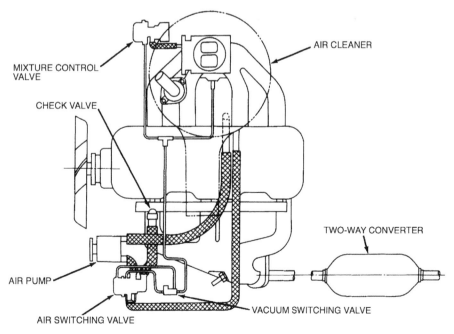

7.1b Typical AIR system and catalytic converter (four cylinder models)

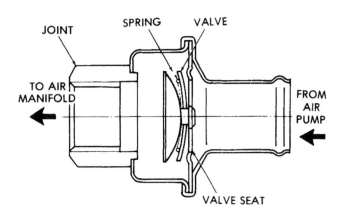

7.7 The check valve should only allow air to pass through in one direction

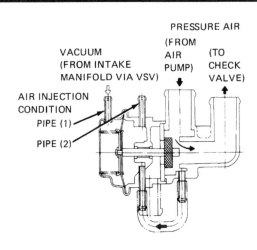

7.9 Typical air switching valve

carbons and carbon monoxide in the exhaust gas by providing conditions favorable for recombustion. On four-cylinder engines, the system is composed of a belt drive air pump, check valve, air switching valve, mixture control valve, air gallery and the associated hoses **(see illustrations)**. On six-cylinder engines the system is made up of the air pump, the Electric Air Control (EAC) valve, an EAC solenoid, a check valve, the ECM and the associated hoses. Because the system is ECM-controlled, the checks that can be made to the system are limited.

2   Air is drawn through the air pump air cleaner, compressed, and directed through the check valve to the air gallery and injection nozzles (four-cylinder engines) in the exhaust manifold. During high speed operation, excessive pump pressure is vented to the atmosphere through the relief valve (four-cylinder models) or to the air cleaner (six-cylinder models).

3   The check valve is installed in the delivery line, just before the exhaust manifold. The function of this valve is to prevent any exhaust gases from passing into the air pump should the manifold pressure be greater than the pump injection pressure. It is designed to close against the exhaust manifold pressure should the air pump fail as a result, for example, of a broken drivebelt.

Checking

Refer to illustrations 7.7, 7.9 and 7.11

4   Check all hoses, air gallery pipes and nozzles for security and condition.

5   Check and adjust the air pump drivebelt tension as described in Chapter 1.

6   With the engine at normal operating temperature, disconnect the hose leading to the check valve.

Check valve

7   Run the engine at approximately 2000 rpm and then let it return to idling speed, while watching for exhaust gas leaks from the valve **(see illustration)**. Where these are evident, replace the valve.

Air pump relief valve

8   Run the engine at a steady 3000 rpm and place your hand on the air outlet of the emergency relief valve. A definite air pressure should be felt. If it is not, replace the valve.

Air switching valve

9   The air switching valve **(see illustration)** is installed near the air pump. If normal, the secondary air continues to blow out from the valve if the accelerator pedal is pressed quickly and then released. Air should blow for not more than 5 seconds. If the valve continues to blow air, replace the valve.

Mixture control valve

10   Locate the mixture control valve near the air cleaner. The air cleaner may have to be removed for access.

11   Remove the air hose from the intake manifold to the mixture control valve and plug it. If the mixture control valve is functioning properly, the secondary air continues to blow from the mixture control valve for only a few seconds when the accelerator pedal is pressed and released quickly. If the valve continues to blow air for more than 5 seconds, then replace the valve **(see illustration)**.

Component replacement

Air pump

12   Remove the air hoses from the pump.

13   Remove the drivebelt from the pump, referring to Chapter 1, if necessary.

14   Remove all hoses from the pump.

15   Remove the mounting and adjusting bolts and lift the pump out.

16   If necessary, remove the pulley from the pump.

17   Installation is the reverse of the removal procedure.

Check valve

18   Remove the air cleaner.

19   Disconnect the air hose from the check valve.

20   Remove the check valve from the air gallery pipe, using two wrenches.

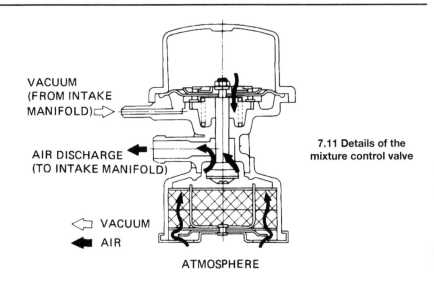

7.11 Details of the mixture control valve

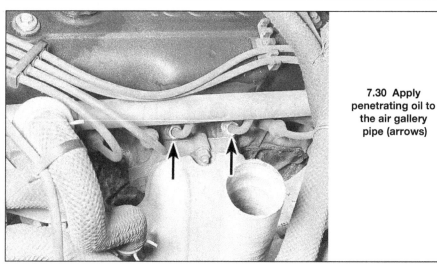

7.30 Apply penetrating oil to the air gallery pipe (arrows)

21   Installation is the reverse of the removal procedure.

Mixture control valve

22   Remove the valve from the clamp bracket and disconnect the hose.
23   Installation is the reverse of removal.

Air switching valve

24   Remove the mounting hardware and hoses from the valve and remove the valve.
25   Installation is the reverse of the removal procedure.

Air gallery pipe and injection nozzles

Refer to illustration 7.30

26   Because of the likelihood of bending or damaging the pipes during removal, the pipes should not be removed unless they are already damaged and need replacing.
27   Remove the air cleaner.
28   Disconnect the hose from the check valve.
29   Disconnect or remove all lines and hoses that will interfere with removal of the air gallery assembly.

30   Apply penetrating oil to the nuts that attach the air gallery to the exhaust manifold **(see illustration)**, then loosen them until the gallery can be lifted out.
31   Lift out the air gallery pipes from the threaded openings in the exhaust manifold.
32   Installation is the reverse of the removal procedure.

8   Coasting Fuel Cut (CFC) device

General information

Refer to illustration 8.1

1   The coasting fuel cutoff system consists principally of a slow cut solenoid valve incorporated in the carburetor, an engine speed sensor, a transmission switch, an accelerator switch and a clutch switch **(see illustration)**.
2   This system acts as an auxiliary fuel system operated by the slow cut solenoid. When the solenoid is de-energized, it pulls the plunger OUT to close the fuel slow passage.
3   The engine speed sensor monitors the engine rpm by sensing the electric pulses from the ignition coil. When the engine speed exceeds 2,000 RPM, the engine speed sensor turns OFF and the coasting fuel cut solenoid becomes de-energized.
4   When the transmission is shifted into any gear, the neutral switch is activated, allowing the coasting richer circuit to energize.

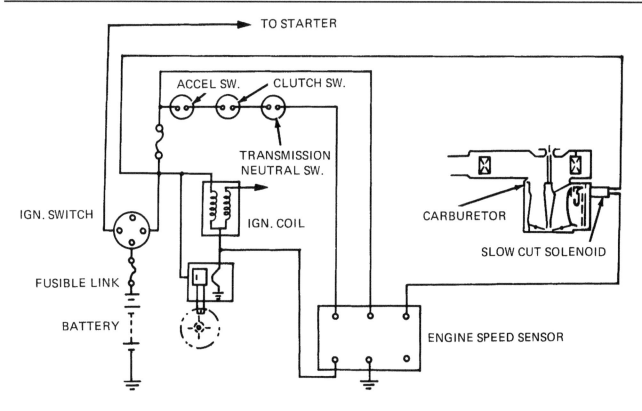

8.1 Schematic of the coasting fuel cutoff system

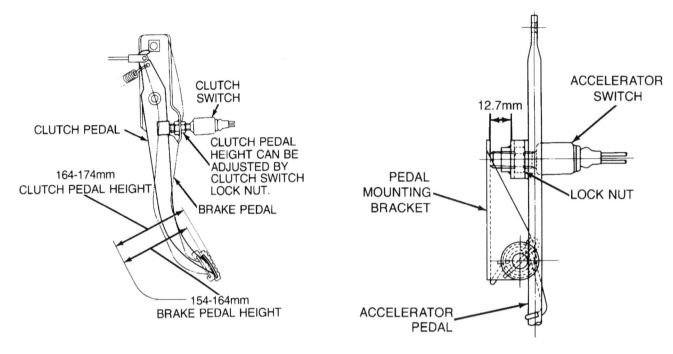

8.8a Clutch switch adjustment details

8.8b Accelerator switch adjustment details

5   The accelerator switch is connected to the accelerator linkage and is turned ON when the pedal is not pressed. When the accelerator is pressed, the switch is turned OFF and the circuit is open.
6   The clutch switch is installed near the clutch pedal and turns OFF when the pedal is pressed, de-energizing the coasting richer circuit.

Checking

Refer to illustrations 8.8a and 8.8b
7   Check the engine speed sensor by connecting the color coded wiring terminals BY and LgW with a suitable jumper wire. Start the engine and check for voltage between Lg and B coded terminals. The engine speed sensor is functioning properly if the voltage is

0 when the engine runs over 2,000 rpm.
8   Check the setting of the clutch switch and the accelerator switch and adjust them if necessary **(see illustrations)**.
9   Check each switch with an ohmmeter. The switches are functioning properly if they turn OFF (open circuit) when the pedal is pressed.

Chapter 7 Part A
Manual transmission

Contents

Specifications

General

Lubricant type...	See Chapter 1

Torque specifications | Ft-lbs

Clutch housing-to-engine bolts..	28 to 30
Shift lever bolts..	10 to 15

1 General information

All vehicles covered in this manual come equipped with either a four or five-speed manual transmission or an automatic transmission. All information on the manual transmission is included in this Part of Chapter 7. Information on the automatic transmission can be found in Part B of this Chapter. Information on the transfer case used on 4WD models is in Part C.

The manual transmission used in these models is a four or five-speed unit with the fifth gear being an overdrive.

Due to the complexity, unavailability of replacement parts and the special tools necessary, internal repair by the home mechanic is not recommended. The information in this Chapter is limited to general information and removal and installation of the transmission.

Depending on the expense involved in having a faulty transmission overhauled, it may be a good idea to replace the unit with either a new or rebuilt one. Your local dealer or transmission shop should be able to supply you with information concerning cost, availability and exchange policy. Regardless of how you decide to remedy a transmission problem, you can still save a lot of money by removing and installing the unit yourself.

2 Oil seal replacement

Refer to illustrations 2.4, 2.8, 2.9a, 2.9b and 2.9c

1 Oil leaks frequently occur due to wear of the extension housing oil seal and/or the speedometer drive gear O-ring. Replacement of these seals is relatively easy, since the repairs can usually be performed without removing the transmission from the vehicle.

2 The extension housing oil seal is located at the extreme rear of the transmission or transfer case, where the driveshaft is attached. If leakage at the seal is suspected, raise the vehicle and support it securely on

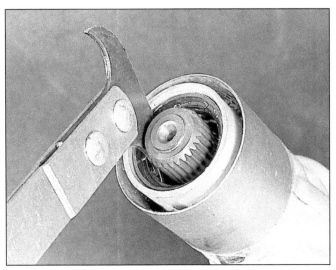

2.4 Use a seal to removal tool (shown) or a large screwdriver to pry the seal out of the transmission extension housing

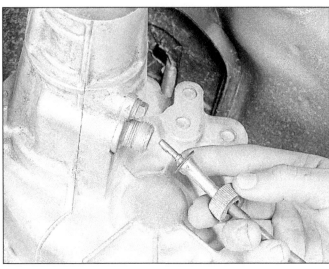

2.8 Unscrew the collar and pull the speedometer cable out of the driven gear housing

2.9a Remove the driven gear housing retaining bolt . . .

jackstands. If the seal is leaking, transmission lubricant will be built up on the front of the driveshaft and may be dripping from the rear of the transmission.

3   Refer to Chapter 8 and remove the driveshaft.

4   Using a screwdriver or oil seal replacement tool, carefully pry the oil seal out of the rear of the transmission or transfer case **(see illustration)**. Do not damage the splines on the transmission output shaft.

5   Using a large section of pipe or a very large deep socket as a drift, install the new oil seal. Drive it into the bore squarely and make sure it's completely seated.

6   Lubricate the splines of the transmission output shaft and the outside of the driveshaft sleeve yoke with lightweight grease, then install the driveshaft. Be careful not to damage the lip of the new seal.

7   The speedometer cable and driven gear

housing is located on the side of the extension housing. Look for transmission oil around the cable housing to determine if the O-ring is leaking.

8   Disconnect the speedometer cable **(see illustration)**.

9   Remove the driven gear, install a new O-ring on the driven gear housing and reinstall the driven gear housing and cable assembly on the extension housing **(see illustrations)**.

3   Shift lever - removal and installation

Refer to illustrations 3.2 and 3.4

1   Remove the carpet and sound deadener sheet.

2   Pull the shift boot up for access **(see illustration)**.

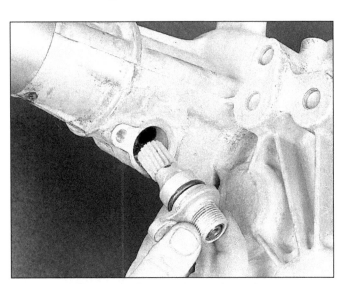

2.9b . . . and withdraw the housing and gear

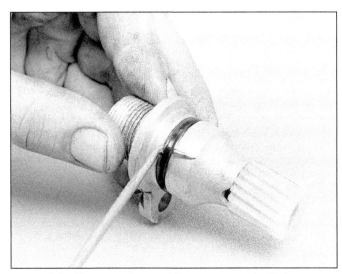

2.9c Use a small screwdriver to remove the O-ring from the groove

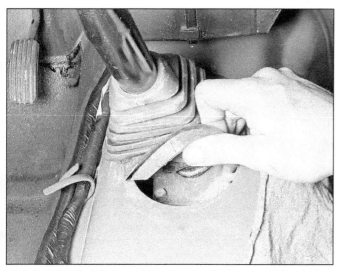

3.2 Pull the shaft boot up for access to the lever retaining bolts

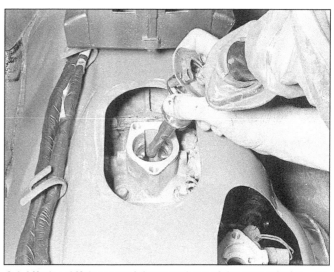

3.4 Lift the shift lever straight up and out of the transmission - be careful not to lose the spring

3   Pry the rubber cover out and remove the three shift lever bolts.
4   Lift the shift lever up and out; it is spring loaded so ease it out or the spring will be disengaged and will fall out **(see illustration)**.
5   Installation is the reverse of removal.

4   Transmission mount - check and replacement

Refer to illustrations 4.2, 4.3 and 4.4

1   Insert a large screwdriver or pry bar into the space between the extension housing and the crossmember and try to pry the transmission up slightly.
2   The transmission should not move away from the insulator much at all **(see illustration)**.
3   To replace the mount, remove the nuts attaching the insulator to the crossmember

4.2 Pry between the crossmember and transmission mount insulator (arrow) - there should be very little movement

(see illustration).
4   Raise the transmission slightly with a jack and remove the bolts from the insulator,

noting which holes are used in the crossmember for proper alignment during installation **(see illustration)**.

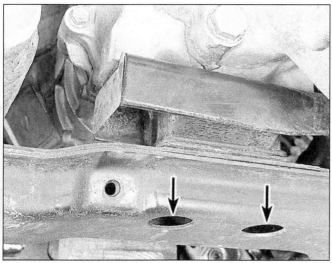

4.3 The insulator nuts are accessible through holes in the crossmember (arrows)

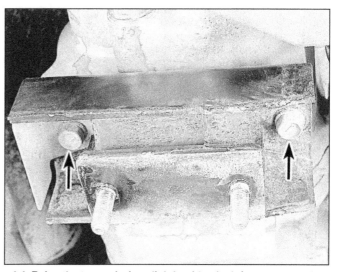

4.4 Raise the transmission slightly with a jack for access to the insulator-to-transmission bolts (arrows)

5   Installation is the reverse of the removal procedure. Be sure to tighten the nuts/bolts securely.

---

5   Manual transmission - removal and installation

Note: *On 4WD models, the transmission and transfer case are removed and installed as a unit.*

Removal

1   Disconnect the negative cable from the battery.
2   Working inside the vehicle, remove the transmission and (if equipped) transfer case shift lever (see Section 3 and Part C of this Chapter).
3   Raise the vehicle and support it securely on jackstands.
4   Drain the transmission and transfer case (4WD models).
5   Disconnect the speedometer cable and wire harness connectors from the transmission/transfer case.
6   Remove the driveshaft(s) (Chapter 8). Use a plastic bag to cover the end of the transmission/transfer case to prevent fluid loss and contamination.
7   Remove the exhaust system components as necessary for clearance (Chapter 4).
8   Support the engine. This can be done from above with an engine hoist, or by placing a jack (with a block of wood as an insulator) under the engine oil pan. The engine should remain supported at all times while the transmission/transfer case is out of the vehicle. On 1987 and later 4WD models, remove the front suspension crossmember.
9   Remove the starter motor and clutch release cylinder.
10   Support the transmission/transfer case with a jack - preferably a special jack made for this purpose. Safety chains will help steady the transmission/transfer case on the jack.
11   Remove the rear transmission/transfer case support-to-crossmember nuts and bolts.
12   Remove the nuts from the crossmember bolts. Raise the transmission/transfer case slightly and remove the crossmember.
13   Remove the bolts securing the transmission clutch housing to the engine block.
14   Make a final check that all wires and hoses have been disconnected from the

transmission/transfer case and then move the transmission/transfer case and jack toward the rear of the vehicle until the transmission input shaft is clear of the clutch housing. Keep the transmission/transfer case level as this is done. On 1981 through 1987 models, rotate the transmission/transfer case 90-degrees in a clockwise direction as it is withdrawn from the engine.
15   Once the input shaft is clear, lower the transmission/transfer case and remove it from under the vehicle. **Caution:** *Do not depress the clutch pedal while the transmission/transfer case is out of the vehicle.*
16   The clutch components can be inspected by removing the clutch housing from the engine (Chapter 8). In most cases, new clutch components should be routinely installed if the transmission is removed.

Installation

17   If removed, install the clutch components (Chapter 8).
18   With the transmission/transfer case secured to the jack as on removal, raise the transmission/transfer case into position behind the clutch housing and then carefully slide it forward, engaging the input shaft with the clutch plate hub. On 1981 through 1987 models, the transmission/transfer case should be placed on the jack with the speedometer opening facing down and then the transmission/transfer case should be rotated counterclockwise 90-degrees as it is moved into place against the engine. Do not use excessive force to install the transmission - if the input shaft does not slide into place, readjust the angle of the transmission/transfer case so it is level and/or turn the input shaft so the splines engage properly with the clutch.
19   Install the transmission clutch housing-to-engine bolts. Tighten the bolts to the torque listed in this Chapter's Specifications.
20   Install the crossmember and transmission/transfer case support. Tighten all nuts and bolts securely.
21   Remove the jacks supporting the transmission/transfer case and the engine.
22   Install the various items removed previously, referring to Chapter 8 for the installation of the driveshaft and Chapter 4 for information regarding the exhaust system components.
23   Make a final check that all wires, hoses and the speedometer cable have been connected and that the transmission and transfer case have been filled with lubricant to the

proper level (Chapter 1). Lower the vehicle.
24   Working inside the vehicle, install the shift lever (see Section 3).
25   Adjust the clutch cable (see Chapter 8).
26   Connect the negative battery cable. Road test the vehicle for proper operation and check for leakage.

---

6   Manual transmission overhaul - general information

Refer to illustrations 6.4a and 6.4b

Overhauling a manual transmission is a difficult job for the do-it-yourselfer. It involves the disassembly and reassembly of many small parts. Numerous clearances must be precisely measured and, if necessary, changed with select fit spacers and snap-rings. As a result, if transmission problems arise, it can be removed and installed by a competent do-it-yourselfer, but overhaul should be left to a transmission repair shop. Rebuilt transmissions may be available - check with your dealer parts department and auto parts stores. At any rate, the time and money involved in an overhaul is almost sure to exceed the cost of a rebuilt unit.

Nevertheless, it's not impossible for an inexperienced mechanic to rebuild a transmission if the special tools are available and the job is done in a deliberate step-by-step manner so nothing is overlooked.

The tools necessary for an overhaul include internal and external snap-ring pliers, a bearing puller, a slide hammer, a set of pin punches, a dial indicator and possibly a hydraulic press. In addition, a large, sturdy workbench and a vise or transmission stand will be required.

During disassembly of the transmission, make careful notes of how each piece comes off, where it fits in relation to other pieces and what holds it in place. Exploded views are included **(see illustrations)** to show where the parts go - but actually noting how they are installed when you remove the parts will make it much easier to get the transmission back together.

Before taking the transmission apart for repair, it will help if you have some idea what area of the transmission is malfunctioning. Certain problems can be closely tied to specific areas in the transmission, which can make component examination and replacement easier. Refer to the Troubleshooting section at the front of this manual for information regarding possible sources of trouble.

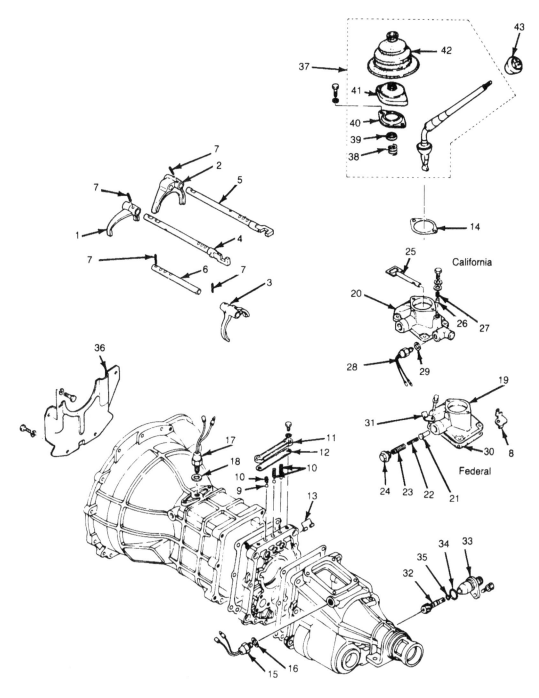

6.4a Typical 4WD transmission shift mechanism components - exploded view

1	3rd/4th shift fork	13	Interlock pin	23	Outer spring	34	Outer bushing O-ring
2	1st/2nd shift fork	14	Interlock plug	24	Reverse stop cap	35	Inner bushing O-ring
3	Reverse shift fork	15	Back-up light switch	25	Neutral switch rod	36	Front transmission case cover
4	3rd/4th gearshift rod	16	Gasket	26	Damper pad	37	Shift lever
5	1st/2nd gearshift rod	17	Switch	27	Spring	38	Spring
6	Reverse gearshift rod	18	Gasket	28	Neutral switch (California model)	39	Spring seat
7	Shift fork pin and spring	19	Shift cover (49-state model)	29	Gasket	40	Cover
8	Bracket	20	Shift cover (California-state model)	30	Gasket	41	Shift lever cover
9	Gearshift detent bolt			31	Bracket	42	Grommet
10	Detent ball spring	21	Revere stop plunger	32	Gear	43	Knob
11	Detent spring plate	22	Inner spring	33	O-ring and bushing		
12	Gasket						

6.4b Typical 4WD transmission and transfer case assembly components - exploded view

1	Case assembly
2	Pin
3	Needle bearing
4	Shift rod plug
5	Stud
6	Stud
7	Plug
8	O-ring
9	Dust cover
10	Mainshaft snap-gear ring
11	Countergear snap-ring
12	Gasket
13	Ball stud
14	Lock washer
15	Flat washer
16	Plug
17	Plug
18	Catch gear shaft
19	Ball bearing
20	Snap-ring
21	Snap-ring
22	Spring
23	Needle bearing
24	Bearing retainer
25	Oil seal
26	Gasket
27	Bolt
28	Transfer case
29	Bushing
30	Oil seal
31	Dust cover
32	Guide pin
33	Stud
34	Center support and transfer case gasket
35	Bolt
36	Nut
37	Lock washer
38	Snap-ring
39	Plug
40	O-ring
41	Breather
42	Bracket
43	Mainshaft
44	Mainshaft snap-ring
45	3rd/4th synchronizer assembly
46	3rd/4th synchronizer hub
47	3rd/4th synchronizer sleeve
48	Synchronizer key
49	Synchronizer spring
50	Blocker ring
51	3rd gear
52	2nd gear
53	1st/2nd gear synchronizer
54	1st/2nd gear synchronizer hub
55	1st/2nd synchronizer sleeve
56	Synchronizer sleeve
57	Synchronizer spring
58	Blocker ring
59	1st gear
60	Needle bearing
61	Needle bearing

62	Needle bearing collar
63	Thrust washer
64	Ball bearing
65	Reverse gear
66	Thrust washer
67	Needle bearing collar
68	Needle bearing
69	Reverse hub
70	Reverse sleeve
71	Transfer hub
72	Transfer input gear
73	Needle bearing
74	Needle bearing collar
75	Mainshaft nut
76	Pilot bearing spacer
77	Pilot bearing
78	Range shift sleeve

79	4-wheel shift sleeve
80	Output transfer shaft
81	Output transfer gear
82	Needle bearing
83	Thrust washer
84	Front output shaft bearing
85	Spacer
86	Speedometer gear spacer
87	Rear output shaft bearing
88	Output shaft nut
89	Speedometer drive gear
90	Key
91	Countergear
92	Bearing

93	Counter reverse gear
94	Washer
95	Lock washer
96	Counter reverse gear nut
97	Transfer case countergear shaft
98	Lock plate
99	Bolt
100	Reverse idler gear
101	Washer
102	Rear thrust washer
103	Thrust washer ball
104	Transfer case countershaft gear
105	Thrust washer
106	Needle bearing
107	O-ring

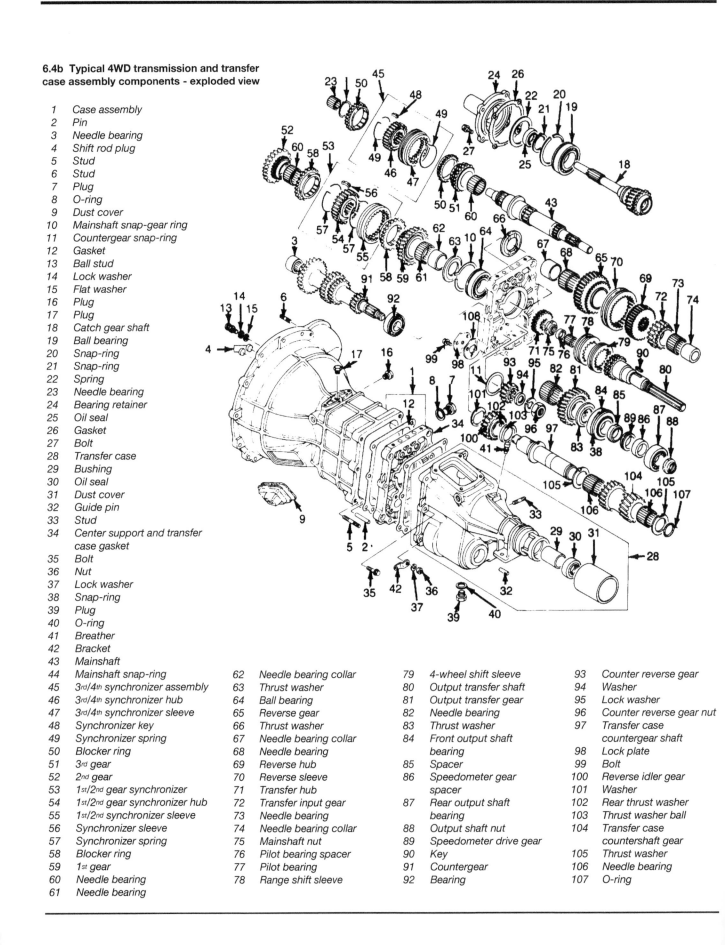

Chapter 7 Part B
Automatic transmission

Contents

Specifications

Fluid type .... See Chapter 1

Throttle valve (TV) cable adjustments*
Boot-to-stopper distance .... 1/32 to 1/16 in (0.8 to 1.5 mm)
Stroke .... 1-5/16 to 1-23/64 in (33 to 34 mm)
* Note: *These adjustments apply to both carburetor-equipped and fuel injected vehicles.*

Torque converter face-to-transmission housing face clearance (2WD models)
1984 through 1987 .... 1-5/16 in (33 mm)
1988-on .... 1-13/64 in (31 mm)

Torque specifications
Ft-lbs

Transmission housing-to-engine bolts
    1984 through 1987 40
    1988-on 47
Torque converter-to-driveplate bolts
    1984 through 1987 24
    1988-on 25

1 General information

All vehicles covered in this manual are equipped with either a four or five-speed manual transmission or a three-speed (1984 through 1986) or four-speed (1987 on) automatic transmission. All information on the automatic transmission is included in this Part of Chapter 7. Information for the manual transmission can be found in Part A. Information on the transfer case used on 4WD models can be found in Part C.

Specialized techniques and equipment are required when working on automatic transmissions, due to their complexity. Consequently, this Chapter addresses only those procedures concerned with routine maintenance, general diagnosis and removal and installation.

If the transmission requires major repair work, it should be left to a dealer service department or an automotive transmission repair shop. You can, however, remove and install the transmission yourself and save the expense, even if the repair work is done by a transmission specialist.

2 Diagnosis - general

Note: *Automatic transmission malfunctions may be caused by five general conditions: poor engine performance, improper adjustments, hydraulic malfunctions, mechanical malfunctions or computer malfunctions. Diagnosis of these problems should always begin with a check of the easily repaired items: fluid level and condition (Chapter 1), and shift linkage adjustment. Next, perform a road test to determine if the problem has been corrected or if more diagnosis is necessary. If the problem persists after the preliminary tests and corrections are completed, additional diagnosis should be done by a dealer service department or transmission repair shop. Refer to the Troubleshooting Section at the front of this manual for information on symptoms of transmission problems.*

Preliminary checks

1   Drive the vehicle to warm the transmission to normal operating temperature.
2   Check the fluid level as described in Chapter 1:

a) *If the fluid level is unusually low, add enough fluid to bring the level within the designated area of the dipstick, then check for external leaks (see below).*

b) *If the fluid level is abnormally high, drain off the excess, then check the drained fluid for contamination by coolant. The presence of engine coolant in the automatic transmission fluid indicates that a failure has occurred in the internal radiator walls that separate the coolant from the transmission fluid (see Chapter 3).*

c) *If the fluid is foaming, drain it and refill the transmission, then check for coolant in the fluid or a high fluid level.*

3   Check the engine idle speed. **Note:** *If the engine is malfunctioning, do not proceed with the preliminary checks until it has been repaired and runs normally.*
4   Inspect the shift linkage (Section 3). Make sure that it's properly adjusted and that the linkage operates smoothly.

Fluid leak diagnosis

5   Most fluid leaks are easy to locate visually. Repair usually consists of replacing a seal or gasket. If a leak is difficult to find, the following procedure may help.
6   Identify the fluid. Make sure it's transmission fluid and not engine oil or brake fluid (automatic transmission fluid is a deep red color).
7   Try to pinpoint the source of the leak. Drive the vehicle several miles, then park it over a large sheet of cardboard. After a minute or two, you should be able to locate the leak by determining the source of the fluid dripping onto the cardboard.
8   Make a careful visual inspection of the suspected component and the area immediately around it. Pay particular attention to gasket mating surfaces. A mirror is often helpful for finding leaks in areas that are hard to see.
9   If the leak still cannot be found, clean the suspected area thoroughly with a degreaser or solvent, then dry it.
10   Drive the vehicle for several miles at normal operating temperature and varying speeds. After driving the vehicle, visually inspect the suspected component again.
11   Once the leak has been located, the cause must be determined before it can be properly repaired. If a gasket is replaced but the sealing flange is bent, the new gasket will not stop the leak. The bent flange must be straightened.
12   Before attempting to repair a leak, check to make sure that the following conditions are corrected or they may cause another leak. **Note:** *Some of the following conditions cannot be fixed without highly specialized tools and expertise. Such problems must be referred to a transmission repair shop or a dealer service department.*

Gasket leaks

13   Check the pan periodically. Make sure the bolts are tight, no bolts are missing, the gasket is in good condition and the pan is flat (dents in the pan may indicate damage to the valve body inside).
14   If the pan gasket is leaking, the fluid level or the fluid pressure may be too high, the vent may be plugged, the pan bolts may be too tight, the pan sealing flange may be warped, the sealing surface of the transmission housing may be damaged, the gasket may be damaged or the transmission casting may be cracked or porous. If sealant instead of gasket material has been used to form a seal between the pan and the transmission housing, it may be the wrong sealant.

Seal leaks

15   If a transmission seal is leaking, the fluid level or pressure may be too high, the vent may be plugged, the seal bore may be damaged, the seal itself may be damaged or improperly installed, the surface of the shaft protruding through the seal may be damaged or a loose bearing may be causing excessive shaft movement.
16   Make sure the dipstick tube seal is in good condition and the tube is properly seated. Periodically check the area around the speedometer gear or sensor for leakage. If transmission fluid is evident, check the O-ring for damage.

Case leaks

17   If the case itself appears to be leaking, the casting is porous and will have to be repaired or replaced.
18   Make sure the oil cooler hose fittings are tight and in good condition.

Fluid comes out vent pipe or fill tube

19   If this condition occurs, the transmission is overfilled, there is coolant in the fluid, the case is porous, the dipstick is incorrect, the vent is plugged or the drain back holes are plugged.

3 Shift linkage - check and adjustment

1   With the engine running and your foot placed firmly on the brake, place the shift lever in each detent and make sure that the transmission shifts into the corresponding gear as the indicator is moved.
2   If adjustment is required, proceed as follows.

Three-speed transmission

Refer to illustration 3.5

3   Raise the vehicle and support it securely on jackstands.
4   Loosen the shift control rod adjusting nuts.
5   Move the shift lever on the transmission all the way in the clockwise direction (when viewed from the left side), back it off to the third stop, then set it in the Neutral position **(see illustration)**.
6   Hold the shift lever in the Neutral position while an assistant places the shift lever in the passenger compartment in Neutral. To remove any play, make sure the shift lever on the transmission and the shift lever in the passenger compartment are both held toward the rear of the Neutral position when tightening the adjusting nuts.

Four-speed transmission

Refer to illustration 3.7

7   Working from under the vehicle, loosen the nut on the shift linkage, push the shift

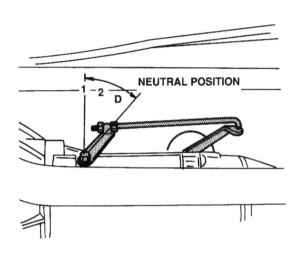

3.5 Three-speed transmission shift lever details

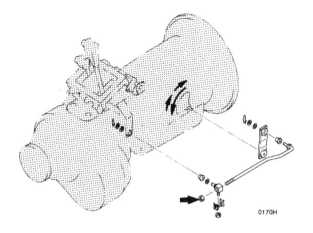

3.7 Loosen the adjusting nut (arrow), then move the shift lever (four-speed transmission)

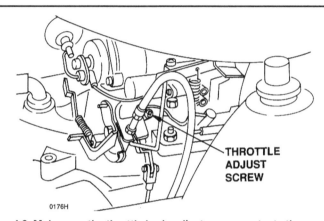

4.2 Make sure the throttle body adjust screw contacts the stopper before adjusting the TV cable

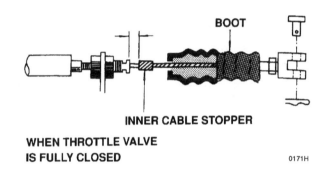

4.5 TV cable adjustment details (carburetor-equipped vehicles)

lever on the transmission all the way to the rear, then return it two notches to the Neutral position **(see illustration)**.

8   Push the lever lightly toward the Reverse side and tighten the shift linkage nut.

All models

9   Check the shift lever operation and position indicator alignment.

4   Throttle valve (TV) cable - adjustment

Carbureted models

Refer to illustrations 4.2, 4.5 and 4.6

1   Loosen the TV cable adjusting nuts.

2   Make sure the throttle adjust screw contacts the stopper **(see illustration)**. If the screw is not in contact with the stopper, the fast idle mechanism is activated and must be de-activated as follows.

3   Disconnect the negative battery cable and remove the air cleaner cover.

4   Have an assistant depress the accelera-

tor pedal, open the choke plate completely, then have the assistant release the pedal. Don't depress the pedal again or the fast idle will engage. Connect the battery cable and install the air cleaner cover.

5   With the fast idle mechanism de-activated, the TV cable may now be adjusted. Pull back the boot on the outer TV cable, then use the adjusting nuts to adjust the inner and outer cable to the distance listed in this Chapter's Specifications **(see illustration)**. Tighten the adjusting nuts.

6   After adjustment, measure the stroke between the open and closed position of the cable and compare this measurement to the distance listed in this Chapter's Specifications **(see illustration)**. Adjust the stroke as necessary, using the adjusting nuts.

7   Move the boot back into position.

Fuel injected models

Refer to illustration 4.9

8   Depress the accelerator pedal fully and make sure the throttle valve opens all the way, adjusting the accelerator linkage as necessary (see Chapter 4).

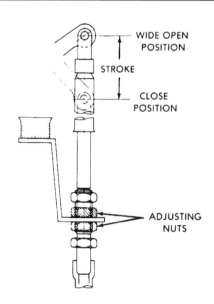

4.6 Measure the TV cable stroke (the distance between the closed and wide open positions)

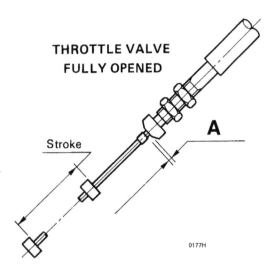

4.9 Fuel injected vehicle TV cable adjustment details - dimension A is the boot-to-stopper distance

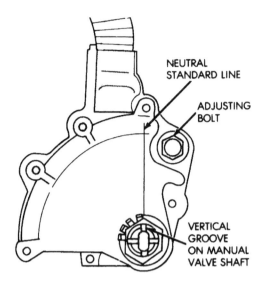

5.7 With the transmission in Neutral, the neutral standard line on the switch must be aligned with the vertical groove in the manual valve shaft

9   Hold the pedal down, loosen the cable adjusting nuts and adjust the cable boot-to-stopper clearance to the figure listed in this Chapter's Specifications **(see illustration)**.
10   After adjusting the distance, check the cable stroke (distance the cable travels between the open and closed position), adjusting as necessary, using the adjusting nuts **(see illustration 4.9)**. Tighten the adjusting nuts.

5   Neutral start switch - check, adjustment and replacement

1   The neutral start switch, located in the shift mechanism on 1984 through 1987 models or on the right-hand side of the transmission on 1988 and later models, prevents the engine from starting with the transmission in gear.

Check

2   Make sure the engine will start only with the selector lever in Park and Neutral. With the key in the On position, make sure the backup lights function when the lever is in Reverse only.
3   If a malfunction is noted, first adjust the switch.

Adjustment

1984 through 1987

4   Remove the shift console cover.
5   Loosen the two screws holding the neutral start switch and adjust the position of the switch until the engine will start only in Neutral and Park.

1988 on

Refer to illustration 5.7

6   Raise the vehicle and support it on jack-

stands. Set the parking brake securely and block the wheels. Place the transmission in Neutral.
7   Loosen the adjusting bolt and rotate the switch to line up the Neutral standard line on the switch with the vertical groove on the manual valve shaft **(see illustration)**.
8   Tighten the adjusting bolt, then recheck the switch operation as described above.

Replacement

9   Disconnect the negative battery cable. On 1988 and later models, raise the vehicle and support it on jackstands. Set the parking brake securely and block the wheels.
10   Shift the transmission into Neutral.

1984 through 1987

11   Unplug the electrical connector, remove the retaining bolts and lift the switch off.
12   Installation is the reverse of removal.

1988 on

13   Unplug the electrical connector, remove the shift shaft nut and the adjusting bolt and lift the switch off.
14   Installation is the reverse of removal. Do not tighten the adjusting bolt fully until the switch has been adjusted (see above).

All models

15   Connect the battery negative cable.
16   Start the engine in both Park and Neutral to verify that the switch is properly adjusted.

6   Automatic transmission - removal and installation

Refer to illustrations 6.7 and 6.23
Note: *On 4WD models, the transmission and*

transfer case are removed and installed as a unit.

Removal

1   Disconnect the negative cable from the battery.
2   Raise the vehicle and support it securely on jackstands.
3   Remove the lower splash shield (if equipped) and drain the transmission fluid and transfer case lubricant (4WD models) (Chapter 1).
4   Remove the starter motor (Chapter 5).
5   Remove the torque converter cover (if equipped).
6   Mark the relationship of the torque converter and driveplate with white paint so they can be installed in the same position.
7   Remove the torque converter-to-drive-plate bolts (and nuts, if equipped). Turn the crankshaft for access to each bolt. On 1984 through 1987 models, work through the torque converter opening and remove the dust cover in the housing for access to the nuts behind the bolts **(see illustration)**. On 1988 and later models, work through the starter motor opening. Turn the crankshaft in a clockwise direction only (as viewed from the front).
8   Remove the driveshaft(s) (Chapter 8).
9   Disconnect the speedometer cable.
10   Detach the wire harness connectors from the transmission and transfer case (4WD models).
11   Remove any exhaust components which will interfere with transmission/transfer case removal (Chapter 4).
12   Disconnect the TV linkage cable.
13   Disconnect the shift linkage from the transmission and transfer case (4WD models).
14   Support the engine with a jack. Use a block of wood under the oil pan to spread the load.

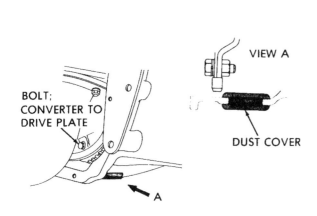

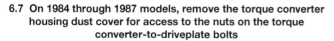

6.7 On 1984 through 1987 models, remove the torque converter housing dust cover for access to the nuts on the torque converter-to-driveplate bolts

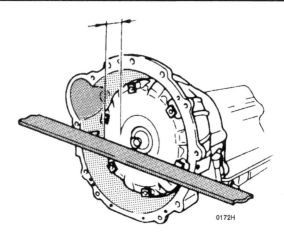

6.23 The distance from the face of the torque converter to the face of the transmission housing must be as listed in this Chapter's Specifications (2WD models only)

15 Support the transmission/transfer case with a jack - preferably a jack made for this purpose. Safety chains will help steady the transmission/transfer case on the jack.

16 Remove the rear mount to crossmember bolts and the crossmember-to-frame bolts.

17 Remove the engine rear support-to-extension housing bolts.

18 Raise the transmission/transfer case enough to allow removal of the cross- member.

19 Remove the bolts securing the transmission to the engine.

20 Lower the transmission/transfer case slightly and disconnect and plug the transmission fluid cooler lines.

21 Remove the transmission dipstick tube.

22 Move the transmission/transfer case to the rear to disengage it from the engine block dowel pins and make sure the torque converter is detached from the driveplate. Secure the torque converter to the transmission so it won't fall out during removal.

Installation

23 Prior to installation, make sure the torque converter hub is securely engaged in the pump. On 2WD models make sure the face of the torque converter is the specified distance from the transmission housing face **(see illustration)**.

24 With the transmission/transfer case secured to the jack, raise it into position. Be sure to keep it level so the torque converter does not slide forward. Connect the transmission fluid cooler lines.

25 Turn the torque converter to line up the holes with the holes in the driveplate. The white paint mark on the torque converter and the flywheel made in Step 6 must line up.

26 Move the transmission/transfer case forward carefully until the dowel pins and torque converter are engaged.

27 Install the transmission housing-to-engine bolts. Tighten the bolts securely.

28 Install the torque converter-to-driveplate bolts. On 1984 through 1987 models, use new bolts with thread-locking compound on

the threads. Tighten the bolts to the specified torque.

29 Install the transmission/transfer case mount crossmember and bolts. Tighten the bolts and nuts securely.

30 Remove the jacks supporting the transmission/transfer case and the engine.

31 Install the dipstick tube.

32 Install the starter motor (Chapter 5).

33 Connect the shift and TV linkages.

34 Plug in the transmission/transfer case wire harness connectors.

35 Install the torque converter cover.

36 Install the driveshaft(s).

37 Connect the speedometer cable.

38 Adjust the transmission and transfer case (if required) shift linkage (see Section 3 and Part C of this Chapter).

39 Install any exhaust system components that were removed or disconnected.

40 Lower the vehicle.

41 Fill the transmission and transfer case (if equipped) with the specified fluid (Chapter 1), run the engine and check for fluid leaks.

Notes

Chapter 7 Part C
Transfer case

Contents

1 General information

Four-wheel drive models are equipped with a transfer case mounted on the transmission housing. Drive is passed through the transmission and transfer case to the front and rear wheels by the driveshafts.

On these models the transfer case is combined with the transmission to form one unit. The transfer case cannot be removed separately; the transmission/transfer case must be disassembled as a unit. Because of the special tools and techniques required, disassembly and overhaul of the transfer case should be left to a dealer or properly equipped shop. You can, however, remove and install the transfer case/transmission yourself and save the expense, even if the repair work is done by a specialist.

2 Shift lever (manual transmission) - removal and installation

Refer to illustrations 2.2a, 2.2b, 2.3 and 2.4

1   Remove the carpet and sound deadener sheet.

2   Unscrew the knob and pull the shift boot up for access **(see illustrations)**.

2.2a Unscrew the shift knob . . .

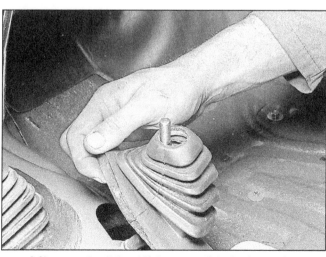

2.2b . . . and pull the shift boot out of the body opening

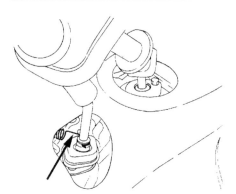

2.3 Unhook the return spring (arrow) from the shift lever before removing the bolts

3   On 1981 through 1987 models, detach the return spring from the shift lever **(see illustration)**.
4   Pull up the rubber cover to expose the shift lever mounting bolts **(see illustration)**.
5   Remove the shift lever mounting bolts.
6   Lift the shift lever up and out of the transfer case. On some models the lever is spring loaded, so ease it out of the case or the spring will be disengaged and will fall out.
7   Installation is the reverse of removal.

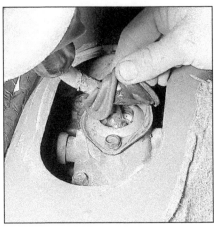

2.4 Pull the rubber cover up to expose the shift lever mounting bolts

3   Shift linkage and position switch (automatic transmission) - adjustment

Shift linkage

1   Place the shift lever in the 2H position.
2   Raise the vehicle and support it on jack-stands.

3   Disconnect the shift linkage from the boss on the control cable.
4   Place the transfer case position switch in the 2H position, then connect the linkage to the cable boss.

Position switch

5   Place the shift lever in the 2H position.
6   Loosen the retaining bolt and rotate the switch to line up the 4H line on the switch with the groove in the shift lever shaft stud.
7   Tighten the retaining bolt.

4   Transfer case overhaul - general information

On these models the transmission/transfer case assembly is designed to be overhauled as a unit. Consequently, overhaul should be left to a repair shop specializing in both transmissions and transfer cases. Rebuilt units may be available - check with your dealer parts department and auto parts stores. At any rate, some cost savings can be realized by removing the transmission/transfer case unit and taking it to the shop (see Part A or B of this Chapter).

Chapter 8
Clutch and driveline

Contents

Specifications

Clutch
Hydraulic system fluid type	See Chapter 1
Clutch disc minimum lining thickness	0.008 in (0.2 mm)
Freeplay	See Chapter 1

Front axle (4WD)
Hub bearing preload	3.31 lbs (measured with spring scale)

Torque specifications
Ft-lbs

Clutch
Pressure plate-to-flywheel bolts	12 to 14

Driveline
Front driveshaft U-joint flange bolts	19 to 22
Rear driveshaft U-joint flange bolts	22 to 25
Center bearing pinion locknut	87
Center bearing bracket bolts	43 to 46

Rear axle
Rear axle bearing retainer/brake backing plate-to-axle housing nuts	55
Rear axle housing-to-leaf spring U-bolt nuts	
1981 through 1987	36 to 43
1988 and on	49
Rear spring shackle nuts	72

Front axle (4WD)
Front axle-to-frame bolts	29 to 51

1   General information

The information in this Chapter deals with the components from the rear of the engine to the rear wheels and to the front wheels on 4WD models, except for the transmission and transfer case (4WD models), which are dealt with in the previous Chapter. For the purposes of this Chapter, these components are grouped into four categories: clutch, driveshaft, front axle and rear axle. Separate Sections within this Chapter offer general descriptions and checking procedures for each of these groups.

Since nearly all the procedures covered in this Chapter involve working under the vehicle, make sure it's securely supported on sturdy jackstands or on a hoist where the vehicle can be easily raised and lowered.

2   Clutch - description and check

1   All models equipped with a manual transmission feature a single dry plate, diaphragm spring-type clutch. The actuation is through a cable (early models) or hydraulic system (later models).

2   When the clutch pedal is depressed on early models, the clutch cable moves the release bearing into contact with the pressure plate release fingers, disengaging the clutch disc. On later models hydraulic fluid (under pressure from the clutch master cylinder) flows into the release cylinder. Because the release cylinder is connected to the clutch fork, the fork moves the release bearing into contact with the pressure plate release fingers, disengaging the clutch disc.

3   Terminology can be a problem regarding the clutch components because common names have in some cases changed from that used by the manufacturer. For example, the driven plate is also called the clutch plate or disc, the clutch release bearing is sometimes called a throwout bearing, the release cylinder is sometimes called the operating or slave cylinder.

4   Due to the slow wearing qualities of the clutch, it is not easy to decide when to go to the trouble of removing the transmission in order to check the wear on the friction lining. The only positive indication that something should be done is when it starts to slip or when squealing noises during engagement indicate that the friction lining has worn down to the rivets. In such instances it can only be hoped that the friction surfaces on the flywheel and pressure plate have not been badly worn or scored.

5   A clutch will wear according to the way in which it is used. Much intentional slipping of the clutch while driving - rather than the correct selection of gears - will accelerate wear. It is best to assume, however, that the disc will need replacement at about 40,000 miles (64,000 km).

6   Because of the clutch's location between the engine and transmission, it cannot be worked on without removing either the engine or transmission. If repairs which would require removal of the engine are not needed, the quickest way to gain access to the clutch is by removing the transmission as described in Chapter 7.

7   Other than to replace components with obvious damage, some preliminary checks should be performed to diagnose a clutch system failure.

a)   *The first check should be of the fluid level in the clutch master cylinder (models with a hydraulic release system). If the fluid level is low, add fluid as necessary and re-test. If the master cylinder runs dry, or if any of the hydraulic components are serviced, bleed the hydraulic system as described in Section 9.*

b)   *To check "clutch spin down time", run the engine at normal idle speed with the transmission in Neutral (clutch pedal up - engaged). Disengage the clutch (pedal down), wait nine seconds and shift the transmission into Reverse. No grinding noise should be heard. A grinding noise would indicate component failure in the pressure plate assembly or the clutch disc.*

c)   *To check for complete clutch release, run the engine (with the brake on to prevent movement) and hold the clutch pedal approximately 1/2-inch from the floor mat. Shift the transmission between 1st gear and Reverse several times. If the shift is not smooth, component failure is indicated. Measure the hydraulic release cylinder pushrod travel. With the clutch pedal completely depressed the release cylinder pushrod should extend substantially. If the pushrod will not extend very far or not at all, check the fluid level in the clutch master cylinder.*

d)   *Visually inspect the clutch pedal bushing at the top of the clutch pedal to make sure there is no sticking or excessive wear.*

e)   *Under the vehicle, check that the release lever is solidly mounted on the ball stud.*

Note: *Because access to the clutch components is an involved process, any time either the engine or transmission is removed, the clutch disc, pressure plate assembly and release bearing should be carefully inspected and, if necessary, replaced with new parts. Since the clutch disc is normally the item of highest wear, it should be replaced as a matter of course if there is any question about its condition.*

3   Clutch components - removal, inspection and installation

Refer to illustrations 3.4, 3.6, 3.10, 3.12a, 3.12b and 3.14

Warning: *Dust produced by clutch wear and deposited on clutch components contains asbestos, which is hazardous to your health. DO NOT blow it out with compressed air and DO NOT inhale it. DO NOT use gasoline or petroleum-based solvents to remove the dust. Brake system cleaner should be used to flush the dust into a drain pan. After the clutch components are wiped clean with a rag, dispose of the contaminated rags and cleaner in a covered container.*

Removal

1   Access to the clutch components is normally accomplished by removing the transmission, leaving the engine in the vehicle. If, of course, the engine is being removed for major overhaul, then the opportunity should always be taken to check the clutch for wear and replace worn components as necessary. The following procedures assume that the engine will stay in place.

2   Remove the release cylinder (if equipped) (see Section 8).

3   Referring to Chapter 7 Part A, remove the transmission from the vehicle. Support the engine while the transmission is out. Preferably, an engine hoist should be used to support it from above. However, if a jack is used underneath the engine, make sure a piece of wood is used between the jack and oil pan to spread the load. **Caution:** *The pickup for the oil pump is very close to the bottom of the oil pan. If the pan is bent or distorted in any way, engine oil starvation could occur.*

4   To support the clutch disc during removal, install a clutch alignment tool through the clutch disc hub **(see illustration)**.

5   Carefully inspect the flywheel and pressure plate for indexing marks. The marks are

3.4 Inexpensive clutch alignment tools are available at most auto parts stores

3.6 Remove the clutch pressure plate and disc (driven plate)

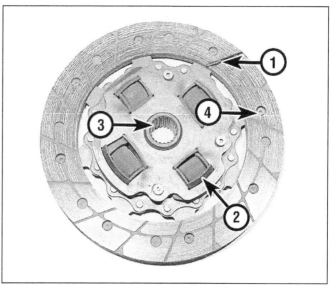

3.10 Typical clutch disc

1   **Lining** - this will wear down in use
2   **Springs or dampers** - check for cracking and deformation
3   **Splined hub** - the splines must not be worn and should slide smoothly on the transmission input shaft splines
4   **Rivets** - the rivets secure the lining and will damage the flywheel or pressure plate if allowed to contact the surfaces

NORMAL FINGER WEAR

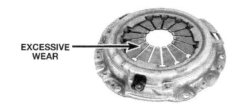

EXCESSIVE WEAR

EXCESSIVE FINGER WEAR

BROKEN OR BENT FINGERS

3.12a Replace the pressure plate if the diaphragm spring fingers exhibit these signs of wear

3.12b Examine the pressure plate friction surface for score marks, cracks and evidence of overheating

usually an X, an O or a white letter. If they cannot be found, apply marks yourself so the pressure plate and the flywheel will be in the same alignment during installation.

6   Turning each bolt only 1/2-turn at a time, slowly loosen the pressure plate-to-flywheel bolts. Work in a diagonal pattern and loosen each bolt a little at a time until all spring pressure is relieved. Then hold the pressure plate securely and completely remove the bolts, followed by the pressure plate and clutch disc **(see illustration)**.

Inspection

7   Ordinarily, when a problem occurs in the clutch, it can be attributed to wear of the clutch disc assembly. However, all components should be inspected at this time.

8   Inspect the flywheel for cracks, heat checking, grooves or other signs of obvious defects. If the imperfections are slight, a machine shop can machine the surface flat and smooth, which is highly recommended regardless of the surface appearance. Refer to Chapter 2 for the flywheel removal and installation procedure.

9   Inspect the pilot bearing (Section 6).

10   Inspect the lining on the clutch disc **(see illustration)**. There should be at least 1mm of lining above the rivet heads. Check for loose rivets, warpage, cracks, distorted springs or damper bushings and other obvious damage.

As mentioned above, ordinarily the clutch disc is replaced as a matter of course, so if in doubt about the condition, replace it with a new one.

11   Ordinarily, the release bearing is also replaced along with the clutch disc (see Section 4).

12   Check the machined surfaces of the pressure plate **(see illustrations)**. If the surface is grooved or otherwise damaged, replace the pressure plate. Also check for obvious damage, distortion, cracking, etc. Light glazing can be removed with medium grit emery cloth. If a new pressure plate is indicated, new or factory-rebuilt units are available.

3.14 The flywheel side of the disc is usually marked

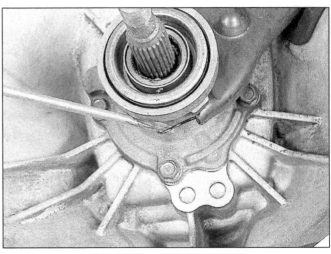

4.3 Use a screwdriver to detach the retaining spring, then slide the release bearing off the shaft

Installation

13   Before installation, carefully wipe the fly-wheel and pressure plate machined surfaces clean with a rubbing-alcohol dampened rag. It's important that no oil or grease is on these surfaces or the lining of the clutch disc. Handle these parts only with clean hands.

14   Position the clutch disc and pressure plate with the clutch held in place with an alignment tool. Make sure it's installed properly (most replacement clutch discs will be marked "flywheel side" or something similar - if not marked, install the clutch with the damper springs or bushings toward the transmission) **(see illustration)**.

15   Tighten the pressure plate-to-flywheel bolts only finger tight, working around the pressure plate.

16   Center the clutch disc by ensuring the alignment tool **(see illustration 3.4)** is through the splined hub and into the pilot bearing in the crankshaft. Wiggle the tool up, down or side-to-side, as needed, to bottom the tool in the pilot bearing. Tighten the pressure plate-to-flywheel bolts a little at a time,

working in a criss-cross pattern to prevent distorting the cover. After all of the bolts are snug, tighten them to the specified torque. Remove the alignment tool.

17   Using high temperature grease, lubricate the inner groove of the release bearing (refer to Section 4). Also place grease on the fork fingers.

18   Install the clutch release bearing as described in Section 4.

19   Install the transmission, release cylinder and all components removed previously, tightening all fasteners to the proper torque specifications.

---

4   Clutch release bearing - removal and installation

Refer to illustrations 4.3 and 4.5

Removal

1   Disconnect the negative cable from the battery.

2   Remove the transmission (Chapter 7).

3   Detach the retaining spring holding it to the release fork, then slide the release bearing off the transmission input shaft **(see illustration)**.

4   Remove the clutch release fork from the ball stud by pulling it straight off.

5   Hold the bearing and turn the inner portion. If the bearing doesn't turn smoothly or if it's noisy, replace it with a new one. Wipe the bearing with a clean rag and inspect it for damage, wear and cracks. Don't immerse the bearing in solvent - it's sealed for life and to do so would ruin it **(see illustration)**.

Installation

Refer to illustrations 4.8a and 4.8b

6   Lubricate the clutch fork ends where they contact the bearing lightly with molybdenum disulphide grease. Apply a thin coat of the same grease to the inner diameter of the bearing and also to the transmission input shaft bearing retainer.

7   Install the release bearing on the clutch fork so that both of the fork ends fit into the

4.5 Check the clutch release (throwout) bearing for excessive wear or damage

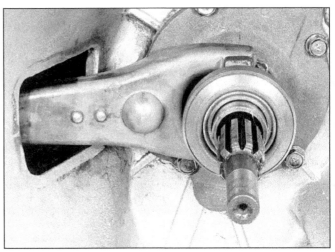

4.8a Install the release fork over the ball stud

4.8b The release fork, showing its spring clip in position over the ball stud

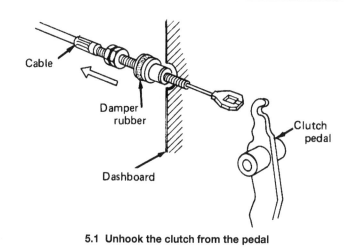

5.1 Unhook the clutch from the pedal

bearing tabs. Make sure the spring clip seats securely.

8   Lubricate the clutch release fork ball socket with molybdenum disulphide grease and push the fork onto the ball stud until it's firmly seated **(see illustrations)**. Check to see that the bearing slides back and forth smoothly on the input shaft bearing retainer.

9   The remainder of the installation is the reverse of the removal procedure, tightening all bolts to the specified torque.

5   Clutch cable - removal and installation

Refer to illustrations 5.1 and 5.3

1   Loosen the cable adjusting and locknuts to provide enough slack to unhook it from the clutch pedal **(see illustration)**.

2   Raise the vehicle and support it securely on jackstands.

3   Unhook the return spring from the shift fork **(see illustration)**.

4   Disconnect the cable from the shift fork, pull it forward through the bracket, detach any retaining clips and remove it from the vehicle.

5   Installation is the reverse of removal, taking care there are no kinks in the cable which will cause it to bind.

6   After installation, adjust the freeplay (Chapter 1).

6   Pilot bearing - inspection, removal and installation

1   The clutch pilot bearing is a ball-type bearing which is pressed into the rear of the crankshaft. Its primary purpose is to support the front of the transmission input shaft. The pilot bearing should be inspected whenever the clutch components are removed from the engine. Due to its inaccessibility, if you are in doubt as to its condition, replace it with a

new one. **Note:** *If the engine has been removed from the vehicle, disregard the following steps which do not apply.*

2   Remove the transmission (refer to Chapter 7 Part A).

3   Remove the clutch components (Section 3).

4   Using a clean rag, wipe the bearing clean and inspect for any excessive wear, scoring or obvious damage. A flashlight will be helpful to direct light into the recess.

5   Check to make sure the pilot bearing turns smoothly and quietly. If the transmission input shaft contact surface is worn or damaged, replace the bearing with a new one.

6   Removal can be accomplished with a special puller but an alternative method also works very well.

7   Find a solid steel bar which is slightly smaller in diameter than the bearing. Alternatives to a solid bar would be a wood dowel or a socket with a bolt fixed in place to make it solid.

8   Check the bar for fit - it should just slip into the bearing with very little clearance.

9   Pack the bearing and the area behind it (in the crankshaft recess) with heavy grease.

Pack it tightly to eliminate as much air as possible.

10   Insert the bar into the bearing bore and lightly hammer on the bar, which will force the grease to the backside of the bearing and push it out. Remove the bearing and clean all grease from the crankshaft recess.

11   To install the new bearing, lubricate the outside surface with oil then drive it into the recess with a hammer and a socket with an outside diameter that matches the bearing outer race.

12   Pack the bearing with lithium base grease. Wipe off all excess grease so the clutch lining will not become contaminated.

13   Install the clutch components, transmission and all other components removed to gain access to the pilot bearing.

7   Clutch master cylinder - removal, overhaul and installation

Refer to illustrations 7.3, 7.4 and 7.5

Caution: *Do not allow brake fluid to contact any painted surfaces of the vehicle, as damage to the finish may result.*

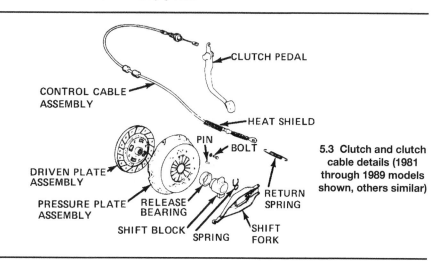

5.3 Clutch and clutch cable details (1981 through 1989 models shown, others similar)

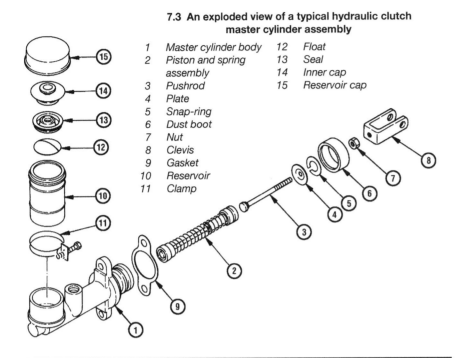

7.3 An exploded view of a typical hydraulic clutch master cylinder assembly

1	Master cylinder body	12	Float
2	Piston and spring assembly	13	Seal
		14	Inner cap
3	Pushrod	15	Reservoir cap
4	Plate		
5	Snap-ring		
6	Dust boot		
7	Nut		
8	Clevis		
9	Gasket		
10	Reservoir		
11	Clamp		

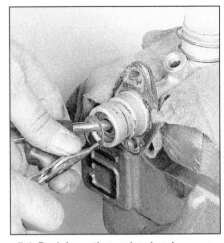

7.4 Push in on the pushrod and remove the snap-ring with a pair of snap-ring pliers

11   Bleed the clutch hydraulic system following the procedure in Section 9, then check the pedal height and freeplay as described in Chapter 1.

Removal

1   Disconnect the hydraulic line from the master cylinder and drain the fluid into a suitable container.
2   Remove the master cylinder flange mounting nuts and withdraw the unit from the engine compartment.

Overhaul

3   Remove the retaining clamp and pull off the reservoir **(see illustration)**.
4   Push the piston down and remove the snap-ring with a pair of snap-ring pliers or a small screwdriver **(see illustration)**.
5   Pull out the piston assembly and spring **(see illustration)**.
6   Examine the inner surface of the cylinder bore. If it is scored or exhibits bright wear areas, the entire master cylinder should be replaced.
7   If the cylinder bore is in good condition, obtain a clutch master cylinder rebuild kit, which will contain all of the necessary replacement parts.
8   Prior to installing any parts, first dip them in brake fluid to lubricate them.
9   Installation of the parts in the cylinder is the reverse of removal.

Installation

10   Position the clutch master cylinder against the firewall, inserting the pedal pushrod into the piston. Install the nuts, tightening them securely.

8   Clutch release cylinder - removal, overhaul and installation

Refer to illustrations 8.4 and 8.5

Removal

1   The clutch release cylinder is located on the side of the transmission bellhousing.
2   Raise the vehicle and support it securely on jackstands.
3   Disconnect the hydraulic line from the release cylinder. On some models, this is done by removing the bolt from the banjo fitting on the cylinder body. On other models, unscrew the fitting from the cylinder body, using a flare-nut wrench, if available.
4   Remove the two bolts and pull off the release cylinder **(see illustration)**.

7.5 Remove the piston and spring from the bore of the master cylinder; note that the seal lip faces forward, toward the front of the master cylinder

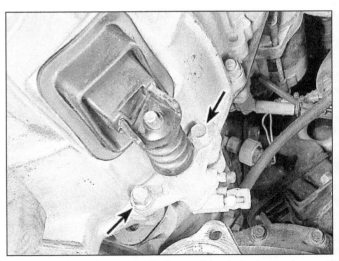

8.4 Clutch release cylinder retaining bolts (arrows)

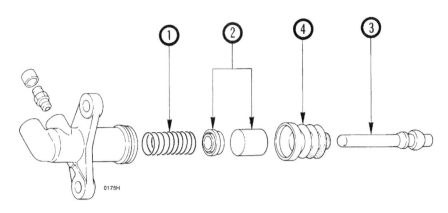

8.5 Exploded view of the clutch release cylinder

1	*Spring*	3	*Pushrod*
2	*Piston and cup assembly*	4	*Bolt*

Overhaul

5   Pull off the dust boot and pushrod and then tap the cylinder gently on a block of wood to extract the piston and spring **(see illustration)**.

6   Unscrew and remove the bleeder screw.

7   Examine the surfaces of the piston and cylinder bore for scoring or bright wear areas. If any are found, discard the cylinder and purchase a new one.

8   If the components are in good condition, wash them in clean brake fluid. Remove the seal and discard it, noting carefully which way the seal lips face.

9   Obtain a repair kit which will contain all the necessary new items.

10   Install the new seal using your fingers only to manipulate it into position. Be sure the lips face in the proper direction.

11   Dip the piston assembly in clean brake fluid before installing it and the spring into the cylinder.

12   Reinstall the bleeder.

13   Complete the reassembly by installing the pushrod and the dust cover. Be sure the dust cover is secure on the cylinder housing.

Installation

14   Installation is the reverse of the removal procedure. After the cylinder has been installed, bleed the clutch hydraulic system as described in Section 9.

9   Clutch hydraulic system - bleeding

Caution: *Do not allow the brake fluid to contact any painted surface of the vehicle, as damage to the finish will result.*

1   Bleeding will be required whenever the hydraulic system has been dismantled and reassembled and air has entered the system.

2   First fill the fluid reservoir with clean brake fluid which has been stored in an airtight container. Never use fluid which has

drained from the system or has bled out previously, as it may contain grit and moisture.

3   Attach a rubber or plastic bleed tube to the bleeder screw on the release cylinder and immerse the open end of the tube in a glass jar containing an inch or two of fluid.

4   Open the bleeder screw about half a turn and have an assistant quickly depress the clutch pedal completely. Tighten the screw and then have the clutch pedal slowly released with the foot completely removed. Repeat this sequence of operations until air bubbles are no longer ejected from the open end of the tube beneath the fluid in the jar.

5   After two or three strokes of the pedal, make sure the fluid level in the reservoir has not fallen too low. Keep it full of fresh fluid, otherwise air will be drawn into the system.

6   Tighten the bleeder screw on a pedal down stroke (do not overtighten it), remove the bleed tube and jar, top-up the reservoir and install the cap.

7   If an assistant is not available, alternative 'one-man' bleeding operations can be carried out using a bleed tube equipped with a one-way valve or a pressure bleed kit, both of which should be used in accordance with the manufacturer's instructions.

10   Driveshafts, differentials and axles - general information

Three different driveshaft assemblies are used on the vehicles covered in this manual. Some use a one-piece driveshaft which incorporates two universal joints, one at either end of the shaft.

Others use a two-piece driveshaft which incorporates a center bearing at the rear of the front shaft. This driveshaft uses three universal joints; one at the transmission end, one behind the center bearing and one at the differential flange.

The 4WD vehicles use two driveshafts; the primary shaft runs between the transfer case and the front differential and the rear

driveshaft runs between the transfer case and the rear differential.

All universal joints are of the solid type and can be replaced separately from the driveshaft.

The driveshafts are finely balanced during production and whenever they are removed or disassembled, they must be reassembled and reinstalled in the exact manner and positions they were originally in, to avoid excessive vibration.

The rear axle is of the semi-floating type, having a 'banjo' design axle housing, which is held in proper alignment with the body by the rear suspension.

Mounted in the center of the rear axle is the differential, which transfers the turning force of the driveshaft to the rear axleshafts, on which the rear wheels are mounted.

The axleshafts are splined at their inner ends to fit into the splines in the differential gears; outer support for the shaft is provided by the rear wheel bearing.

The front axle on 4WD vehicles consists of a frame-mounted differential assembly and two driveaxles. The driveaxles incorporate two constant velocity (CV) joints each, enabling them to transmit power at various suspension angles independent from each other.

Because of the complexity and critical nature of the differential adjustments, as well as the special equipment needed to perform the operations, we recommend any disassembly of the differential be done by a dealer service department or other repair shop.

11   Driveline inspection

1   Raise the rear of the vehicle and support it securely on jackstands.

2   Slide under the vehicle and visually inspect the condition of the driveshaft. Look for any dents or cracks in the tubing. If any are found, the driveshaft must be replaced.

3   Check for any oil leakage at the front and rear of the driveshaft. Leakage where the driveshaft enters the transmission indicates a defective rear transmission seal. Leakage where the driveshaft enters the differential indicates a defective pinion seal.

4   While still under the vehicle, have an assistant turn the rear wheel so the driveshaft will rotate. As it does, make sure that the universal joints are operating properly without binding, noise or looseness. On long bed models, listen for any noise from the center bearing, indicating it is worn or damaged. Also check the rubber portion of the center bearing for cracking or separation, which will necessitate replacement.

5   The universal joint can also be checked with the driveshaft motionless, by gripping your hands on either side of the joint and attempting to twist the joint. Any movement at all in the joint is a sign of considerable wear. Lifting up on the shaft will also indicate movement in the universal joints.

12.1 Mark the direction the driveshaft faces

12.3 Lock the driveshaft so it can't turn, and remove the bolts

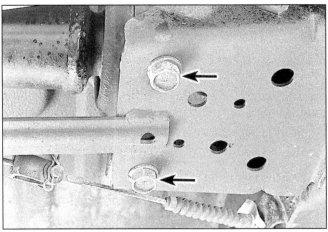

12.8 Center bearing bracket retaining bolts (arrows)

12.9 Mark the rear driveshaft-to -flange relationship

6   Finally, check the driveshaft mounting bolts at the ends to make sure they are tight.

7   On 4WD models, the above driveshaft checks should be repeated on all driveshafts. In addition, check for grease leakage around the sleeve yoke, indicating failure of the yoke seal.

8   Check for leakage at each connection of the driveshafts to the transfer case and front differential. Leakage indicates worn oil seals.

9   At the same time, check for looseness in the joints of the front driveaxles.

12   Driveshafts - removal and installation

Front driveshaft (4WD)

Refer to illustrations 12.1 and 12.3

1   Raise the front of the vehicle and place it on jackstands. Mark the relationship of the front driveshaft flange to the front differential companion flange so they can be realigned upon installation. Also mark the direction the driveshaft faces **(see illustration)**.

2   Mark the relationship of the rear drive-shaft flange to the transfer case companion flange.

3   Lock the driveshaft from turning with a large screwdriver or prybar, then remove the four nuts and bolts from the front flange **(see illustration)**.

4   Again lock the driveshaft from turning and remove the four nuts and bolts from the rear flange. On later models, it may be necessary to support the transmission with a jack and remove the crossmember for access to the bolts and nuts.

5   Slide the front portion of the driveshaft toward the rear, so it telescopes along the splined portion of the shaft. Lower the drive-shaft from the vehicle.

6   Installation is the reverse of removal. Be sure to align all marks and tighten the flange bolts to the specified torque.

Rear driveshaft

Refer to illustrations 12.8, 12.9, 12.10 and 12.13

7   Raise the rear of the vehicle and support it on jackstands.

8   Remove the bolts holding the center support bearing bracket to the frame (3-joint

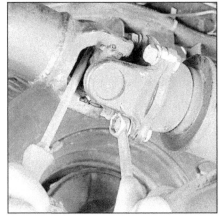

12.10 Use a large screwdriver to keep the driveshaft from turning while unscrewing the bolts

type) **(see illustration)**.

9   Mark the edges of the driveshaft rear flange and the differential companion flange so they can be realigned upon installation **(see illustration)**.

10   Remove the four nuts and bolts **(see illustration)**.

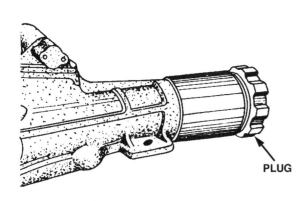

12.13 Insert a plug into the transmission or transfer case to keep oil from running out (a plastic bag and rubber bands can also be used)

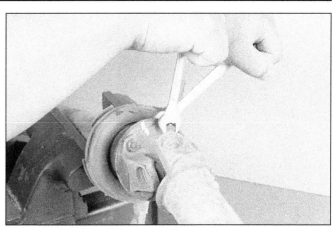

13.1 Mark the relationship of the center U-joint to the rear driveshaft flange, then remove the nuts and bolts that attach the U-joint to the flange

13.3a Mark the relationship of the flange to the front driveshaft, then unstake the nut and remove it

13.3b Remove the flange with a puller; if you don't have a suitable puller, have it pressed off at an automotive machine shop

11 Push the shaft forward slightly to disconnect the rear flange.

12 Pull the yoke from the transmission transfer case while supporting the driveshaft (2WD models) or driveshafts and center bearing assembly as a unit (4WD models) with your hands.

13 While the driveshafts are removed, insert a plug in the transmission/transfer case to prevent lubricant leakage **(see illustration)**.

14 Installation is the reverse of the removal procedure. During installation, make sure all flange marks line up. When connecting the center bearing support to the frame, first finger-tighten the two mounting bolts, then make sure that the bearing bracket is at right angles to the driveshaft. Tighten all nuts and bolts to the specified torque.

13 Center bearing - replacement

Refer to illustrations 13.1, 13.3a and 13.3b

1 Mark the relative flange positions, remove the bolts and nuts **(see illustration)** and detach the rear driveshaft from the cen-

ter bearing assembly.

2 Support the front driveshaft in a vise with padded jaws, with the center bearing assembly facing out.

3 Remove the pinion locknut **(see illustration)**, then remove the pinion flange from the driveshaft with a puller **(see illustration)**.

4 Remove the two bolts retaining the rubber cushion retainer and remove the retainer.

5 Remove the support ring from the rubber cushion, then remove the cushion from the bearing.

6 Remove the bearing assembly from the shaft.

7 To begin installation, install the rubber support cushion and the support ring on the shaft and push them onto the shaft so they will clear the bearing face.

8 Install the cushion support ring, aligning the pin in the ring with the hole in the cushion.

9 After packing the new bearing with grease, install it on the shaft.

10 The spacer ring and pinion flange can now be installed by driving them on the shaft with a hammer and suitable diameter socket or piece of pipe.

11 Install the washer and a new locknut.

Tighten the locknut to the specified torque and then stake the outer face of the nut so that it engages the slot in the shaft.

12 Install the rubber cushion, making sure to align the stop groove in the inner face of the cushion with the bearing projection.

13 The remainder of installation is the reverse of removal.

14 Universal joints - replacement

Refer to illustrations 14.1, 14.2a, 14.2b, 14.2c, 14.3 and 14.5

Note: *Selective fit snap-rings are used to retain the universal joint spiders in the yokes. In order to maintain the driveshaft balance, you must use replacement snap-rings of the same size as originally used.*

1 Clean away all dirt from the ends of the bearings on the yokes so the snap-rings can be removed with a pair of long-nose pliers or a screwdriver **(see illustration)**. If they are very tight, tap the end of the bearing with a hammer to relieve the pressure.

2 Once the snap-rings are removed, tap the universal joints at the yoke with a soft-

14.1 Use needle-nose pliers to remove the U-joint snap-rings

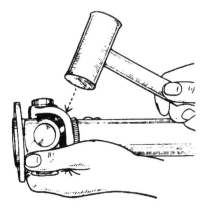

14.2a Tapping the driveshaft yoke with a hammer will cause the bearings to be dislodged . . .

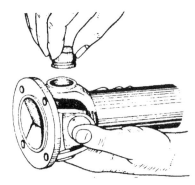

14.2b . . . so they can be grasped and lifted out

face hammer; the bearings will come out of the housing and can be removed easily **(see illustrations)**. If the bearings *don't* come out, proceed as follows. Place a piece of pipe or a large socket with the same inside diameter over one end of the bearing caps. Position a socket of slightly smaller diameter than the

14.2c To press the U-joint out of the driveshaft yoke, set it up in a vise with a small socket pushing the joint and bearing cap into a large socket

cap on the opposite bearing cap **(see illustration)** and use the vise or press to force out the cap (into the pipe or socket), stopping just before it comes completely out of the yoke. Use the vise or large pliers to work the cap the rest of the way out.

3   Once the bearings are removed from each opposite journal yoke the journal can be disengaged **(see illustration)**.

4   In cases of extreme wear or neglect, it is possible that the bearing housings in the driveshaft, sleeve or flange will be worn so much that the bearings are a loose fit in them. In such a case, it will also be necessary to replace the worn component as well.

5   Lubricate the inside of the bearing cap with moly-base grease and install the bearing needle rollers and secure them with the seal **(see illustration)**.

6   Installation is the reverse of the removal procedure. Lubricate the inner surface of the bearings prior to installing them. If the U-joints are equipped with grease fittings, fill each joint with grease after assembly is completed. Also be sure to install the snap-rings in their proper positions. When assembled, the U-joint should have no play in it and should be reasonably stiff, but should still be able to be flexed easily by hand.

15   Rear axle assembly - removal and installation

1   Loosen the rear wheel lug nuts, raise the vehicle and support it securely on jackstands placed underneath the frame. Remove the wheels.

2   Support the rear axle assembly with a floor jack placed underneath the differential.

3   Remove the shock absorber lower mounting nuts and compress the shocks to get them out of the way (Chapter 10).

4   Disconnect the driveshaft from the differential companion flange and hang it with a piece of wire from the underbody (Section 12).

5   Disconnect the parking brake cables from the parking brake levers.

6   Disconnect the rear flexible brake hose from the brake line above the rear axle housing. Plug the ends of the line and hose or wrap plastic bags tightly around them to prevent excessive fluid loss and contamination.

7   Support the rear axle assembly with a jack.

8   Remove the U-bolt nuts from the leaf spring seats and raise the axle assembly slightly with the jack.

9   Remove the rear spring shackle bolts and lower the rear of each leaf spring to the floor. Carefully lower the axle assembly to the floor with the jack, then remove it from under the vehicle. It would be a good idea to have

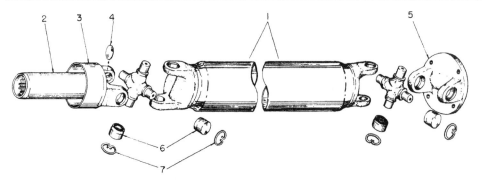

14.3 Typical driveshaft – exploded view

1	Driveshaft	3	Cover	5	Flange	7	Snap-ring
2	Spline yoke	4	Plate plug	6	Bearing caps		

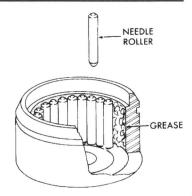

14.5 Pack the bearing cup with grease before installing the needle rollers

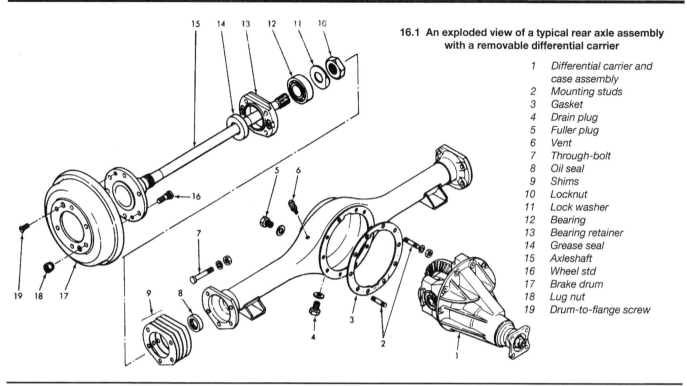

16.1 An exploded view of a typical rear axle assembly with a removable differential carrier

1   Differential carrier and
    case assembly
2   Mounting studs
3   Gasket
4   Drain plug
5   Fuller plug
6   Vent
7   Through-bolt
8   Oil seal
9   Shims
10 Locknut
11 Lock washer
12 Bearing
13 Bearing retainer
14 Grease seal
15 Axleshaft
16 Wheel std
17 Brake drum
18 Lug nut
19 Drum-to-flange screw

an assistant on hand, as the assembly is very heavy.

10   Installation is the reverse of the removal procedure. Be sure to tighten the U-bolt nuts and the driveshaft companion flange bolts to the torques listed in this Chapter's Specifications. Tighten the spring shackle bolts to the torque listed in Chapter 10. Bleed the brakes (see Chapter 9).

16   Rear axleshaft, bearing and seal - replacement

Models with a removable differential carrier

Refer to illustrations 16.1, 16.7, 16.8 and 16.9
1   The axleshafts can be removed without

disturbing the differential assembly. They must be removed in order to replace the oil seals and when removing the differential carrier from the rear axle housing **(see illustration)**. **Note:** *Read this entire procedure before starting work.*

2   Raise the rear of the vehicle and support it securely on jackstands. Block the front wheels to keep the vehicle from rolling.
3   Remove the rear wheels and release the parking brake.
4   Remove the drain plug and drain the differential oil into a suitable container. When the draining is complete, finger-tighten the drain plug in place.
5   On drum brake-equipped models, remove the drum and disconnect the parking brake cable from the actuator lever. Then disconnect the brake line from the wheel cylin-

der (see Chapter 9 for details).
6   On disc brake-equipped models, remove the caliper and hang it out of the way on a piece of wire, then remove the brake disc (see Chapter 9).
7   Remove the four axle bearing retainer nuts **(see illustration)**.
8   The axleshaft, complete with the bearing assembly and adjusting shims (if equipped) can now be pulled out from the rear axle **(see illustration)**. If the old bearings are worn, take the axleshaft assembly to an automotive machine shop and have them pressed off and new bearings pressed on.
9   Use a large screwdriver or oil seal removal tool (available at auto parts stores) to pry the oil seal out of the axle housing **(see illustration)**.
10   Install the axleshaft assembly into the

16.7 Rear axleshaft bearing retaining nuts (arrow)

16.8 Withdraw the axleshaft from the housing, complete with the bearing assembly

16.9 Pry the seal out of the housing, using a screwdriver or seal remover tool (shown)

16.16 Remove the lock screw and pull the pinion shaft out of the differential far enough to clear the inner end of the axleshaft

16.17 Push the axleshaft in and remove the C-lock

16.24 Installing a rear axleshaft oil seal with a large socket

axle housing with the shims (if equipped) in place on the bearing retainer bolts. Install the nuts, tightening them to the specified torque.
11   The remaining installation procedures are the reverse of those of removal.
12   Following installation, tighten the drain plug and fill the differential with the proper grade of lubricant as specified in Chapter 1.
13   On drum brake-equipped models, bleed the brakes, following the procedure in Chapter 9.

Models with a non-removable differential carrier

Axleshaft - removal and installation

Refer to illustrations 16.16 and 16.17

14   Raise the rear of the vehicle, support it securely and remove the wheel and brake drum (see Chapter 9).
15   Unscrew and remove the pressed steel cover from the differential housing and allow the lubricant to drain into a container.
16   Remove the lock bolt from the differential pinion shaft **(see illustration)**. Remove the pinion shaft.
17   Push the outer (flanged) end of the axleshaft in and remove the C-lock from the inner end of the shaft **(see illustration)**.
18   Withdraw the axleshaft, taking care not to damage the oil seal in the end of the axle housing as the splined end of the axleshaft passes through it.
19   Installation is the reverse of removal. Tighten the lock bolt to the torque listed in this Chapter's Specifications. **Note:** *The manufacturer recommends using a new lock bolt.*
20   Always use a new cover gasket and tighten the cover bolts to the torque listed in this Chapter's Specifications.
21   Refill the differential with the correct quantity and grade of lubricant (see Chapter 1).

Axleshaft oil seal - replacement

Refer to illustration 16.24

22   Remove the axleshaft (see Steps 14 through 18).

16.27 Bearing puller and slide hammer being used to remove a rear axleshaft bearing

23   Pry out the old oil seal from the end of the axle housing, using a large screwdriver or the inner end of the axleshaft itself as a lever.
24   Apply high melting point grease to the oil seal recess and tap the seal into position **(see illustration)** so that the lips are facing in and the metal face is visible from the end of the axle housing. When correctly installed, the face of the oil seal should be flush with the end of the axle housing.
25   Install the axleshaft.

Axleshaft bearing - replacement

Refer to illustration 16.27

26   Remove the axleshaft (see Step 8) and the oil seal (see Step 17).
27   A bearing puller will be required or a tool which will engage behind the bearing will have to be fabricated **(see illustration)**.
28   Attach a slide hammer and pull the bearing from the axle housing.
29   Lubricate the new bearing with gear lubricant. Clean out the bearing recess and drive in the new bearing using a bearing driver. If you don't have a bearing driver, use a piece of pipe or a socket of the proper diameter, applied against the outer race of the bearing. Make sure that the bearing is tapped into the full depth of its recess and that the numbers on the bearing are visible from the outer end of the housing.

30   Install a new seal, then install the axleshaft.

---

17   Front axle assembly (4WD models) - removal and installation

---

Refer to illustration 17.12

1   Loosen the wheel lug nuts, raise the front of the vehicle and support it securely on jackstands. Remove the wheels.
2   Remove the bolts that secure the front driveshaft to the differential (see Section 12), then detach the driveshaft and hang it out of the way on a piece of wire.
3   Remove the brake calipers and hang them out of the way on pieces of wire (see Chapter 9).
4   Remove the skid plate (if equipped).
5   On early models, remove the strut bars (see Chapter 10).
6   Remove the stabilizer bar (see Chapter 10).
7   Disconnect the steering linkage from the spindles (see Chapter 10).
8   Remove the upper control arms (see Chapter 10).
9   Remove the link ends from the lower control arms.

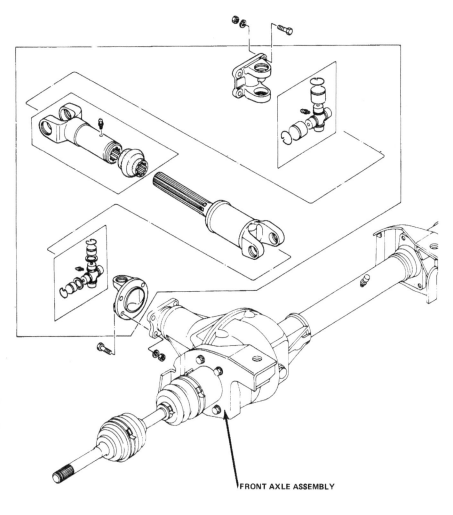

17.12 Front axle details

18.3 Mark the relationship of the hub cap with white paint

18.4a Lock the hub from turning, remove the bolts . . .

10   Remove the lower shock absorber bolts.
11   Remove the front hubs (Section 18).
12   Support the front axle assembly with a floor jack. Remove the four bolts attaching the front axle housing mounting brackets to the frame **(see illustration)**.
13   Carefully lower the assembly and remove it from the vehicle.
14   Installation is the reverse of the removal procedure.

18   Front hub assembly (4WD models) - removal, bearing repack, installation and adjustment

Removal

Refer to illustrations 18.3, 18.4a, 18.4b, 18.5a, 18.5b, 18.6a, 18.6b, 18.7, 18.8, 18.9a, 18.9b, 18.10, 18.11 and 18.12

1   Set the hub control handle to the 2H position and the free-wheeling hub to the Free position.

2   Remove the brake caliper and hang it out of the way on a piece of wire (see Chapter 9).
3   Mark the position of the hub cap so it

can be reinstalled in the same position **(see illustration)**.
4   Remove the six bolts and lift the cap off **(see illustrations)**.
5   Mark the position of the hub housing, then remove it **(see illustrations)**.

18.4b . . . then lift the cap off

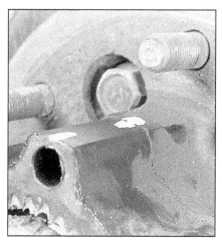

18.5a Mark the relationship of the housing . . .

18.5b . . . then remove it

18.6a A small screwdriver and needle nose pliers can be used to remove the snap-ring from the hub

6   On free-wheeling hubs, use snap-ring pliers to remove the snap-ring which retains the inner assembly **(see illustration)**. Use a screwdriver and needle-nose pliers to remove the snap-ring that retains the drive clutch assembly **(see illustration)**. Be sure to keep track of the number and locations of any shims so they can be reinstalled in the

18.6b Exploded view of a typical free-wheeling hub

1	Knuckle	27	Compression spring
2	Oil seal	28	Snap-ring
3	Bearing	29	Detent and ball spring
4	Washer		
5	Adapter	30	Knob
6	Shield	31	Ring
7	Retainer ring	32	Cover
8	Oil seal	33	Bolt
9	Inner wheel bearing		
10	Dust shield		
11	Bolt		
12	Wheel stud		
13	Bolt		
14	Hub and disc assembly		
15	Outer wheel bearing		
16	Locknut		
17	Lock washer		
18	Shim		
19	Snap-ring		
20	Spacer		
21	Ring		
22	Spacer		
23	Body		
24	Clutch		
25	Retaining spring		
26	Follower		

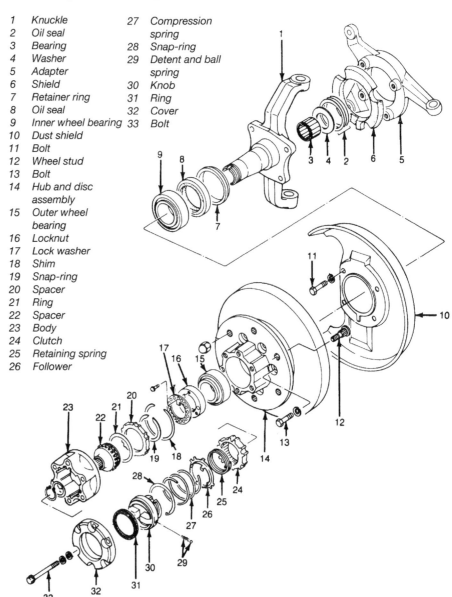

18.7 Lift the drive clutch assembly out of the hub

18.8 Lift the inner cam off

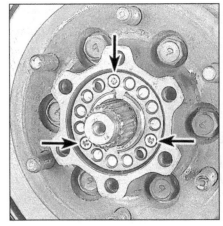

18.9a Remove the retaining screws (arrows) . . .

18.9b ... then lift the lock washer off

18.10 Use a punch or Phillips screwdriver to unscrew the locknut

same positions.

7   Remove the drive clutch assembly **(see illustration)**.

8   Remove the inner cam **(see illustration)**.

9   Remove the screws and lift the lock washer off, using needle-nose pliers **(see illustrations)**.

10   Unscrew the hub nut, using a punch **(see illustration)**.

11   Pull the hub out slightly, then push it back into its original position. This should force the outer wheel bearing off the spindle enough so it can be removed **(see illustration)**.

12   Pull the hub off the spindle **(see illustration)**.

13   Use a screwdriver to pry the grease seal out of the rear of the hub. As this is done, note how the seal is installed.

14   Remove the inner wheel bearing from the hub.

Bearing repack

15   Use solvent to remove all traces of the old grease from the bearings, hub and spindle. A small brush may prove helpful; however make sure no bristles from the brush

embed themselves inside the bearing rollers. Allow the parts to air dry.

16   Carefully inspect the bearings for cracks, heat discoloration, worn rollers, etc. Check the bearing races inside the hub for wear and damage. If the bearing races are defective, the hubs should be taken to a machine shop with the facilities to remove the old races and press new ones in. Note that the bearings and races come as matched sets and old bearings should never be installed on new races.

17   Use high-temperature moly-base front wheel bearing grease to pack the bearings. Work the grease completely into the bearings, forcing it between the rollers, cone and cage from the back side.

18   Apply a thin coat of grease to the spindle at the outer bearing seat, inner bearing seat, shoulder and seal seat.

Installation and adjustment

Refer to illustration 18.22

19   Place the grease-packed inner bearing into the rear of the hub and put a little more grease outside of the bearing.

20   Place a new seal over the inner bearing

and tap the seal evenly into place with a hammer and block of wood until it's flush with the hub.

21   Carefully place the hub assembly onto the spindle and push the grease-packed outer bearing into position.

22   Install hub nut and tighten it only slightly **(see illustration)**.

23   Spin the hub in a forward direction to seat the bearings and remove any grease or burrs which could cause excessive bearing play later.

24   Attach a spring scale to one of the wheel studs at right angles to the hub and measure the amount of pull necessary to rotate the hub. This is the preload. Compare the preload reading to the Specifications at the front of this Chapter.

25   Tighten or loosen the hub nut to achieve the specified preload.

26   Install the lock washer and see if the bolt holes in the washer are aligned with the holes in the hub nut. If they are not, turn the lock washer over. The holes should now line up. If they don't, turn the hub nut just enough to align them. Check the preload to make sure it is still within specification.

27   Install the retaining screws.

18.11 Dislodge the outer bearing, then lift it out of the hub

18.12 Grasp the hub and disc assembly securely and lift it off the spindle

18.22 Use needle-nose pliers to tighten the hub nut

28 Clean the contact surfaces of the lock washer and the inner cam and then slide the inner cam into place. If the inner cam is difficult to install, tap it lightly with a plastic hammer to seat it in the hub.

29 Install the drive clutch assembly and secure it with the snap-ring.

30 The remainder of installation is the reverse of removal. When installing the snap-ring on the drive clutch assembly, thread a bolt into the hub and pull the hub out.

31 Install the caliper (Chapter 9).

32 Install the tire/wheel assembly on the hub and tighten the lug nuts.

33 Lower the vehicle and tighten the lug nuts to the torque specified in Chapter 1.

19 Front driveaxle (4WD) - removal and installation

1 Remove the front axle assembly (Section 17).

2 Remove the four bolts retaining the axle mounting bracket to the axle case.

3 Grasp the driveaxle securely and pull it out of the axle assembly.

4 Installation is the reverse of removal.

20 Front driveaxle boot replacement and constant velocity (CV) joint overhaul (4WD)

Note: *If the CV joints exhibit signs of wear indicating need for an overhaul (usually due to torn boots), explore all options before beginning the job. Complete rebuilt driveaxles are available on an exchange basis, which eliminates much time and work. Whichever route you choose to take, check on the cost and availability of parts before disassembling the vehicle.*

Disassembly and inspection

Refer to illustrations 20.1, 20.2, 20.4, 20.5, 20.6, 20.8, 20.11a and 20.11b

1 Remove the front axle assembly from the vehicle (Section 17) **(see illustration)**.

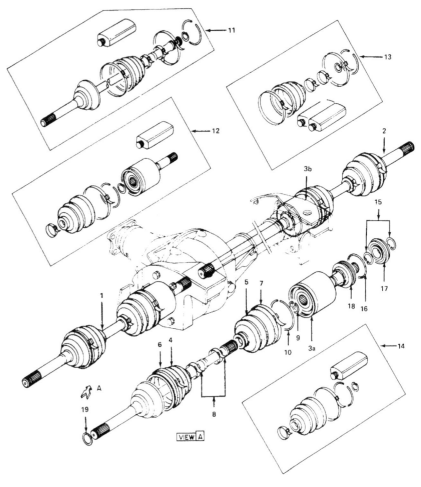

20.1 Exploded view of the driveaxle assemblies

1	Outer right hand CV joint assembly	10	Circlip
2	Outer left hand CV joint assembly	11	Outer CV joint components
3a	Inner right hand CV joint assembly	12	Inner CV joint components
3b	Inner left hand CV joint assembly	13	Boot kit
4	Boot	14	Boot kit
5	Boot	15	Snap-rings
6	Boot band	16	O-ring
7	Boot	17	Ball bearing
8	Boot bands	18	Oil seal
9	Snap-ring	19	Shim

20.2 Use a screwdriver to pry the boot band tab up

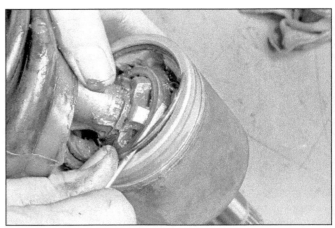

20.4 Pry the circlip from the outer race with a screwdriver

20.5 With the retainer removed, the outer race can be pulled off the bearing assembly

20.6 Pry the balls from the cage with a screwdriver (be careful not to nick or scratch anything)

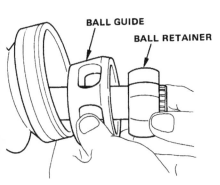

BALL GUIDE

BALL RETAINER

20.8 Remove the snap-ring with snap-ring pliers so the ball retainer can be removed

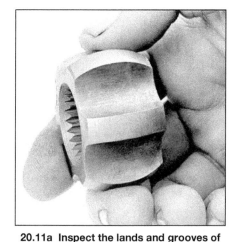

20.11a Inspect the lands and grooves of the ball retainer for pitting and score marks

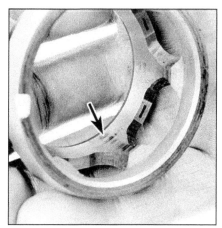

20.11b Inspect the ball guide for cracks, pitting and score marks (shiny spots are normal and don't affect operation)

2   Using a screwdriver or similar tool, loosen both boot bands from the rubber boot covering the double offset joint (inner) and remove them from the inner assembly **(see illustration)**. Since boot bands should not be reused, these should be discarded.

3   Slide the rubber boot off of the inner joint.

4   Using a screwdriver or similar tool, pry out the circlip from inside the inner assembly **(see illustration)**.

5   The driveaxle shaft can now be withdrawn from the driveaxle housing **(see illustration)**.

6   Remove the snap-ring that secures the retainer to the axle shaft **(see illustration)**.

7   Wipe off the grease with a rag and then use a screwdriver to remove the six balls from the joint assembly.

8   With the balls removed, rotate the outer part of the cage so that the ball guide holes line up with the projections in the inner part, the ball retainer. Then separate the two parts by sliding the ball guide along the axle shaft **(see illustration)**.

9   Remove the retainer, ball guide and boot from the shaft.

10   The outer joint assembly is permanently fixed to the driveaxle shaft and cannot be removed or disassembled. To inspect the

joint, remove the two boot bands from the rubber boot. Then slide the boot off the axle shaft. Thoroughly wash the outer joint assembly in clean solvent and blow it dry with compressed air. Bend the joint housing at an angle to the driveaxle to expose the bearings, ball retainer and ball guide.

11   Check all of the disassembled parts for signs of wear or damage and replace them as necessary **(see illustrations)**.

12   Inspect the outer case and the center shaft for any signs of contact. Also check the bottom face of the inner case for signs of contact. If any are found, the steering angles should be checked after reinstallation.

13   Also check the inside face of the inner ball guide for any signs of contact with the retainer. Check also for signs of contact between the spring clip and the balls. Contact of these parts indicate transverse misalignment of the front axle assembly.

Reassembly

Refer to illustrations 20.14, 20.16a, 20.16b, 20.16c and 20.26

14   Carefully slip the outer assembly rubber boot onto the axle shaft, and slide it into its

groove in the shaft. Wrapping the splines on the shaft with tape will prevent them from damaging the new boot as it's slid on **(see illustration)**.

15   Apply sufficient moly-base CV joint

20.14 Wrap the splined area of the axleshaft with tape to prevent damage to the new boot

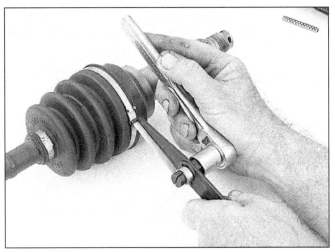

20.16a Secure the boot clamps with a special banding tool such as the one shown here (available at most auto parts stores): install the clamp, thread it onto the tool, pull the clamp tight . . .

20.16b . . . peen over the locking tabs . . .

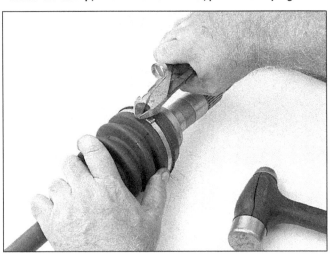

20.16c . . . and cut off the excess

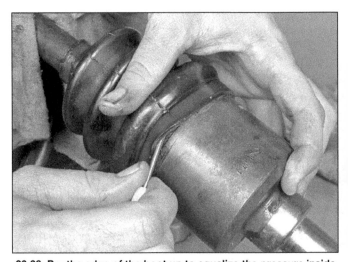

20.26 Pry the edge of the boot up to equalize the pressure inside so the boot won't be distorted when the band is tightened

grease (normally included with the rebuild kit) to the outer assembly so that the interior of the joint is adequately lubricated. Also fill the boot about half-way with grease.

16   Position the boot over the outer case and check that it is not collapsed or distorted. Then fit the large boot band into position on the boot and install it as shown **(see illustrations)**.

17   Next, install the small boot band in the same manner as the large one. Check again that the boot is not distorted.

18   To assemble the inner CV joint assembly, first slide the large boot band onto the axle shaft.

19   Again, tape the splines to protect the new boot. Apply a thin coat of grease to the shaft and then carefully slide the boot onto the shaft. Remove the tape.

20   Slide the ball guide onto the shaft, with the smaller diameter side toward the outer assembly.

21   Fit the ball retainer onto the shaft and secure it by installing the snap-ring.

22   Align the ball guide holes with the protrusions in the retainer and slide them together. Then turn the ball guide one-half pitch, so the track on the retainer is aligned with the cage holes.

23   Install the six balls into position and

pack the joint with moly-base CV joint grease.

24   Fit this assembly into the inner case and secure it by installing the circlip. The open ends of the circlip should be positioned at the inner circumference away from the ball groove.

25   Apply grease to the joint and boot, then slide the boot into position over the case.

26   Insert a screwdriver under the large end to allow the inside and outside air to equalize **(see illustration)**.

27   With the boot free of distortion, install both boot bands in the same manner the outer bands were installed.

Chapter 9 Brakes

Contents

Specifications

General

Brake fluid type	See Chapter 1
Brake booster flange face-to-pushrod end clearance	
1981 through 1983	0.729 to 0.736 in (18.5 to 18.7 mm)
1984-on	0.709 to 0.0717 in (18.0 to 18.2 mm)

Disc brakes

Minimum brake pad thickness	See Chapter 1
Disc standard thickness	
Front	
1981 through 1983	0.492 in (12.5 mm)
1984 through 1987	0.705 in (18 mm)
1988-on	0.866 in (22 mm)
Rear (all)	0.472 in (12 mm)
Disc minimum thickness*	
Front	
1981 through 1983	0.453 in (11.51 mm)
1984 through 1987	0.668 in (16.97 mm)
1988-on	0.826 in (20.97 mm)
Rear (all)	0.432 in (10.20 mm)
Lateral runout (all)	0.005 in (0.13 mm)

* Refer to marks stamped on the disc (they supercede information printed here)

Drum brakes

Minimum brake shoe lining thickness	See Chapter 1
Brake drums (all)	
Standard drum diameter	10.0 in (254 mm)
Maximum drum diameter*	10.39 in (255 mm)

* Refer to marks stamped on the drum (they supercede information printed here)

Torque specifications

	Ft-lbs
Brake caliper mounting bolt	22 to 25
Caliper support bracket bolts	
Front	
1981 through 1987	62 to 66
1988 on	103 to 126
Rear	69 to 84
Caliper inlet fitting bolt	24 to 27
Disc-to-hub bolts	36
Wheel cylinder mounting bolts	6 to 9
Brake backing plate-to-axle housing	50 to 55
Master cylinder mounting nuts	12 to 14
Power brake booster mounting nuts	18 to 21
Wheel lug nuts	
Steel wheels	60 to 85
Aluminum wheels	80 to 94

1   General information

The vehicles covered by this manual are equipped with hydraulically operated front and rear brake systems. The front brakes are disc-type. The rear brakes on all early models are drum-type while some later models use rear disc brakes.

These models are equipped with a dual master cylinder which allows the operation of half of the system if the other half fails. This system also incorporates a proportioning bypass valve which limits pressure to the rear brakes under heavy braking to prevent rear wheel lock-up. Some later Trooper models have a Load Sensing Proportioning Valve (LSPV) mounted to the frame and connected to the rear axle by a link.

All models are equipped with a power brake booster which utilizes engine vacuum to assist in application of the brakes.

The parking brake operates the rear brakes only, through cable actuation.

There are some notes and cautions involving the brake system on this vehicle:

a) *Use only DOT 3 brake fluid in this system.*

b) *The brake pads and linings contain asbestos fibers which are hazardous to your health if inhaled. Whenever you work on the brake system components, carefully clean all parts with brake cleaner. Do not allow the fine asbestos dust to become airborne.*

c) *Safety should be paramount whenever any servicing of the brake components is performed. Do not use parts or fasteners which are not in perfect condition, and be sure that all clearances and torque specifications are adhered to. If you are at all unsure about a certain procedure, seek professional advice. Upon completion of any brake system work, test the brakes carefully in a controlled area before putting the vehicle into normal service. If a problem is suspected in the brake system, do not drive the vehicle until the fault is corrected.*

d) *Tires, load and front end alignment are factors which also affect braking performance.*

2   Disc brake pads - replacement

Refer to illustration 2.3
Warning: *Disc brake pads must be replaced* on both front wheels at the same time - never replace the pads on only one wheel. Also, the dust created by the brake system may contain asbestos, which is harmful to your health. Never blow it out with compressed air and don't inhale any of it. An approved filtering mask should be worn when working on the brakes. Do not, under any circumstances, use petroleum based solvents to clean brake parts. Use brake cleaner or denatured alcohol only!

Note: *When servicing the disc brakes, use only high quality, nationally recognized name brand pads.*

1   Remove the master cylinder reservoir cap and siphon out approximately half of the fluid into a container. Be careful not to spill fluid onto any of the painted surfaces - it will damage the paint.

2   Loosen the wheel lug nuts, raise the vehicle and support it securely on jackstands. Remove the wheels. Work on one brake assembly at a time, using the assembled brake for reference if necessary.

3   Using a large C-clamp, bottom the piston back into the caliper bore. The frame end of the C-clamp should be positioned on the backside of the caliper body and the screw should bear on the outer brake pad **(see illustration)**.

2.3 Use a C-Clamp to bottom the piston in its bore

2.4a Unscrew the caliper bolt with a box end or socket so the bolt head will not be rounded off

2.4b Rotate the caliper up for access to the pads

2.5 Slide the pads away from the disc, out of the caliper bracket

Front brake

Refer to illustrations 2.4a, 2.4b, 2.5, 2.6a and 2.6b

4   Remove the caliper lower mounting bolt and rotate the caliper up to allow removal of the pads **(see illustrations)**.

5   Remove the anti-rattle springs and pull the brake pads and shims from the caliper bracket **(see illustration)**.

6   Remove the retaining clips from the bracket **(see illustrations)**.

7   Clean the mounting surfaces of the caliper and bracket, removing any dirt or corrosion. Install new retaining clips in the bracket.

8   Coat the shims with disc brake grease and install the pads and shims into the caliper bracket.

9   Install the wear indicator, then swing the caliper down over the pads. Install the lower mounting bolt, tightening it to the specified torque.

10   Repeat the procedure on the other wheel.

2.6a Remove the lower . . .

Rear brake

Refer to illustrations 2.11, 2.12a, 2.12b, 2.13, 2.17a and 2.17b

11   Disconnect the parking brake cable

2.6b . . . and upper pad retaining clips from the bracket

from the lever **(see illustration)**.

12   Remove the caliper lower mounting bolt and rotate the caliper up to allow removal of the pads **(see illustrations)**.

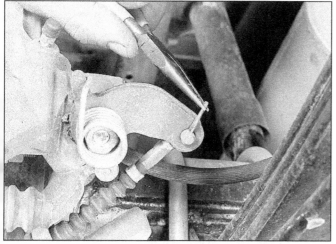

2.11 Use pliers to pull the cotter pin out of the retaining pin, then disconnect the parking brake cable

2.12a Remove the lower bolt on the rear caliper

2.12b Rotate the caliper up to expose the pads

2.13 Lift the pads from the bracket, noting that the wear indicators are positioned at the bottom

13   Rotate the brake pads and shims out of the caliper bracket **(see illustration)**.

14   Remove the retaining clips from the bracket. Clean the mounting surfaces of the caliper and bracket, removing any dirt or corrosion. Lubricate the bracket contact surfaces with brake lube (this should be included with the pads).

15   Install new retaining clips in the bracket.

16   Coat the pad shims with disc brake grease and install the pads and shims into the caliper bracket.

17   Use needle nose pliers to rotate the piston clockwise until bottoms in the bore **(see illustration)**. Make sure that when the piston is bottomed in the bore the notches in the piston line up with center line of the caliper **(see illustration)**. The brake pads must seat in the notches in the piston when they are installed.

18   Swing the caliper down over the pads. Install the lower mounting bolt, tightening it to the specified torque.

19   Repeat the procedure on the other wheel.

All models

20   Firmly depress the brake pedal a few times to bring the pads into contact with the disc. Check the fluid level in the master cylinder, topping it up if necessary.

21   Road test the vehicle carefully before placing it into normal use.

---

3   Disc brake caliper - removal, overhaul and installation

Warning: *Dust created by the brake system may contain asbestos, which is harmful to your health. Never blow it out with compressed air and don't inhale any of it. An approved filtering mask should be worn when working on the brakes. Do not, under any circumstances, use petroleum-based solvents to clean brake parts. Use brake cleaner or denatured alcohol only!*

Note: *If an overhaul is indicated (usually because of fluid leakage) explore all options before beginning the job. New and factory*

rebuilt calipers are available on an exchange basis, which makes this job quite easy. If it's decided to rebuild the calipers, make sure that a rebuild kit is available before proceeding. Always rebuild the calipers in pairs - never rebuild just one of them. Rear brake calipers, because of the special tools required, should not be rebuilt by the home mechanic, therefore this job should be left to a dealer service department or properly equipped shop.

Removal

Refer to illustration 3.5

1   Remove the cap from the brake fluid reservoir, siphon off two-thirds of the fluid into a container and discard it.

2   Loosen the wheel lug nuts, raise the front of the vehicle and support it securely on jackstands. Remove the wheels.

3   On rear disc brakes, disconnect the parking brake cable from the lever on the caliper (see Section 2).

4   Remove the brake hose inlet fitting bolt and detach the hose. Have a rag handy to

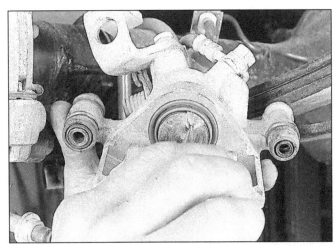

2.17a Rotate the piston clockwise into the bore until it bottoms, using needle-nose pliers

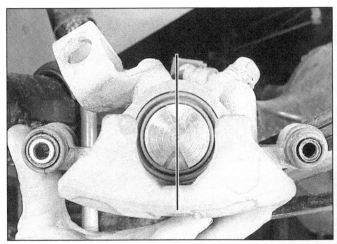

2.17b The notches must line up with the caliper centerline when the piston is bottomed in the bore so the pads can seat in the notches

3.5 Slide the caliper off the pivot pin at the top of the bracket

catch spilled fluid and wrap a plastic bag tightly around the end of the hose to prevent fluid loss and contamination.

5   Remove the caliper (see Section 2), and slide it off the pin at the top of the bracket **(see illustration)**.

Overhaul

Refer to illustrations 3.7, 3.9, 3.10, 3.13, 3.14, 3.15 and 3.16

Note: *The rear caliper cannot be overhauled without special tools - it should be taken to a dealer service department or properly equipped shop for overhaul.*

6   Clean the exterior of the caliper with brake cleaner or denatured alcohol. Never use gasoline, kerosene or petroleum-based cleaning solvents. Place the caliper on a clean workbench.

7   Remove the sleeve, sleeve dust boot and guide pin dust boot **(see illustration)**.

8   Using a small screwdriver, pry the dust seal ring from the caliper. Then pry out the dust seal.

9   Position a wooden block or several shop rags in the caliper as a cushion, then use compressed air to remove the piston from

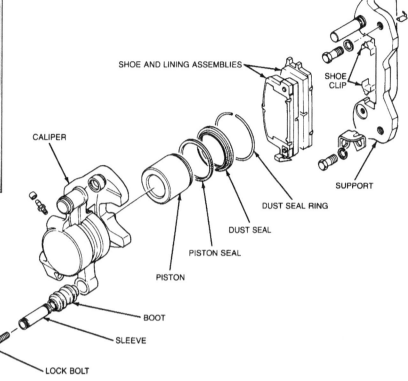

3.7 Typical caliper - exploded view

the caliper **(see illustration)**. Use only enough air pressure to ease the piston out of the bore. If the piston is blown out, even with the cushion in place, it may be damaged. **Warning:** *Never place your fingers in front of the piston in an attempt to catch or protect it when applying compressed air, as serious injury could occur.*

10   Using a wood or plastic tool, remove the piston seal from the groove in the caliper bore **(see illustration)**. Metal tools may cause bore damage.

11   Clean all components in brake cleaner or clean brake fluid and blow dry with compressed air. Check the cylinder bore and pistons for signs of wear, corrosion or surface defects such as scoring and replace if necessary.

12   The dust seal and piston seal must always be replaced when the caliper is overhauled.

13   To reassemble the caliper, lubricate the piston seal and piston with brake lube (usually supplied with the overhaul kit) or brake

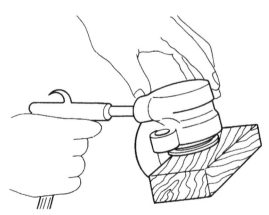

3.9 Apply compressed air to the brake fluid hose connection on the caliper body - position a wood block between the piston and caliper to prevent damage

3.10 The piston seal should be removed with a wood or plastic tool to prevent damage to the bore and seal groove - a pencil will do the job

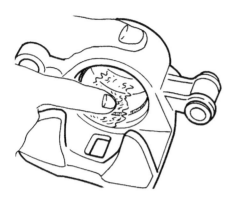

3.13 Apply brake lube to the new seal, press it into its groove, then lubricate the contact surface and the bore itself

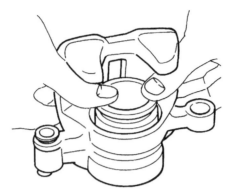

3.14 Press the piston evenly into the caliper bore, using finger pressure only - don't use a hammer

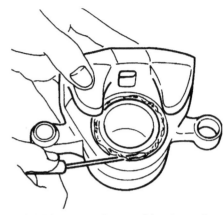

3.15 Use a small screwdriver to apply brake lube around the circumference of the groove before installing the seal

fluid, install it in the groove in the caliper bore, then lubricate the seal and the bore **(see illustration)**.

14 Carefully insert the piston into the caliper using only finger pressure **(see illustration)**.

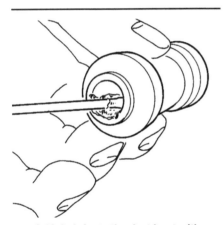

3.16 Lubricate the dust boot with brake lube

15 Apply brake lube to the piston and then install the dust seal in the piston and caliper **(see illustration)**. Insert the seal ring into the dust seal.

16 Apply brake lube into the sleeve and guide pin dust boots **(see illustration)**. Then install the sleeve dust boot to the caliper and insert the sleeve into the dust boot.

17 Apply brake lube into the guide pin hole and then install the guide pin dust boot to the caliper.

Installation

18 Install the caliper on the guide pin then install the dust boot over the guide pin.

19 Install the caliper bolt and tighten it to the specified torque.

20 Install the flexible brake hose to the caliper. Be sure the hose does not interfere with any suspension or steering components.

21 Install the wheel and tire.

22 Bleed the brake system (see Section 9).

23 Lower the vehicle to the ground. Test the brakes carefully before placing the vehicle into normal operation.

4 Brake disc - inspection, removal and installation

Refer to illustrations 4.2, 4.4a, 4.4b, 4.5, 4.7 and 4.8

Inspection

1 Loosen the wheel lug nuts, raise the vehicle and support it securely on jackstands. Remove the wheel.

2 Remove the brake caliper as outlined in Sections 2 and 3. It's not necessary to disconnect the brake hose. After removing the caliper bolt, suspend the caliper out of the way with a piece of wire **(see illustration)**. Don't let the caliper hang by the hose and don't stretch or twist the hose.

3 Visually check the disc surface for score marks and other damage. Light scratches and shallow grooves are normal after use and may not always be detrimental to brake operation, but deep score marks - over 0.015-inch (0.38 mm) - require disc removal and refinishing by an automotive machine shop.

4.2 Hang the caliper from a piece of wire - be careful not to allow the hose to stretch

4.4a Check disc runout with a dial indicator positioned approximately 1/2-inch from the edge of the disc - if the reading exceeds the maximum allowable runout, the disc will have to be resurfaced or replaced

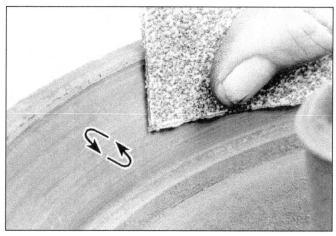

4.4b Using a swirling motion, remove the glaze from the disc with sandpaper or emery cloth

4.5 Use a micrometer to measure the thickness of the disc

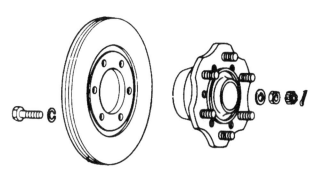

4.7 Remove the bolts from the backside of the disc/hub assembly, then separate the two components (it may be necessary to tap the disc off the hub with a hammer and a block of wood)

4.8 On rear discs, separate the brake disc from the hub by threading a bolt into the hub

Be sure to check both sides of the disc. If pulsating has been noticed during application of the brakes, suspect excessive disc runout.

4   To check disc runout, place a dial indicator at a point about 1/2-inch from the outer edge of the disc **(see illustration)**. Set the indicator to zero and turn the disc. The indicator reading should not exceed the specified allowable runout limit. If it does, the disc should be refinished by an automotive machine shop. **Note:** *Professionals recommend resurfacing of brake discs regardless of the dial indicator reading (to produce a smooth, flat surface, that will eliminate brake pedal pulsations and other undesirable symptoms related to questionable discs). At the very least, if you elect not to have the discs resurfaced, deglaze the brake pad surface with medium-grit emery cloth (use a swirling motion to ensure a non-directional finish)* **(see illustration)**.

5   The disc must not be machined to a thickness less than the specified minimum refinish thickness. The minimum wear thickness is cast into the inside of the disc. The disc thickness can be checked with a micrometer **(see illustration)**.

Removal

6   Remove the caliper bracket bolts and lift the bracket off.

7   On front disc brakes, remove the front hub/disc assembly, referring to Chapter 1 for 2WD models or Chapter 8 for 4WD models. Unbolt the disc from the hub **(see illustration)**.

8   On rear disc brakes, detach the disc by screwing a bolt into the threaded hole in the hub, then lift the disc off **(see illustration)**.

Installation

9   Install the front disc to the hub, tightening the bolts to the specified torque in a crisscross pattern. Install the rear disc by sliding it on the hub over the wheel studs.

10   Install the front disc and hub assembly and adjust the wheel bearing (see Chapter 1).

11   Install the caliper/bracket, tightening the mounting bolts to the specified torque. Position the pads in the bracket and install the caliper (refer to Section 3 for the caliper installation procedure, if necessary). Tighten the caliper bolts to the specified torque.

12   Install the wheel, then lower the vehicle

to the ground. Depress the brake pedal a few times to bring the brake pads into contact with the disc. Bleeding of the system will not be necessary unless the brake hose was disconnected from the caliper. Check the operation of the brakes carefully before placing the vehicle into normal service.

5   Drum brake shoes - replacement

Refer to illustration 5.2

Warning: *Drum brake shoes must be replaced on both rear wheels at the same time - never replace the shoes on only one wheel. Also, the dust created by the brake system may contain asbestos, which is harmful to your health. Never blow it out with compressed air and don't inhale any of it. An approved filtering mask should be worn when working on the brakes. Do not, under any circumstances, use petroleum based solvents to clean brake parts. Use brake cleaner or denatured alcohol only!*

Caution: *Whenever the brake shoes are replaced, the retractor and hold-down*

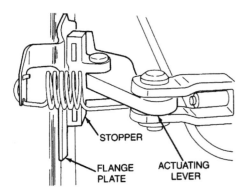

5.2 On 1981 through 1983 models, remove the stopper from the parking brake lever and push it against the flange plate

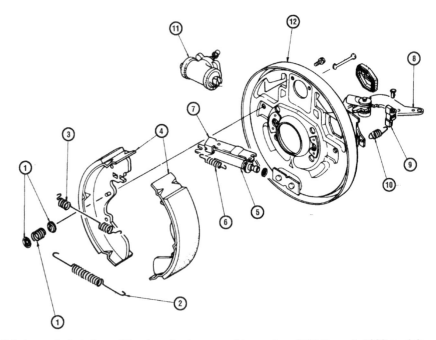

5.4 An exploded view of the drum brake assembly used on 1981 through 1983 models

1	Hold-down spring and retainers	7	Actuator bracket
2	Lower return spring	8	Actuating lever
3	Upper return spring	9	Stopper
4	Brake shoes	10	Actuating lever return spring
5	Actuator strut and adjuster assembly	11	Wheel cylinder
6	Actuating spring	12	Backing plate

springs should also be replaced. Due to the continuous heating/cooling cycle that the springs are subjected to, they lose their tension over a period of time and may allow the shoes to drag on the drum and wear at a much faster rate than normal. When replacing the brake shoes, use only high quality nationally recognized brand-name parts.

1   Loosen the wheel lug nuts, raise the vehicle and support it securely on jackstands. Remove the wheel.

2   Remove the brake drum attaching screws and pull the drum off the axle flange. On 1981 through 1983 models it may be necessary to retract the brake shoes away from the inside of the drum. This is accomplished by removing the stopper from the parking brake lever and then bringing the lever in contact with the flange plate **(see illustration)**.

3   Before removing anything, clean the brake assembly with brake cleaner - DO NOT use compressed air to blow the dust from the brake assembly!

1981 through 1983 models

Refer to illustration 5.4

4   Remove the leading brake shoe hold-down spring and pin **(see illustration)**. This is accomplished by grasping the spring with a pair of pliers, pushing down and rotating it 90-degrees, disengaging it from the pin. Place a finger behind the pin to prevent it from being pushed out.

5   Use pliers to remove the lower pull-back spring.

6   Expand the brake shoes until they disengage from the actuator assembly, then remove both shoes together with the upper pull-back spring.

7   If it is necessary to remove the actuator assembly, disconnect the actuating spring, then remove the actuating strut and lever. Check the wheel cylinders for signs of leaking fluid, replacing or rebuilding them if necessary (see Section 6). Check the brake drum for hard spots, cracks, score marks and grooves. Hard spots will appear as small dis-

colored areas. If they can't be removed with emery cloth or if any of the other conditions listed above exist, the drum must be taken to an automotive machine shop to have it turned (machined on a lathe). **Note:** *Professional mechanics recommend resurfacing the drums whenever a brake job is performed. Resurfacing will eliminate the possibility of out-of-round drums.*

8   Prior to installing the shoes, lubricate the contact surfaces of the backing plate and wheel cylinder clevis' with high temperature grease. Be careful not to get any lubricant on the brake shoe surfaces.

9   Installation is the reverse of removal.

10   Prior to installing the brake drum, check the operation of the self-adjuster by moving the actuating lever.

11   Install the brake drums and move the parking brake actuating lever until the adjuster stops clicking. Connect the parking brake cable, install the wheels and lower the vehicle. Make the final parking brake adjustment by operating the parking brake repeatedly while depressing the brake pedal until sufficient brake pedal stroke is achieved.

1984 and later models

Refer to illustration 5.12

12   Remove the leading brake shoe hold-down spring and pin **(see illustration)**. This is accomplished by grasping the spring with a pair of pliers, pushing down and rotating it 90-degrees, disengaging it from the pin.

Place a finger behind the pin to prevent it from being pushed out.

13   Remove the lower and upper brake shoe return springs.

14   Remove the leading brake shoe.

15   Remove the trailing shoe hold-down spring and pin.

16   Remove the adjuster assembly retainer, pin and washer and lift off the adjuster assembly.

17   Remove the trailing brake shoe from the backing plate.

18   Check the wheel cylinders for signs of leaking fluid, replacing or rebuilding them if necessary (see Section 6).

19   Check the brake drum for hard spots, cracks, score marks and grooves. Hard spots will appear as small discolored areas. If they can't be removed with emery cloth or if any of the other conditions listed above exist, the drum must be taken to an automotive machine shop to have it turned (machined on a lathe). **Note:** *Professional mechanics recommend resurfacing the drums whenever a brake job is performed. Resurfacing will eliminate the possibility of out-of-round drums.*

20   Unscrew the brake adjuster, clean it and apply a little high temperature grease to the adjuster screw threads. Assemble the adjuster screw, turning it in completely.

21   Connect the adjuster assembly to the lever and position the trailing shoe against the backing plate. Install the hold-down pin and spring.

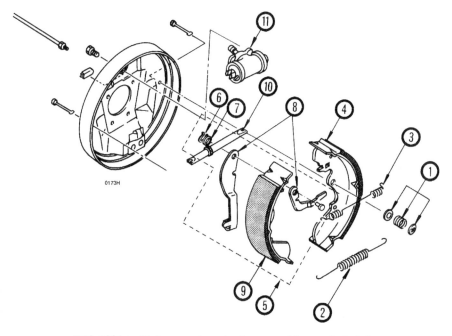

5.12 1984 and later rear drum brake assembly - exploded view

1	Hold-down spring and pin	7	Washer
2	Lower return spring	8	Adjuster lever
3	Upper return spring	9	Trailing brake shoe
4	Leading brake shoe	10	Adjuster screw assembly
5	Shoe and lever assembly	11	Wheel cylinder assembly
6	Adjuster assembly retainer		

Overhaul

7   Remove the bleeder screw, piston cups, pistons, boots and expander spring assembly from the wheel cylinder body (see illustration).

8   Clean the wheel cylinder with brake fluid, denatured alcohol or brake system cleaner. **Warning:** *Do not, under any circumstances, use petroleum based solvents to clean brake parts!*

9   Use compressed air to remove excess fluid from the wheel cylinder and to blow out the passages.

10   Check the cylinder bore for corrosion and score marks. Crocus cloth can be used to remove light corrosion and stains, but the cylinder must be replaced with a new one if the defects cannot be removed easily, or if the bore is scored.

11   Lubricate the new seals with brake fluid.

12   Assemble the brake cylinder components. Make sure the seal lips face in.

Installation

13   Place the wheel cylinder in position and install the bolts.

14   Connect the brake line and install the brake shoe assembly.

15   Bleed the brakes (see Section 9).

---

7   Master cylinder - removal, overhaul and installation

Refer to illustration 7.11
Note: *Before deciding to overhaul the master cylinder, check on the availability and cost of a new or factory rebuilt unit and also the availability of a rebuild kit.*

Removal

1   The master cylinder is located in the engine compartment, mounted to the power brake booster.

2   Remove as much fluid as you can from the reservoir with a syringe.

3   Place rags under the fluid fittings and prepare caps or plastic bags to cover the

---

22   Install the adjuster assembly and place the leading shoe against the backing plate. Install the hold-down pin and spring.

23   Install the return springs.

24   Wiggle the brake shoe assembly to center it on the backing plate. Make sure the tops of the shoes are seated in the slots of the wheel cylinder pushrods and the bottoms of the shoes are resting under the tabs of the anchor plate.

25   Install the brake drum and wheel. Don't forget to tighten the lug nuts to the specified torque. Repeat the operation on the other wheel, then road test the vehicle carefully before placing it into normal service.

vehicle from rolling.

2   Remove the brake shoe assembly (see Section 5).

3   Remove all dirt and foreign material from around the wheel cylinder.

4   Disconnect the brake line using a flare nut wrench. Don't pull the brake line away from the wheel cylinder.

5   Remove the wheel cylinder mounting bolts.

6   Detach the wheel cylinder from the brake backing plate and place it on a clean workbench. Immediately plug the brake line to prevent fluid loss and contamination.

---

6   Wheel cylinder - removal, overhaul and installation

Refer to illustration 6.7
Note: *If an overhaul is indicated (usually because of fluid leakage or sticky operation) explore all options before beginning the job. New wheel cylinders are available, which make this job quite easy. If it's decided to rebuild the wheel cylinder, make sure that a rebuild kit is available before proceeding. Never overhaul only one wheel cylinder - always rebuild both of them at the same time.*

Removal

1   Raise the vehicle and support it securely on jackstands. Block the wheels to keep the

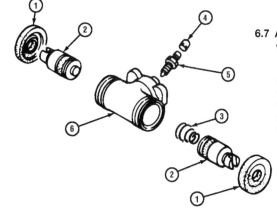

6.7 An exploded view of a typical wheel cylinder assembly

1	Dust boot
2	Piston and cup
3	Expander spring
4	Cap
5	Bleeder screw
6	Wheel cylinder

1   *master cylinder housing*
2   *Primary piston assembly*
3   *Secondary piston assembly*
4   *Primary spring*
5   *Secondary spring*
6   *Snap-ring*
7   *Stop bolt*
8   *Sealing washer*
9   *Reservoir assembly*
10   *Filter*
11   *Screw*
12   *Cap*
13   *Gasket*
14   *Grommets*
15   *Lower seal*
16   *Cap seal*

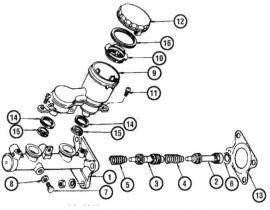

7.11 Exploded view of a typical hydraulic brake master cylinder

ends of the lines once they are disconnected. **Caution:** *Brake fluid will damage paint. Cover all body parts and be careful not to spill fluid during this procedure.*

4   Loosen the tube nuts at the ends of the brake lines where they enter the master cylinder. To prevent rounding off the flats on these nuts, the use of a flare nut wrench, which wraps around the nut, is preferred.

5   Pull the brake lines slightly away from the master cylinder and plug the ends to prevent contamination.

6   Pull the master cylinder off the studs and out of the engine compartment. Again, be careful not to spill the fluid as this is done.

Overhaul

7   Before attempting the overhaul of the master cylinder, obtain the proper rebuild kit, which will contain the necessary replacement parts and also any instructions which may be specific to your model.

8   Inspect the reservoir or inlet grommet(s) for indications of leakage near the base of the reservoir. Remove the reservoir.

9   Place the cylinder in a vise and use a punch or Phillips screwdriver to fully depress the pistons until they bottom against the other end of the master cylinder. Hold the pistons in this position and remove the stop bolt on the side of the master cylinder. Remove the two outlet plugs and the copper gaskets.

10   Depress the pistons once again, then carefully remove the snap-ring at the end of the master cylinder.

11   The internal components can now be removed from the cylinder bore **(see illustration)**. Make a note of the proper order of the components so they can be returned to their original locations. **Note:** *The two springs are of different tension, so pay particular attention to their order.*

12   Carefully inspect the bore of the master cylinder. Any deep scoring or other damage will mean a new master cylinder is required.

13   Replace all parts included in the rebuild kit, following any instructions in the kit. Clean all reused parts with clean brake fluid or denatured alcohol. Do not use any

petroleum-based cleaners. During assembly, lubricate all parts liberally with clean brake fluid. Be sure to tighten all fittings and connections to the specified torque.

14   Push the assembled components into the bore, bottoming them against the end of the master cylinder and install the stop bolt.

15   Install the new snap-ring, making sure it is seated properly in the groove.

16   Before installing the new master cylinder it should be bench bled. Because it will be necessary to apply pressure to the master cylinder piston and, at the same time, control flow from the brake line outlets, it is recommended that the master cylinder be mounted in a vise, with the jaws of the vise clamping on the mounting flange.

17   Insert threaded plugs into the brake line outlet holes and snug them down so that there will be no air leakage past them, but not so tight that they cannot be easily loosened.

18   Fill the reservoir with brake fluid of the recommended type (see Chapter 1).

19   Remove one plug and push the piston assembly into the master cylinder bore to expel the air from the master cylinder. A large Phillips screwdriver can be used to push on the piston assembly.

20   To prevent air from being drawn back into the master cylinder the plug must be replaced and snugged down before releasing the pressure on the piston assembly.

21   Repeat the procedure until only brake fluid is expelled from the brake line outlet hole. When only brake fluid is expelled, repeat the procedure with the other outlet hole and plug. Be sure to keep the master cylinder reservoir filled with brake fluid to prevent the introduction of air into the system.

22   Since high pressure is not involved in the bench bleeding procedure, an alternative to the removal and replacement of the plugs with each stroke of the piston assembly is available. Before pushing in on the piston assembly, remove the plug as described in Step 19. Before releasing the piston, however, instead of replacing the plug, simply put your finger tightly over the hole to keep air from being drawn back into the master cylinder. Wait several seconds for brake fluid to

be drawn from the reservoir into the piston bore, then depress the piston again, removing your finger as brake fluid is expelled. Be sure to put your finger back over the hole each time before releasing the piston, and when the bleeding procedure is complete for that outlet, replace the plug and snug it before going on to the other port.

Installation

23   Install the master cylinder over the studs on the power brake booster and tighten the attaching nuts only finger tight at this time.

24   Thread the brake line fittings into the master cylinder. Since the master cylinder is still a bit loose, it can be moved slightly in order for the fittings to thread in easily. Do not strip the threads as the fittings are tightened.

25   Fully tighten the mounting nuts and the brake fittings.

26   Fill the master cylinder reservoir with fluid, then bleed the master cylinder (only if the cylinder has not been bench bled) and the brake system as described in Section 9. To bleed the cylinder on the vehicle, have an assistant pump the brake pedal several times and then hold the pedal to the floor. Loosen the fitting nut to allow air and fluid to escape. Repeat this procedure on both fittings until the fluid is clear of air bubbles. Test the operation of the brake system carefully before placing the vehicle in normal service.

8   Brake lines and hoses - inspection and replacement

1   About every six months the flexible hoses which connect the steel brake lines with the front and rear brakes should be inspected for cracks, chafing of the outer cover, leaks, blisters, and other damage (Chapter 1).

2   Replacement steel and flexible brake lines are commonly available from dealer parts departments and auto parts stores. Do not, under any circumstances, use anything other than genuine steel lines or approved flexible brake hoses as replacement items.

3   When installing the brake line, leave at least 0.75 in (19 mm) clearance between the line and any moving or vibrating parts.

4   When disconnecting a hose and line, first remove the spring clip. Then, using a normal wrench to hold the hose and a flare-nut wrench to hold the tube, make the disconnection. Use the wrenches in the same manner when making a connection, then install a new clip. **Note:** *Make sure the tube passes through the center of its grommet.*

5   When disconnecting two hoses, use normal wrenches on the hose fittings. When connecting two hoses, make sure they are not bent, twisted or strained.

6   Steel brake lines are usually retained along their span with clips. Always remove these clips completely before removing a fixed brake line. Always reinstall these clips, or new ones if the old ones are damaged, when replacing a brake line, as they provide

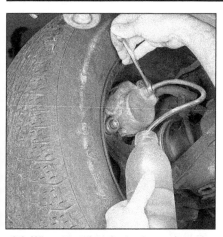

9.8 When bleeding the brakes, a hose is connected to the bleeder valve at the caliper or wheel cylinder and then submerged in brake fluid - air will be seen as bubbles in the tube and container when the valve is opened - all air must be expelled before moving to the next wheel

support and keep the lines from vibrating, which can eventually break them.

7   Remember to bleed the hydraulic system after replacing a hose or line.

9   Brake system - bleeding

Refer to illustration 9.8

Warning: *Wear eye protection when bleeding the brake system. If the fluid comes in contact with your eyes, immediately rinse them with water and seek medical attention.*

Note: *Bleeding the hydraulic system is nec-*

essary to remove any air that manages to find its way into the system when it's been opened during removal and installation of a hose, line, caliper or master cylinder.

1   It will probably be necessary to bleed the system at all four brakes if air has entered the system due to low fluid level, or if the brake lines have been disconnected at the master cylinder.

2   If a brake line was disconnected only at a wheel, then only that caliper or wheel cylinder must be bled.

3   If a brake line is disconnected at a fitting located between the master cylinder and any of the brakes, that part of the system served by the disconnected line must be bled.

4   Remove any residual vacuum from the brake power booster by applying the brake several times with the engine off.

5   Remove the master cylinder reservoir cover and fill the reservoir with brake fluid. Reinstall the cover. **Note:** *Check the fluid level often during the bleeding operation and add fluid as necessary to prevent the fluid level from falling low enough to allow air bubbles into the master cylinder.*

6   Have an assistant on hand, as well as a supply of new brake fluid, a clear container partially filled with clean brake fluid, a length of 3/16-inch plastic, rubber or vinyl tubing to fit over the bleeder valve and a wrench to open and close the bleeder valve.

7   Beginning at the right rear wheel, loosen the bleeder valve slightly, then tighten it to a point where it is snug but can still be loosened quickly and easily.

8   Place one end of the tubing over the bleeder valve and submerge the other end in brake fluid in the container **(see illustration)**.

9   Have the assistant pump the brakes

slowly a few times to get pressure in the system, then hold the pedal firmly depressed.

10   While the pedal is held depressed, open the bleeder valve just enough to allow a flow of fluid to leave the valve. Watch for air bubbles to exit the submerged end of the tube. When the fluid flow slows after a couple of seconds, close the valve and have your assistant release the pedal.

11   Repeat Steps 9 and 10 until no more air is seen leaving the tube, then tighten the bleeder valve and proceed to the left rear wheel, the right front wheel and the left front wheel, in that order, and perform the same procedure. Be sure to check the fluid in the master cylinder reservoir frequently.

12   Never use old brake fluid. It contains moisture which will lower the boiling point of the brake fluid and also deteriorate the brake system rubber components.

13   Refill the master cylinder with fluid at the end of the operation.

10   Power brake booster - check, removal and installation

Refer to illustrations 10.7 and 10.14

Operating check

1   Depress the brake pedal several times with the engine off and make sure that there is no change in the pedal reserve distance.

2   Depress the pedal and start the engine. If the pedal goes down slightly, operation is normal.

Air tightness check

3   Start the engine and turn it off after one or two minutes. Depress the brake pedal several times slowly. If the pedal goes down farther the first time but gradually rises after the second or third depression, the booster is air tight.

4   Depress the brake pedal while the engine is running, then stop the engine with the pedal depressed. If there is no change in the pedal reserve travel after holding the pedal for 30 seconds, the booster is air tight.

Removal

5   Power brake booster units should not be disassembled. They require special tools not normally found in most service stations or shops. They are fairly complex and because of their critical relationship to brake performance it is best to replace a defective booster unit with a new or rebuilt one.

6   To remove the booster, first remove the brake master cylinder as described in Section 7. On some vehicles it is not necessary to disconnect the brake lines from the master cylinder, as there is enough room to reposition the cylinder to allow booster removal.

7   Locate the pushrod clevis connecting the booster to the brake pedal **(see illustration)**. This is accessible from the interior in front of the driver's seat.

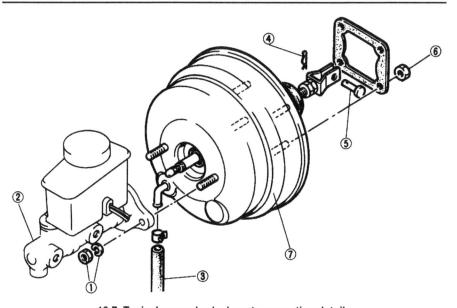

10.7 Typical power brake booster mounting details

1	Nut and washer	4	Cotter pin	6	Nut
2	Master cylinder	5	Clevis pin	7	Power booster
3	Vacuum hose				

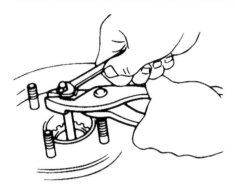

10.14 To adjust the length of the booster pushrod, hold the serrated portion of the rod with a pair of pliers and turn the adjusting screw in or out, as necessary, to achieve the desired setting

8   Remove the clevis pin retaining clip with pliers and pull out the pin.
9   Holding the clevis with pliers, disconnect the clevis locknut with a wrench. The clevis is now loose.
10   Disconnect the hose leading from the engine to the booster. Be careful not to damage the hose when removing it from the booster fitting.
11   Remove the four nuts and washers holding the brake booster to the firewall. You may need a light to see these, as they are up under the dash area.
12   Slide the booster straight out from the firewall until the studs clear the holes and pull the booster, brackets and gaskets from the engine compartment area.

Installation

13   Installation procedures are basically the reverse of those for removal. Tighten the clevis locknut and booster mounting nuts to the specified torque figures.
14   If the power booster unit is being replaced, the pushrod protrusion must be checked and, if necessary, adjusted. Connect a hand-held vacuum pump to the fitting on the booster and apply 20 in-Hg of vacuum. Measure the distance from the master cylinder flange face on the power booster to the end of the pushrod, comparing your measurement with the value listed in this Chapter's Specifications. Turn the end of the pushrod in or out to achieve the desired setting **(see illustration)**.
15   Install the master cylinder.
16   After the final installation of the master cylinder and brake hoses and lines, the brake pedal height and free play must be adjusted (see Chapter 1) and the system must be bled (see Section 9).

11   Parking brake cable - adjustment

Refer to illustrations 11.2a and 11.2b

1   If the parking brake doesn't keep the vehicle from rolling when the handle is applied 12 to 14 clicks on 1987 and earlier models, or 9 to 11 clicks on later models, adjust the cable.
2   Locate the cable adjusting nut for your particular model. On 1983 and earlier models, the adjusting nut is located between the intermediate cable and on later models at the cable equalizer (where the front and rear cables meet) **(see illustrations)**.
3   Turn the adjusting nut to remove all slack from the parking brake cables. After adjustment apply the parking brake several times and check that handle travel is 12 to 14 clicks on 1987 and earlier models or 9 to 11 clicks on later models with drum rear brakes. On models with rear disc brakes there should be a clearance of 0.08-inch (2 mm) between the parking brake lever at the caliper and the stop pin. Make sure the rear brakes don't drag when the parking brake is released. Check to see that the parking brake light on the dash glows when the handle is applied.
4   Lower the vehicle and verify that the parking brake will hold the vehicle on a moderate incline. If it still won't keep the vehicle from rolling, inspect the rear brakes as described in Chapter 1.

12   Parking brake cable(s) - replacement

1981 through 1983 models

Refer to illustration 12.1

Rear cable

1   Remove the cotter pins, plain washers, waved washers and pins that attach the rear cables to the actuating levers at each rear wheel **(see illustration)**.
2   Remove the same attaching components that secure the rear cables to the rear relay lever and lift out the left rear cable.
3   Raise the cable guides on the rear axle case to release them from the guide bracket, then disconnect the right rear cable.
4   Installation is the reverse of the removal procedure.
5   Following installation, adjust the parking brake (see Section 11).

First intermediate cable

6   If necessary for clearance, raise the front of the vehicle and support it on jackstands.
7   Disconnect the front end of the front cable from the front relay lever.
8   Pry out the one (2-wheel drive models) or two (4-wheel drive models) cable guides from the frame crossmember located under the driver's compartment.
9   Remove the return spring from the adjuster at the rear end of the first intermediate cable.
10   Loosen the adjuster locknut. Then separate the first and second intermediate cables by turning the second cable out from the adjuster.
11   Installation is the reverse of the removal procedure.
12   Following installation, adjust the parking brake (see Section 11).

Second intermediate cable

13   If necessary for clearance, raise the rear of the vehicle and support it on jackstands.

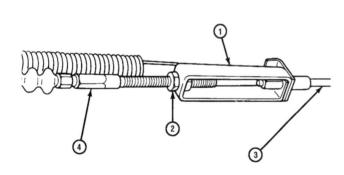

11.2a On 1981 through 1983 models, loosen the locknut and turn the adjuster to adjust the cable

1 Parking brake cable adjuster	*3 Forward intermediate cable*
2 Locknut	*4 Rear intermediate cable*

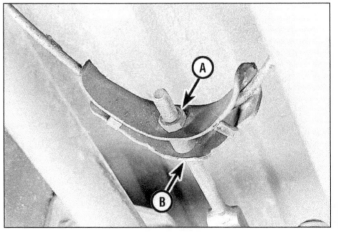

11.2b On 1984 and later models, loosen the locknut (A) and turn the adjusting nut (B) to adjust the parking brake cable

14   Remove the return spring from the adjuster located at the rear of the first intermediate cable.

15   Loosen the adjuster locknut. Then turn the second intermediate cable out of the adjuster until the two are separated.

16   Remove the clip attaching the front of the cable to the frame bracket and remove the cable from the bracket.

17   Remove the cotter pin, plain washer, waved washer and pin that attach the cable to the rear relay lever.

18   Remove the plate that attaches the cable to the rear axle case and lift out the cable.

19   Installation is the reverse of the removal procedure.

20   Lower the vehicle to the ground.

1984 and later models

Refer to illustration 12.22

Front lower cable

21   If necessary for clearance, raise the vehicle at the front and support it on jackstands.

22   Loosen the locknut, then completely unscrew the adjust nut securing the equalizer bracket to the 2nd relay lever assembly **(see illustration)**.

23   Remove the cotter pin, plain washer, waved washer and pin that secure the front lower cable to the 2nd relay lever assembly.

24   Disconnect the front lower cable from the 1st relay lever assembly and remove the cable.

25   Installation is the reverse of the removal procedure.

26   Following installation, adjust the parking brake (see Section 11).

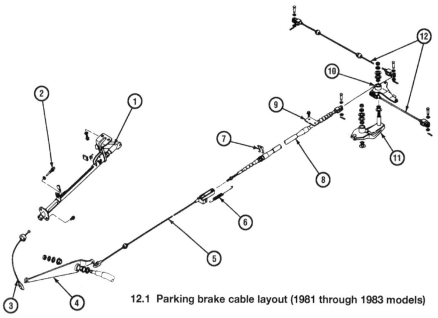

12.1 Parking brake cable layout (1981 through 1983 models)

1	Parking brake lever assembly	7	Clip
2	Parking brake switch	8	Rear intermediate cable
3	Front cable	9	Fixing plate
4	Front relay lever	10	Rear relay lever
5	Forward intermediate cable	11	Relay lever bracket
6	Return spring	12	Rear cables

Intermediate cable and rear cable assembly

27   If necessary for clearance, raise the vehicle at the front and support it on jackstands.

28   Remove the cotter pins, plain washers, waved washers and pins that secure the rear cables to the rear wheels **(see illustration 12.22)**.

29   Loosen the locknut, then completely unscrew the adjust nut securing the equalizer bracket to the 2nd relay lever assembly.

30   Remove the intermediate cable and cable guides from the frame brackets.

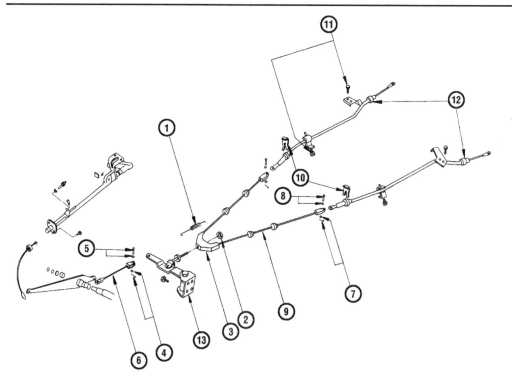

12.22 Parking brake cable layout (1984 and later models)

1	Return spring
2	Locknut
3	Equalizer bracket
4	Cotter pin and washer
5	Pin and curved washer
6	Front lower cable
7	Cotter pin and washer
8	Pin and curved washer
9	Rear intermediate cable
10	Clip
11	Fixing bracket clip
12	Rear cable assembly
13	Relay lever assembly

31 Remove the clips and bolts attaching the intermediate cable and rear cable assembly to the frame brackets and remove the cables from the brackets and lift out the cable.

32 Installation is the reverse of the removal procedure.

33 Following installation, adjust the parking brake (see Section 11).

13 Anti-lock Brake System (ABS) - general information

Some 1991 and later models are equipped with a rear-wheel ABS system, sometimes referred to as RWAL (Rear Wheel Anti-Lock). The system is designed to maintain vehicle maneuverability, directional stability and optimum deceleration under severe braking conditions on most road surfaces. It does so by monitoring the rotational speed of the wheels and controlling the brake line pressure to the rear wheels during braking. This prevents the rear wheels from locking up prematurely during hard braking.

Anti-lock valve assembly

The anti-lock valve assembly consists of a dump valve and an isolation valve. The valve assembly operates by changing the rear brake fluid pressure in response to signals from the control module. The valve is mounted to the right frame rail, just behind the cab area.

Control module

The control module is mounted in the cab, under the seat and is the "brain" of the system. It accepts and processes information received from the speed sensor and brake light switch to control the hydraulic line pressure, avoiding wheel lock up. The control module also constantly monitors the system, even under normal driving conditions, to find faults within the system. If a problem develops within the system, the RWAL light will glow on the dashboard. A diagnostic code will also be stored, which, when retrieved, will indicate the problem area or component.

Speed sensor

A speed sensor is located on top of the rear axle housing. The speed sensor sends a signal to the control module, indicating rear wheel rotational speed.

Brake light switch

The brake light switch signals the control module when the driver steps on the brake pedal. Without this signal the anti-lock system won't activate.

Diagnosis and repair

If the RWAL warning light on the dashboard comes on and stays on, make sure the parking brake is not applied and there's no problem with the brake hydraulic system. If neither of these is the cause, the RWAL system is probably malfunctioning. The home mechanic can perform a few preliminary checks before taking the vehicle to a dealer service department.

a) Make sure the brakes and wheel cylinders are in good condition.
b) Check the electrical connectors at the control module assembly.
c) Check the fuses.
d) Retrieve the diagnostic code and follow the wiring harness to the indicated component (speed sensor, anti-lock valve, brake light switch, etc.) and make sure all connections are secure and the wiring or component isn't damaged. If the above preliminary checks don't rectify the problem, the vehicle should be diagnosed by a dealer service department.

Self-diagnostic codes - retrieving

Refer to illustration 13.1

To retrieve self- diagnostic information from the ECM memory, you must turn the

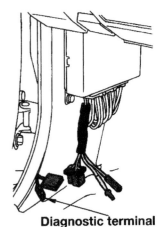

Diagnostic terminal

13.1 Diagnostic terminal location

ignition switch to on and ground the diagnostic terminal. The diagnostic terminal is located in the corner of the floor on the drivers side under the ECM **(see illustration)**. Grounding the diagnostic terminal with the rear anti-lock light on will cause the light to flash a code if a failure has been detected and is in memory. Count the number of flashes, starting wit h the long flash, and including the long flash as a count. The light will continue to flash, repeating the code, as long as the terminal is grounded. If there is more than one failure, only the first recognized code will be retained in memory and flashed. Removing battery voltage for 30 seconds will clear stored trouble codes. Trouble codes should be cleared after repairs have been completed. **Caution:** *To prevent damage to the ECM, the ignition must be off before disconnecting power to the ECM. The following is a list of the typical codes which may be encountered while diagnosing the computerized system. If the problem persists after checks have been made, more detailed service procedures will have to be done by a dealer service department.*

Diagnosis codes

Code 2	Open isolation valve circuit or malfunctioning ECM
Code 3	Open dump valve circuit or malfunctioning ECM
Code 4	Closed rear wheel antilock valve switch
Code 5	System dumps too many times, condition occurs while making normal or hard stops, rear brakes may lock
Code 6	Erratic speed sensor signal while driving
Code 7	Isolation valve output signal missing or antilock valve wiring shorted to ground
Code 8	Dump valve output signal missing or antilock valve wiring shorted to ground
Code 9	High speed sensor resistance
Code 10	Low speed sensor resistance
Code 11	Stop light switch circuit defective, condition indicated only when driving over 35 mph
Code 13	Speed processor failure (ECM)
Code 14	Program check failure (ECM)
Code 15	Memory failure (ECM)

Note: *Component replacement may not cure the problem in all cases. For this reason, you may want to seek professional advise before purchasing replacement parts.*

Chapter 10
Suspension and steering systems

Contents

Specifications

Torque specifications

	Ft-lbs
Front suspension	
Upper balljoint-to-steering knuckle nut	75 to 80
Upper balljoint-to-upper arm bolt/nut 1987 and earlier	19
1988 and later	24
Lower balljoint-to-steering knuckle nut	
1987 and earlier	75
1988 and later	
2WD	108
4WD	94
Lower balljoint-to-lower arm bolts	
1987 and earlier	50
1988 and later	
2WD	50
4WD	76
Upper arm-to-frame bolts	
1987 and earlier	75 to 80
1988 and later	
2WD	75 to 80
4WD	112
Lower arm-to-frame (pivot) bolts/nuts	
1987 and earlier	95
1988 and later	
2WD	93
4WD	
Front	116
Rear	145
Strut bar-to-lower arm bolts	51
Strut bar-to-frame outer nut	
1987 and earlier	60
1988 and later	94
Stabilizer bar mounting bracket bolts	
1983 and earlier	55
1984 through 1986 (all models) and 1987 pick-up	Not available
1987 and 1988 Trooper	7
1988 and later pick-up and 1989 and later Trooper	
2WD	14
4WD	20

1 General information

Refer to illustrations 1.1 and 1.2

The front suspension on the vehicles covered by this manual is an independent type, made up of upper and lower arms, torsion bars, balljoint mounted steering knuckles and shock absorbers **(see illustration)**. Some models are equipped with a stabilizer bar to limit body roll during cornering. Four wheel drive models use the same basic arrangement, but with the necessary modifications to accommodate the driveaxles.

The rear suspension consists of the rear axle housing, leaf springs and shock absorbers **(see illustration)**. Information regarding axles and the rear axle housing can be found in Chapter 8.

The steering system is composed of a steering column, steering gear, Pitman arm, idler arm, a steering damper and two tie-rod assemblies. Some models feature power assisted steering, which includes a belt-driven pump and associated hoses to provide hydraulic pressure to the steering gear.

Frequently, when working on the suspension or steering system components, you may come across fasteners which seem impossible to loosen. These fasteners on the underside of the vehicle are continually subjected to water, road grime, mud, etc., and can become rusted

or "frozen," making them extremely difficult to remove. In order to unscrew these stubborn fasteners without damaging them (or other components), be sure to use lots of penetrating oil and allow it to soak in for a while. Using a wire brush to clean exposed threads will also ease removal of the nut or bolt and prevent damage to the threads. Sometimes a sharp blow with a hammer and punch is effective in breaking the bond between a nut and bolt threads, but care must be taken to prevent the punch from slipping off the fastener and ruining the threads. Heating the stuck fastener and surrounding area with a torch sometimes helps too, but isn't recommended because of the obvious dangers associated with fire. Long breaker bars and extension, or "cheater," pipes will increase leverage, but never use an extension pipe on a ratchet - the ratcheting mechanism could be damaged. Sometimes, turning the nut or bolt in the tightening (clockwise) direction first will help to break it loose. Fasteners that require drastic measures to unscrew should always be replaced with new ones.

Since most of the procedures that are dealt with in this Chapter involve jacking up the vehicle and working underneath it, a good pair of jackstands will be needed. A hydraulic floor jack is the preferred type of jack to lift the vehicle, and it can also be used to support certain components during various oper-

ations. **Warning:** *Never, under any circumstances, rely on a jack to support the vehicle while working on it. Whenever any of the suspension or steering fasteners are loosened or removed they must be inspected and, if necessary, replaced with new ones of the same part number or of original equipment quality and design. Torque specifications must be followed for proper reassembly and component retention. Never attempt to heat or straighten any suspension or steering component. Always replace them with new ones.*

2 Front stabilizer bar - removal and installation

Refer to illustrations 2.2a, 2.2b and 2.2c
Warning: *Whenever any of the suspension or steering fasteners are loosened or removed, they must be inspected and, if necessary, replaced with new ones of the same part number or of original equipment quality and design. Torque specifications must be followed for proper reassembly and component retention.*

Removal

1   Raise the vehicle and support it securely on jackstands. Apply the parking brake.
2   Remove the stabilizer bar-to-lower arm

1.1 Front suspension and steering components

1	Stabilizer bar	3	Shock absorber	5	Torsion bar	7	Tie-rod end
2	Steering gear	4	Lower suspension bar	6	Lower balljoint	8	Upper suspension arm

1.2 Rear suspension components

1	Shock absorber	3	Spring hanger pivot bolt
2	Leaf spring	4	U-bolt nut

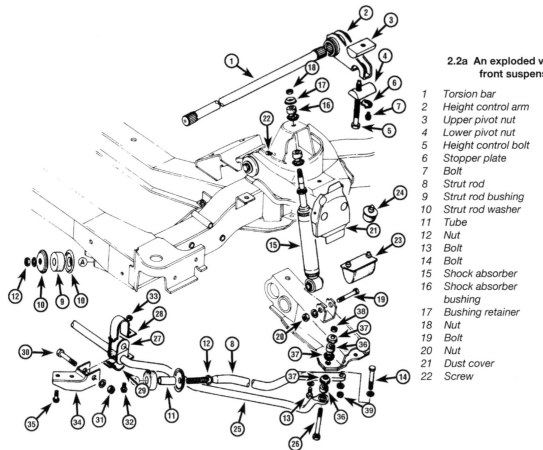

2.2a An exploded view of a typical 2WD front suspension assembly

1	Torsion bar	23	Lower control arm
2	Height control arm		bumper
3	Upper pivot nut	24	Upper control arm
4	Lower pivot nut		bumper
5	Height control bolt	25	Stabilizer bar
6	Stopper plate	26	Bolt
7	Bolt	27	Stabilizer bar
8	Strut rod		bushing
9	Strut rod bushing	28	Stabilizer bar
10	Strut rod washer		upper clamp
11	Tube	29	Stabilizer bar lower
12	Nut		clamp
13	Bolt	30	Bolt
14	Bolt	31	Nut
15	Shock absorber	32	Bolt
16	Shock absorber	33	Nut
	bushing	34	Stabilizer bar-to-
17	Bushing retainer		frame bracket
18	Nut	35	Bolt
19	Bolt	36	Bushing
20	Nut	37	Washer
21	Dust cover	38	Nut
22	Screw	39	Nut

2.2b An exploded view of a typical 4WD front suspension assembly (1981 through 1989 models)

1	Adjustment bolt and seat
2	Height control arm
3	Torsion bar
4	Rubber seat
5	Nut and washer
6	Rubber bushing and washers
7	Bolt, nut and washer
8	Strut bar
9	Rubber bushing and washer
10	Bolt
11	Bracket and rubber bushing
12	Bolt, nut and washer
13	Rubber bushing and washer
14	Stabilizer bar
15	Nut
16	Rubber bushing and washer
17	Bracket
18	Rubber bushing and washer
19	Nut
20	Rubber bushing and washer
21	Bolt, nut and washer
22	Shock absorber
23	Rubber bushing and washer
24	Lower control arm bumper

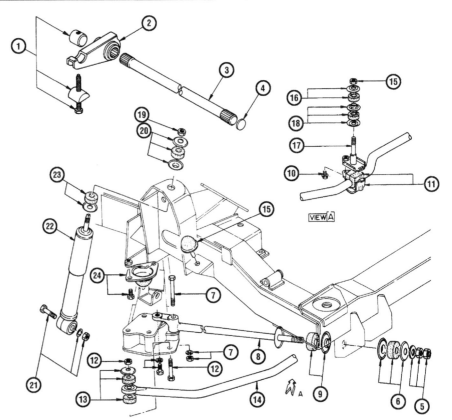

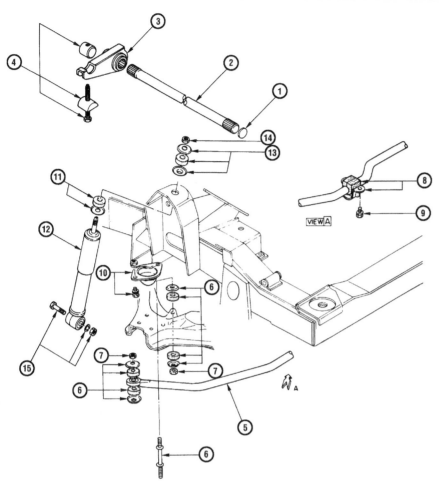

2.2c An exploded view of a typical 4WD front suspension assembly
(1990 and later models)

1	Rubber seat	9	Bolt
2	Torsion bar	10	Lower link bumper
3	Height control arm	11	Rubber bushing and washer
4	Adjusting bolt and seat	12	Shock absorber
5	Stabilizer bar	13	Rubber bushing and washer
6	Rubber bushing, bolt and washer	14	Nut
7	Nut	15	Bolt, nut and washer
8	Bracket		

3.3 The upper end of the front shock
absorber is held in place by a nut (arrow)

3.5 The lower end of the front shock
absorber is secured by a bolt and
nut (arrow)

nuts and bolts, noting how the spacers, washers and bushings are positioned **(see illustrations)**.
3   Remove the stabilizer bar bracket bolts and detach the bar from the vehicle.
4   Pull the brackets off the stabilizer bar and inspect the bushings for cracks, hardening and other signs of deterioration. If the bushings are damaged, replace them.

Installation

5   Position the stabilizer bar bushings on the bar. Push the brackets over the bushings and raise the bar up to the frame. Install the bracket bolts but don't tighten them completely at this time.
6   Install the stabilizer bar-to-lower arm bolts, washers, spacers and rubber bushings and tighten the nuts securely.
7   Tighten the bracket bolts.

3   Front shock absorber - removal and installation

Refer to illustrations 3.3 and 3.5
Warning: *Whenever any of the suspension or steering fasteners are loosened or removed, they must be inspected and, if necessary, replaced with new ones of the same part number or of original equipment quality and design. Torque specifications must be followed for proper reassembly and component retention.*
1   Jack up the front of the vehicle and place it securely on jackstands.
2   Remove the front wheels.
3   Remove the nut holding the shock absorber to the mount **(see illustration)**. Clamp a pair of locking pliers to the flats at the top of the shock rod to prevent it from turning.

4   Remove the upper washer and rubber cushion from the shaft of the shock absorber.
5   Disconnect the shock from the lower arm by removing the single through bolt **(see illustration)**.
6   Fully compress the shock absorber.
7   Tilt the shock forward, turn it 90-degrees so the bushing is at right angles to the vehicle and pull it out.
8   Installation is the reverse of the removal procedure. Make sure the washers and bushings are assembled in the proper order. Tighten the bolts and nuts securely.

4   Front strut bar - removal and installation

Warning: *Whenever any of the suspension or steering fasteners are loosened or removed, they must be inspected and, if necessary, replaced with new ones of the same part number or of original equipment quality and design. Torque specifications must be followed for proper reassembly and component retention.*
1   Raise the front of the vehicle and support it securely on jackstands. Apply the parking brake.

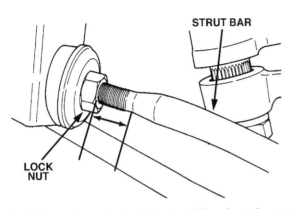

4.2 On 2WD models, mark the locknut and threads on the strut bar to ensure correct alignment during installation

4.3 To disconnect the strut bar from the frame, remove this nut (arrow), washer and rubber bushing

4.4 To disconnect the strut bar from the lower control arm, remove these bolts (arrows)

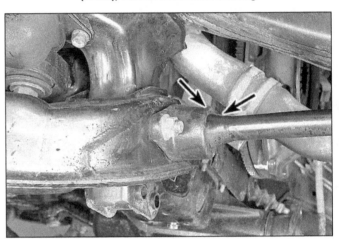

5.4 Mark the lower arm and torsion bar relationship (arrows) before removing the torsion bar

2WD models

Refer to illustrations 4.2, 4.3 and 4.4

2   Place alignment marks on the threaded portion of the strut bar and the inner locknut **(see illustration)**. If the position of the inner locknut changes, the caster setting will be affected.

3   Remove the outer nut from the strut bar **(see illustration)**.

4   Remove the bolts holding the strut bar to the lower arm and remove the strut bar **(see illustration)**.

5   Inspect the strut bar and components for wear or damage. Replace parts as necessary with new ones.

6   Installation is the reverse of the removal procedure. Make sure to first thread the inner locknut onto the strut bar and align the marks made before removal. Tighten the strut bar-to-lower arm bolts to the torque listed in this Chapter's Specifications. Lower the vehicle and tighten the outer nut to the torque listed in this Chapter's Specifications.

4WD models (1981 through 1989 models only)

7   Remove the outer nuts, bushings and

washers securing the strut bar to the frame **(see illustration 2.2b)**.

8   Remove the bolts holding the strut bar to the lower arm and remove the strut bar.

9   Inspect the strut bar and components for wear or damage. Replace parts as necessary with new ones.

10   Installation is the reverse of the removal procedure. Tighten the strut bar-to-lower arm bolts according to this Chapter's Specifications.

11   Lower the vehicle and tighten the outer nuts to the torque listed in this Chapter's Specifications.

---

5   Torsion bar - removal, installation and adjustment

Refer to illustration 5.4

Warning: *Whenever any of the suspension or steering fasteners are loosened or removed, they must be inspected and, if necessary, replaced with new ones of the same part number or of original equipment quality and design. Torque specifications must be followed for proper reassembly and component retention.*

Removal

1   Loosen the wheel lug nuts, raise the vehicle and support it securely on jackstands. Remove the wheel.

2   Use white paint to place alignment marks on the torsion bar and the height control arm **(see illustration 2.2a or 2.2b)**.

3   Place alignment marks on the threads of the height control arm seat and bolt, then remove the bolt and height control arm.

4   Mark the torsion bar-to-lower arm relationship, withdraw the torsion bar from the arm and remove the torsion bar assembly from the vehicle **(see illustration)**.

5   Inspect the torsion bar, control arm and height control arm splines for damage and wear. Inspect the bar for distortion, cracks and nicks, which will cause the bar to fail prematurely.

Installation and adjustment

6   Apply a coat of multi-purpose grease to the splines of the torsion bar.

7   Slide the height control arm onto the torsion bar, aligning the marks made earlier.

8   Raise the lower arm as necessary to line up the match marks from the torsion bar to

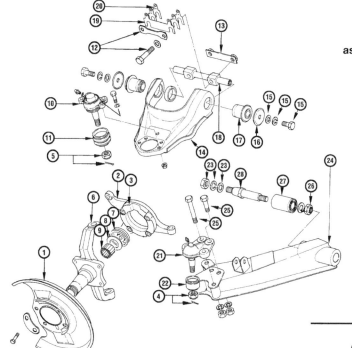

6.4a An exploded view of a typical 4WD front suspension assembly (1981 through 1987 pick-up models and 1984 through 1986 Trooper models)

1	Brake backing plate	16	Plate
2	Steering knuckle arm	17	Bushing
3	Dust shield	18	Upper control arm
4	Lower balljoint nut and		pivot shaft
	cotter pin	19	Camber shims
5	Upper balljoint nut and	20	Caster shims
	cotter pin	21	Lower balljoint
6	Steering knuckle	22	Balljoint dust boot
7	Oil seal	23	Nut and washers
8	Washer	24	Lower control arm
9	Needle bearing	25	Balljoint-to-lower
10	Upper balljoint		control arm bolts
11	Balljoint dust boot	26	Nut and washer
12	Bolt and plate	27	Bushing
13	Pivot bolt nut	28	Lower control arm
14	Upper control arm		pivot bolt
15	Bolt and washers		

the lower arm and insert the bar into the arm. Install the height control arm and bolt, tightening it until the alignment marks are matched.

6 Front suspension upper arm - removal and installation

Refer to illustrations 6.4a, 6.4b and 6.4c

Warning: *Whenever any of the suspension or steering fasteners are loosened or removed, they must be inspected and, if necessary, replaced with new ones of the same part number or of original equipment quality and design. Torque specifications must be followed for proper reassembly and component retention.*

Removal

1 Loosen the wheel lug nuts, raise the front of the vehicle and support it securely on jackstands. Apply the parking brake. Remove the wheel.

2 Support the lower arm with a jack or jackstand. The support point must be as close to the balljoint as possible to give maximum leverage on the lower arm.

3 Remove the cotter pin and nut and separate the upper balljoint from the steering knuckle, but be careful not to let the steering knuckle/brake caliper assembly fall outward, as this may damage the brake hose (see Section 8).

4 Remove the upper arm-to-frame bolts, noting the positions of any alignment shims. They must be reinstalled in the same locations to maintain wheel alignment **(see illustrations)**.

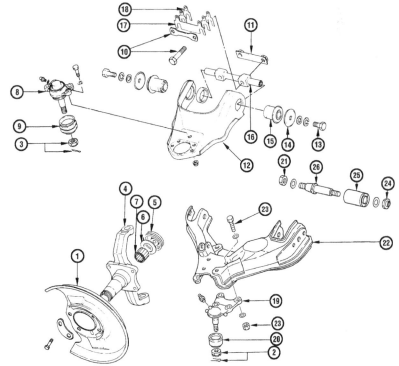

6.4b An exploded view of a typical 4WD front suspension assembly (1987 and later Trooper models and 1988 and later pick-up models)

1	Brake backing plate	10	Bolt and plate	20	Balljoint dust boot
2	Lower balljoint nut	11	Pivot bolt nut	21	Nut and washer
	and cotter pin	12	Upper control arm	22	Lower control arm
3	Upper balljoint nut	13	Bolt and washers	23	Balljoint-to-lower
	and cotter pin	14	Plate		control arm bolts and
4	Steering knuckle	15	Bushing		nuts (4)
5	Oil seal	16	Upper control arm	24	Nut and washer
6	Washer		pivot shaft	25	Bushing
7	Needle bearing	17	Camber shims	26	Lower control arm
8	Upper balljoint	18	Caster shims		pivot bolt
9	Balljoint dust boot	19	Lower balljoint		

5   Detach the upper arm from the vehicle. Inspect the bushings for deterioration, cracks and other damage, replacing them, if necessary.

Installation

6   Position the arm on the frame and install the bolts. Install any alignment shims that were removed. Tighten the bolts to the torque listed in this Chapter's Specifications.
7   Connect the balljoint to the steering knuckle and tighten the nut to the torque listed in this Chapter's Specifications.
8   Install the wheel and lug nuts and lower the vehicle. Tighten the lug nuts to the torque specified in Chapter 1.
9   Drive the vehicle to an alignment shop to have the front end alignment checked and, if necessary, adjusted.

---

7   Front suspension lower arm - removal and installation

Warning: *Whenever any of the suspension or steering fasteners are loosened or removed, they must be inspected and, if necessary, replaced with new ones of the same part number or of original equipment quality and design. Torque specifications must be followed for proper reassembly and component retention.*

Removal

1   Loosen the wheel lug nuts, raise the front of the vehicle and support it securely on jackstands. Remove the wheel.
2   Disconnect the shock absorber from the lower arm (Section 3).
3   Disconnect the stabilizer bar from the lower arm (Section 2).
4   Remove the strut rod (Section 4), if so equipped.
5   Remove the torsion bar (Section 5).
6   Loosen, but do not remove the lower arm pivot bolt(s).
7   Disconnect the balljoint from the lower arm by removing the nuts and bolts.
8   To remove the pivot bolt, remove the nut, then pull the bolt out or drive it out with a hammer and drift punch **(see illustration 6.4a, 6.4b or 6.4c)**.

Installation

9   Inspect the lower arm bushing for deterioration, cracking and other damage. If the bushing is in need of replacement, take the lower arm to an automotive machine shop to have the old bushing pressed out and a new one pressed in.
10   Connect the balljoint to the lower arm and install the bolts and nuts, tightening them to the specified torque. Raise the lower arm to simulate the normal ride height, then tighten the pivot bolt nut(s) to the specified torque. Install the torsion bar (Section 6).
11   Install the strut rod (Section 4) and shock absorber (Section 3) and connect the stabilizer bar to the lower arm (Section 2).

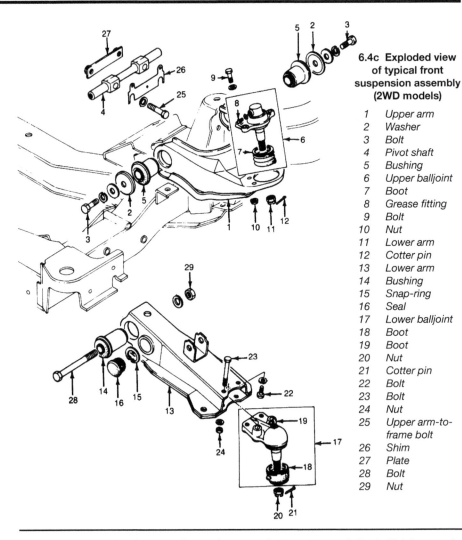

6.4c Exploded view of typical front suspension assembly (2WD models)

1   Upper arm
2   Washer
3   Bolt
4   Pivot shaft
5   Bushing
6   Upper balljoint
7   Boot
8   Grease fitting
9   Bolt
10   Nut
11   Lower arm
12   Cotter pin
13   Lower arm
14   Bushing
15   Snap-ring
16   Seal
17   Lower balljoint
18   Boot
19   Boot
20   Nut
21   Cotter pin
22   Bolt
23   Bolt
24   Nut
25   Upper arm-to-frame bolt
26   Shim
27   Plate
28   Bolt
29   Nut

12   Install the wheel and lug nuts. Lower the vehicle and tighten the lug nuts to the specified torque.
13   Drive the vehicle to an alignment shop to have the front end alignment checked and, if necessary, adjusted.

---

8   Balljoints - check and replacement

Warning: *Whenever any of the suspension or steering fasteners are loosened or removed, they must be inspected and, if necessary, replaced with new ones of the same part number or of original equipment quality and design. Torque specifications must be followed for proper reassembly and component retention.*

Check

1   Raise the vehicle and support it securely on jackstands.
2   Visually inspect the rubber seal for cuts, tears or leaking grease. If any of these conditions are noticed, the balljoint should be replaced.
3   Place a large pry bar under the balljoint

and attempt to push the balljoint upwards. Next, position the pry bar between the steering knuckle and the arm and apply downward pressure. If any movement is seen or felt during either of these checks, a worn out balljoint is indicated.
4   Have an assistant grasp the tire at the top and bottom and shake the top of the tire in an in-and-out motion. Touch the balljoint stud castellated nut. If any looseness is felt, suspect a worn out balljoint stud or a widened hole in the steering knuckle boss. If the latter problem exists, the steering knuckle should be replaced as well as the balljoint.

Replacement

Refer to illustrations 8.6a and 8.6b

5   With the vehicle still raised and supported, position a floor jack under the lower arm - it must stay there throughout the entire operation. Remove the wheel.
6   Remove the cotter pin and loosen the castellated nut a couple of turns, but don't remove it (it will prevent the balljoint and steering knuckle from separating violently) **(see illustrations)**.
7   Using a balljoint separating tool, separate the balljoint from the steering knuckle. There are several types of balljoint tools avail-

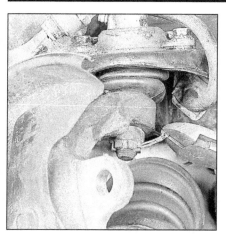

8.6a Use wire cutters or pliers to remove the cotter pin

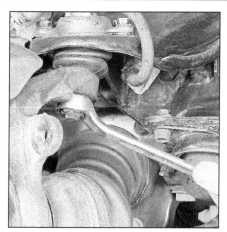

8.6b Unscrew the balljoint nut two turns

10.3 The lower mounting nut on the rear shock absorber

able, but the kind that pushes the balljoint stud out of the knuckle boss works the best. The wedge, or "pickle fork" type works fairly well, but it tends to damage the balljoint seal. Some two-jaw pullers will do the job, also.

8   Remove the castle nut and disconnect the balljoint from the steering knuckle. If the upper arm balljoint is being worked on, be careful not to let the steering knuckle/hub/brake assembly fall outward, because the brake hose might be damaged. If necessary, wire the assembly to the frame to prevent this from happening.

9   On most models, the balljoint is retained to the suspension arm by bolts. Remove the bolts (and nuts on some models), then detach the balljoint from the arm. On some 4WD models, the balljoint is attached to the upper suspension arm by a large nut. Unscrew the nut and remove the balljoint. Take note of how the balljoint is positioned on the arm - the new one must be installed the same way **(see illustrations 6.4a and 6.4b).**

10   Position the new balljoint on the arm and install the nut or bolts, tightening them to the specified torque.

11   Insert the balljoint stud into the steering knuckle boss, install the castellated nut and tighten it to the specified torque. Install a new cotter pin. If necessary, tighten the nut an additional amount to line up the slots in the nut with the hole in the balljoint stud (never loosen the nut to align the hole).

12   Install the wheel and lug nuts. Lower the vehicle and tighten the lug nuts to the specified torque.

9   Steering knuckle - removal and installation

Warning: *Whenever any of the suspension or steering fasteners are loosened or removed, they must be inspected and, if necessary, replaced with new ones of the same part number or of original equipment quality and design. Torque specifications must be followed for proper reassembly and component retention.*

Removal

1   Loosen the front wheel lug nuts, raise the front of the vehicle and support it securely on jackstands placed under the front suspension lower arms. Apply the front brake. Remove the front wheels.

2   Remove the brake caliper and place it on top of the upper arm or wire it up out of the way. Remove the caliper mounting bracket from the steering knuckle (see Chapter 9 if necessary).

3   Remove the front hub assembly and brake disc (see Chapter 1 or 8).

4   Remove the splash shield from the steering knuckle.

5   Separate the tie-rod end from the knuckle arm (see Section 13).

6   Remove the cotter pins from the upper and lower balljoint studs and back off the nuts two turns each (see Section 8).

7   Break the balljoints loose from the steering knuckle (see Section 8).

8   Remove the nuts from the balljoint studs, separate the suspension arms from the steering knuckle and remove the knuckle from the vehicle.

Installation

9   Place the knuckle between the upper and lower suspension arms and insert the balljoint studs into the knuckle, beginning with the lower balljoint. Install the nuts and tighten them to the specified torque. Install new cotter pins, tightening the nuts slightly to align the slots in the nuts with the holes in the balljoint studs, if necessary.

10   Install the splash shield.

11   Connect the tie-rod end to the knuckle arm and tighten the nut to the specified torque. Be sure to use a new cotter pin.

12   Install the hub and brake disc assembly (Chapter 1 or 8).

13   Install the brake caliper mounting bracket and caliper (Chapter 9).

14   Install the wheel and lug nuts. Lower the vehicle to the ground and tighten the nuts to the specified torque.

10   Rear shock absorber - removal and installation

Refer to illustration 10.3

Warning: *Whenever any of the suspension or steering fasteners are loosened or removed, they must be inspected and, if necessary, replaced with new ones of the same part number or of original equipment quality and design. Torque specifications must be followed for proper reassembly and component retention.*

1   If the shock absorber is to be replaced with a new one, it is recommended that both shocks on the rear of the vehicle be replaced at the same time.

2   Raise and support the rear of the vehicle according to the jacking and towing procedures in this manual. Use a jack to raise the differential until the tires clear the ground, then place jackstands under the axle housing. Do not attempt to remove the shock absorbers with the vehicle raised and the axle unsupported.

3   Unscrew and remove the lower shock absorber mounting nut to disconnect it at the spring seat **(see illustration).**

4   Unscrew and remove the upper mounting nut at the frame and remove the shock absorber.

5   If the unit is defective, it must be replaced with a new one. Replace worn rubber bushings with new ones.

6   Install the shocks in the reverse order of removal, but let the vehicle be free standing before tightening the mounting bolts and nuts.

7   Bounce the rear of the vehicle a couple of times to settle the bushings into place, then tighten the nuts securely.

11   Rear leaf spring - removal and installation

Refer to illustration 11.6

Warning: *Whenever any of the suspension or steering fasteners are loosened or removed,*

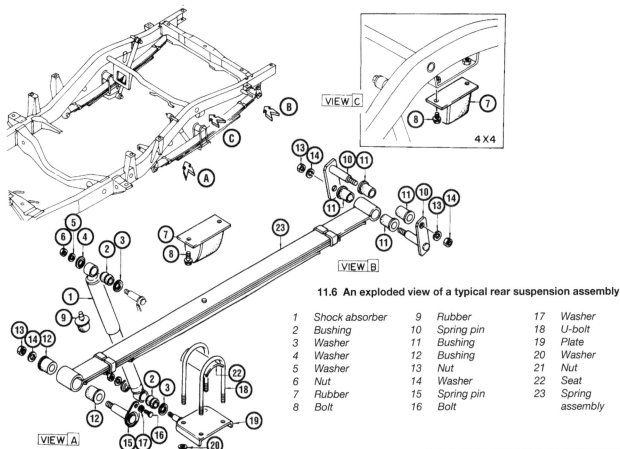

11.6 An exploded view of a typical rear suspension assembly

1	Shock absorber	9	Rubber	17	Washer
2	Bushing	10	Spring pin	18	U-bolt
3	Washer	11	Bushing	19	Plate
4	Washer	12	Bushing	20	Washer
5	Washer	13	Nut	21	Nut
6	Nut	14	Washer	22	Seat
7	Rubber	15	Spring pin	23	Spring
8	Bolt	16	Bolt		assembly

they must be inspected and, if necessary, replaced with new ones of the same part number or of original equipment quality and design. Torque specifications must be followed for proper reassembly and component retention.

1   Jack up the vehicle and support the frame securely on jackstands.
2   Remove the rear wheels and tires.
3   Place a jack under the rear differential housing.
4   Lower the axle housing until the leaf spring tension is relieved, and lock the jack in this position.
5   Disconnect the shock absorber from the spring seat (see Section 10).
6   Remove the U-bolt mounting nuts **(see illustration)**.
7   Remove the U-bolt nuts and spring seat.
8   Unbolt and remove the shackle pin assemblies and lower the spring from the vehicle.
9   If the bushings in the spring or frame are deteriorated, replace them. If you need to replace the bushings in the spring, take the spring to an automotive machine shop to have the old bushings pressed out and new ones pressed in.
10   Installation is the reverse of the removal procedure. Lubricate the shackle pins and bushings with lithium base grease. When installing the U-bolt and shackle pin nuts, tighten them to the specified torque values.

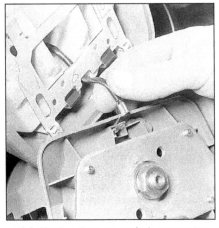

12.2 Unplug the horn switch connector

12   Steering wheel - removal and installation

Refer to illustrations 12.2, 12.3 and 12.4
Warning: Whenever any of the suspension or steering fasteners are loosened or removed, they must be inspected and, if necessary, replaced with new ones of the same part number or of original equipment quality and design. Torque specifications must be followed for proper reassembly and component retention.

12.3 Mark the steering wheel hub-to-shaft relationship (arrows)

Removal

1   Disconnect the cable from the negative battery terminal.
2   Remove the horn button or center pad and unplug the horn switch connector **(see illustration)**.
3   Unscrew the steering wheel nut and mark the steering wheel hub and the column shaft with paint to ensure correct repositioning during reassembly **(see illustration)**.
4   Remove the wheel using a puller **(see illustration). Caution:** Do not hammer on the wheel or the shaft to separate them.

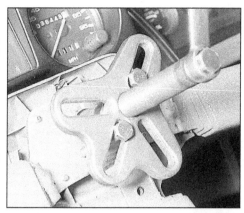

12.4 Remove the steering wheel from the shaft with a puller – DO NOT hammer on the shaft!

Installation

5   Realign the steering wheel and the column shaft using the match marks. Install and tighten the retaining nut to the specified torque, then attach the horn button or center pad.
6   Hook up the negative battery cable.

---

13   Steering linkage - removal and installation

Warning: *Whenever any of the suspension or steering fasteners are loosened or removed, they must be inspected and, if necessary, replaced with new ones of the same part number or of original equipment quality and design. Torque specifications must be followed for proper reassembly and component retention.*

1   All steering linkage removal and installation procedures should be performed with the front end of the vehicle raised and placed securely on jackstands.
2   Before removing any steering linkage components, obtain a balljoint separator. It may be a screw-type puller or a wedge-type (pickle fork) tool, although the wedge-type tool tends to damage the balljoint seals. It is possible to jar a balljoint taper pin free from

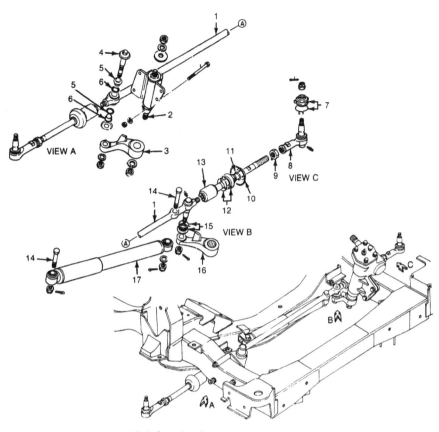

13.4 Steering linkage components

1	Center link	7	Boot	13	Center link
2	Idler arm pivot shaft	8	Tie-rod end	14	Bolt
3	Idler arm	9	Nut	15	Boot
4	Pin	10	Cover	16	Pitman arm
5	Bushing	11	Clamp ring	17	Steering damper
6	O-ring	12	Boot		

its eye by striking opposite sides of the eye simultaneously with two large hammers, but the space available to do so is usually very limited.
3   After installing any of the steering linkage components, the front wheel alignment should be checked by a reputable front end alignment shop.

Pitman arm

Refer to illustration 13.4
4   Remove the nut securing the Pitman arm to the steering gear sector shaft **(see illustration)**.
5   Scribe or paint match marks on the arm and shaft.
6   Using a puller, disconnect the Pitman arm from the shaft splines.
7   Remove the cotter pin and castle nut securing the Pitman arm to the center link.
8   Using a puller, disconnect the Pitman arm from the center link.
9   Installation is the reverse of the removal procedure. Be sure to tighten the nuts to the specified torque and install a new cotter pin.

Tie-rod

Refer to illustrations 13.10a, 13.10b and 13.11
10   Remove the cotter pins and castle nuts securing the tie-rod to the center link and knuckle arm **(see illustrations)**.
11   Separate the tie-rod from the center link

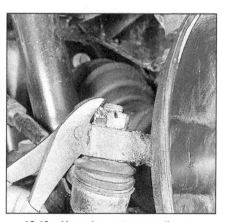

13.10a Use wire cutters or pliers to remove the cotter pin

13.10b Unscrew the tie-rod end nut at the steering knuckle

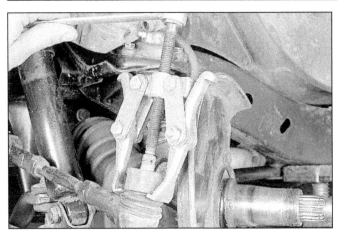

13.11 Use a puller to separate the tie-rod end from the steering knuckle

15.6 When tightening the adjusting screw locknut, be sure to hold the screw to keep it from turning

and knuckle arm with a puller **(see illustration)**.

12   If a tie-rod end is to be replaced, loosen the jam nut and mark the relationship of the tie-rod end to the tie-rod with white paint. When installing the new rod end, thread it onto the tie-rod until it reaches the mark, then turn the jam nut until it contacts the rod end. Don't tighten if fully at this time.

13   Turn the tie-rod ends so they are at approximately 90-degree angles to each other, then tighten the jam nuts to lock the ends in position.

14   The remaining installation steps are the reverse of those for removal. Make sure to tighten the castle nuts to the specified torque.

Center link

15   Remove the cotter pins and castle nuts securing the tie-rod ends to the center link.

16   Remove the cotter pin and castle nut securing the center link to the Pitman arm.

17   Remove the cotter pin and castle nut securing the center link to the idler arm.

18   Using a puller, separate the center link from the tie-rod ends, Pitman arm and idler arm.

19   Installation is the reverse of the removal procedure. Be sure to tighten all nuts to the specified torque.

Steering damper

20   Remove the cotter pin, nut and washer from the bolt that attaches the damper to the center link.

21   Remove the cotter pin, nut and washer attaching the damper to the cross- member (2WD) or idler arm bracket.

22   Lower the steering damper from the vehicle.

23   Installation is the reverse of removal.

14   Steering gear - removal and installation

Warning: *Whenever any of the suspension or steering fasteners are loosened or removed,*

they must be inspected and, if necessary, replaced with new ones of the same part number or of original equipment quality and design. Torque specifications must be followed for proper reassembly and component retention.

Note: *If you find that the steering gear is defective, it is not recommended that you overhaul it. Because of the special tools needed to do the job, it is best to let your dealer service department overhaul it for you (or replace it with a factory rebuilt unit). However, you can remove and install it yourself by following the procedure outlined here. The removal and installation procedures for manual steering and power steering gear housings are identical except that the inlet and outlet lines must be removed from the steering gear housing on power steering-equipped models before the housing can be removed. The steering system should be filled and power steering systems should be bled after the gear housing is reinstalled (see Section 17).*

1   Raise the front of the vehicle and place it securely on jackstands. Apply the parking brake.

2   Place an alignment mark on the steering coupling and the gear housing worm shaft to assure correct reassembly, then remove the coupling bolt.

3   Remove the Pitman arm (Section 13).

4   Remove the bolts securing the gear housing to the chassis and pull the gear housing from the coupling **(see illustration 13.4)**.

5   Installation is the reverse of the removal procedure except that the Pitman arm nut should be tightened before the Pitman arm is connected to the center link. Be sure to tighten all nuts and bolts to the specified torque.

15   Steering freeplay - adjustment

Refer to illustration 15.6

1   Raise the vehicle with a jack so that the front wheels are off the ground and place the vehicle securely on jackstands.

2   Point the wheels straight ahead.

3   Using a wrench, loosen the locknut on the steering gear.

4   Turn the adjusting screw clockwise to decrease wheel freeplay and counterclockwise to increase it. **Note:** *Turn the adjusting screw in small increments, checking the steering wheel freeplay between them.*

5   Turn the steering wheel halfway around in both directions, checking that the freeplay is correct and that the steering is smooth.

6   Hold the adjusting screw so that it will not turn and tighten the locknut **(see illustration)**.

7   Remove the jackstands and lower the vehicle.

16   Power steering pump - removal and installation

Refer to illustration 16.1

Warning: *Whenever any of the suspension or steering fasteners are loosened or removed, they must be inspected and, if necessary, replaced with new ones of the same part number or of original equipment quality and design. Torque specifications must be followed for proper reassembly and component retention.*

Note: *If you find that the steering pump is defective, it is not recommended that you overhaul it. Because of the special tools needed to do the job, it is best to let your dealer service department overhaul it for you (or replace it with a factory rebuilt unit). However, you can remove it yourself using the procedure which follows.*

Removal

1   Position a container to catch the pump fluid. Disconnect the hoses from the pump body. As each is disconnected, cap or tape over the hose opening and then secure the end in a raised position to prevent leakage and contamination. Cover or plug the pump openings to so dirt won't enter the unit **(see illustration)**.

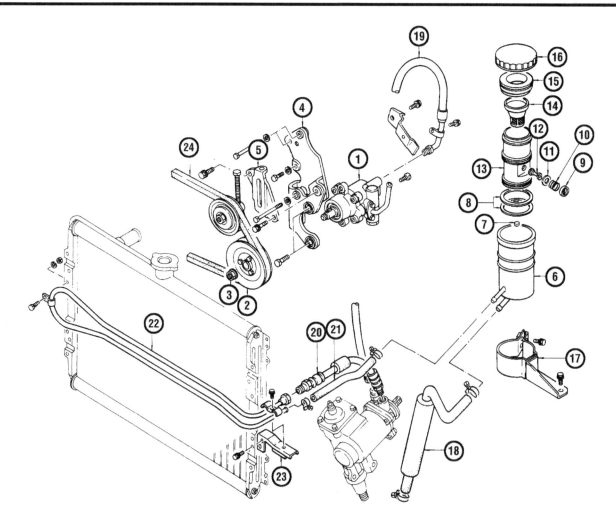

16.1 An exploded view of a typical power steering system

1	Power steering pump	9	Spring cover	17	Reservoir bracket
2	Pump pulley	10	Spring	18	Supply hose
3	Pump pulley nut	11	Ring	19	Pressure hose
4	Pump bracket	12	Holder	20	Return hose
5	Adjuster bracket	13	Strainer	21	Return hose
6	Reservoir	14	Filter	22	Return pipe
7	Magnet	15	Cap gasket	23	Pipe bracket
8	Seal	16	Cap	24	Drivebelt

2   Loosen the idler pulley locknut and turn the adjusting bolt until the drivebelt can be removed from the pump pulley.

3   Remove the pump mounting bolts and lift the pump out of the engine compartment.

Installation

4   Installation is the reverse of the removal procedure. When installing the pressure line, make sure there is sufficient clearance between the line and the exhaust manifold.

5   To adjust the drivebelt tension, see Chapter 1.

6   Fill the power steering fluid reservoir with the specified fluid and bleed the power steering system (see Section 17).

7   Check for fluid leaks.

17 Power steering system - bleeding

1   Check the fluid in the reservoir and add fluid of the specified type if it is low.

2   Jack up the front of the vehicle and place it securely on jackstands.

3   With the engine off, turn the steering wheel fully in both directions two or three times.

4   Recheck the fluid in the reservoir and add more fluid if necessary.

5   Start the engine and turn the steering wheel fully in both directions two or three times. The engine should be running at 1000 rpm or less.

6   Remove the jackstands and lower the vehicle completely.

7   With the engine running at 1000 rpm or less, turn the steering wheel fully in both directions two or three times.

8   Return the steering wheel to the center position.

9   Check that the fluid is not foamy or cloudy.

10   Measure the fluid level with the engine running.

11   Turn the engine and again measure the fluid level. It should rise no more than 0.20 in (5 mm) when the engine is turned off.

12   If a problem is encountered, repeat Steps 7 through 11.

13   If the problem persists, remove the pump (see Section 16), and have it repaired by a dealer service department.

18 Front end alignment - general information

Refer to illustration 18.1

A front end alignment refers to the adjustments made to the front wheels so they are in proper angular relationship to the suspension and the ground. Front wheels that are out of proper alignment not only affect steering control, but also increase tire wear. The front end adjustments normally required are camber, caster and toe-in **(see illustration)**.

Getting the proper front wheel alignment is a very exacting process, one in which complicated and expensive machines are necessary to perform the job properly. Because of this, you should have a technician with the proper equipment perform these tasks. We will, however, use this space to give you a basic idea of what is involved with front end alignment so you can better understand the process and deal intelligently with the shop that does the work.

Toe-in is the turning in of the front wheels. The purpose of a toe specification is to ensure parallel rolling of the front wheels. In a vehicle with zero toe-in, the distance between the front edges of the wheels will be the same as the distance between the rear edges of the wheels. The actual amount of toe-in is normally only a fraction of an inch. Toe-in adjustment is controlled by the tie-rod end position on the inner tie-rod. Incorrect toe-in will cause the tires to wear improperly by making them scrub against the road surface.

Camber is the tilting of the front wheels from the vertical when viewed from the front of the vehicle. When the wheels tilt out at the top, the camber is said to be positive (+). When the wheels tilt in at the top the camber is negative (-). The amount of tilt is measured in degrees from the vertical and this measurement is called the camber angle. This angle affects the amount of tire tread which contacts the road and compensates for changes in the suspension geometry when the vehicle is cornering or travelling over an undulating surface.

Caster is the tilting of the front steering axis from the vertical. A tilt toward the rear is positive caster and a tilt toward the front is negative caster. Caster angle affects the self-centering action of the steering, which governs straight-line stability.

Caster is adjusted by moving shims from one end of the upper arm mount to the other.

19 Wheels and tires - general information

Wheels can be damaged by an impact with a curb or other solid object. If the wheels are bent, the result is a hazardous condition which must be corrected. To check the

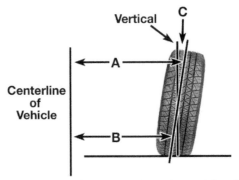

CAMBER ANGLE (FRONT VIEW)

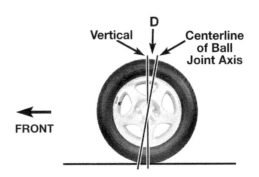

CASTER ANGLE (SIDE VIEW)

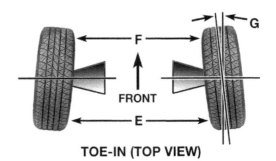

TOE-IN (TOP VIEW)

18.1 Front end alignment details

A minus B = C (degrees camber)
D = caster (expressed in degrees)
E minus F = toe-in (measured in inches)
G = toe-in (expressed in degrees)

wheels, raise the vehicle and set it on jackstands. Visually inspect the wheels for obvious signs of damage such as cracks and deformation.

Tire and wheel balance is very important to the overall handling, braking and ride performance of the vehicle. Whenever a tire is dismounted for repair or replacement, the tire and wheel assembly should be balanced before being installed on the vehicle.

Wheels should be periodically cleaned, especially on the inside, where mud and road salts accumulate and eventually cause rust and, ultimately, possible wheel failure.

Tires are extremely important from a safety standpoint. The tread should be checked periodically to see that the tires have not worn excessively, a condition which can be dangerous, especially in wet weather.

To equalize wear and add life to a set of tires, it is recommended that they be rotated periodically. When rotating, check for signs of abnormal wear and foreign objects in the tread or sidewalls (refer to Chapter 1, Routine Maintenance).

Proper tire inflation is essential for maximum life of the tread and for proper handling and braking.

Tires that are wearing in an abnormal way are an indication that their inflation is incorrect or that the front end components are not adjusted properly. Take the vehicle to a reputable front end alignment and repair shop to correct the situation.

Chapter 11   Body

Contents

1   General information

These models feature a separate boxed steel frame and body. Certain components are particularly vulnerable to accident damage and can be unbolted and repaired or replaced. Among these parts are the body moldings, doors, bumpers, hood, tailgate and all glass.

Only general body maintenance practices and body panel repair procedures within the scope of the do-it-yourselfer are included in this Chapter.

2   Body - maintenance

1   The condition of your vehicle's body is very important, because the resale value depends a great deal on it. It's much more difficult to repair a neglected or damaged body than it is to repair mechanical components. The hidden areas of the body, such as the wheel wells, the frame and the engine compartment, are equally important, although they don't require as frequent attention as the rest of the body.
2   Once a year, or every 12,000 miles, it's a good idea to have the underside of the body steam cleaned. All traces of dirt and oil will be removed and the area can then be inspected carefully for rust, damaged brake lines, frayed electrical wires, damaged cables and other problems. The front suspension components should be greased after completion of this job.
3   At the same time, clean the engine and the engine compartment with a steam cleaner or water soluble degreaser.
4   The wheel wells should be given close attention, since undercoating can peel away and stones and dirt thrown up by the tires can cause the paint to chip and flake, allowing rust to set in. If rust is found, clean down to the bare metal and apply an anti-rust paint.
5   The body should be washed about once a week. Wet the vehicle thoroughly to soften the dirt, then wash it down with a soft sponge and plenty of clean soapy water. If the surplus dirt is not washed off very carefully, it can wear down the paint.
6   Spots of tar or asphalt thrown up from the road should be removed with a cloth soaked in solvent.
7   Once every six months, wax the body and chrome trim. If a chrome cleaner is used to remove rust from any of the vehicle's plated parts, remember that the cleaner also removes part of the chrome, so use it sparingly.

3   Vinyl trim - maintenance

1   Don't clean vinyl trim with detergents, caustic soap or petroleum based cleaners. Plain soap and water works just fine, with a soft brush to clean dirt that may be ingrained. Wash the vinyl as frequently as the rest of the vehicle.
2   After cleaning, application of a high quality rubber and vinyl protectant will help prevent oxidation and cracks. The protectant can also be applied to weather-stripping, vacuum lines and rubber hoses, which often fail as a result of chemical degradation, and to the tires.

4   Upholstery and carpets - maintenance

1   Every three months remove the carpets or mats and clean the interior of the vehicle (more frequently if necessary). Vacuum the upholstery and carpets to remove loose dirt and dust.
2   Leather upholstery requires special care. Stains should be removed with warm water and a very mild soap solution. Use a clean, damp cloth to remove the soap, then wipe again with a dry cloth. Never use alcohol, gasoline, nail polish remover or thinner to clean leather upholstery.
3   After cleaning, regularly treat leather upholstery with a leather wax. Never use car wax on leather upholstery.
4   In areas where the interior of the vehicle is subject to bright sunlight, cover leather seats with a sheet if the vehicle is to be left out for any length of time.

5   Body repair - minor damage

See photo sequence

Repair of scratches

1   If the scratch is superficial and does not penetrate to the metal of the body, repair is very simple. Lightly rub the scratched area with a fine rubbing compound to remove loose paint and built up wax. Rinse the area with clean water.
2   Apply touch-up paint to the scratch, using a small brush. Continue to apply thin layers of paint until the surface of the paint in

These photos illustrate a method of repairing simple dents. They are intended to supplement *Body repair - minor damage* in this Chapter and should not be used as the sole instructions for body repair on these vehicles.

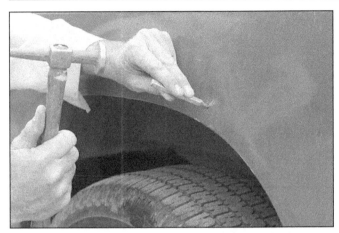

1 If you can't access the backside of the body panel to hammer out the dent, pull it out with a slide-hammer-type dent puller. In the deepest portion of the dent or along the crease line, drill or punch hole(s) at least one inch apart . . .

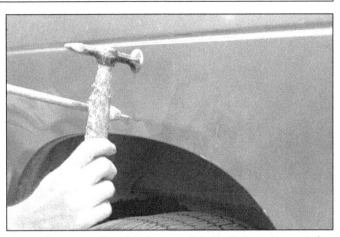

2 . . . then screw the slide-hammer into the hole and operate it. Tap with a hammer near the edge of the dent to help 'pop' the metal back to its original shape. When you're finished, the dent area should be close to its original contour and about 1/8-inch below the surface of the surrounding metal

3 Using coarse-grit sandpaper, remove the paint down to the bare metal. Hand sanding works fine, but the disc sander shown here makes the job faster. Use finer (about 320-grit) sandpaper to feather-edge the paint at least one inch around the dent area

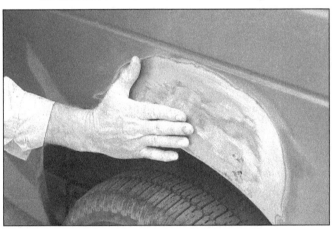

4 When the paint is removed, touch will probably be more helpful than sight for telling if the metal is straight. Hammer down the high spots or raise the low spots as necessary. Clean the repair area with wax/silicone remover

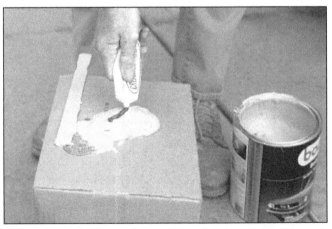

5 Following label instructions, mix up a batch of plastic filler and hardener. The ratio of filler to hardener is critical, and, if you mix it incorrectly, it will either not cure properly or cure too quickly (you won't have time to file and sand it into shape)

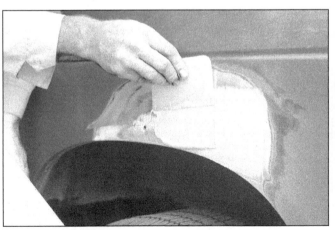

6 Working quickly so the filler doesn't harden, use a plastic applicator to press the body filler firmly into the metal, assuring it bonds completely. Work the filler until it matches the original contour and is slightly above the surrounding metal

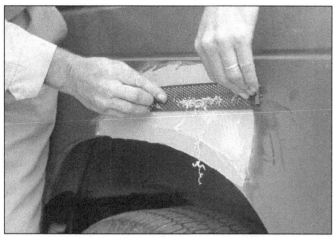

7 Let the filler harden until you can just dent it with your fingernail. Use a body file or Surform tool (shown here) to rough-shape the filler

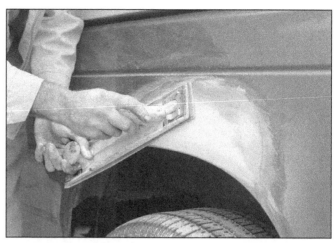

8 Use coarse-grit sandpaper and a sanding board or block to work the filler down until it's smooth and even. Work down to finer grits of sandpaper - always using a board or block - ending up with 360 or 400 grit

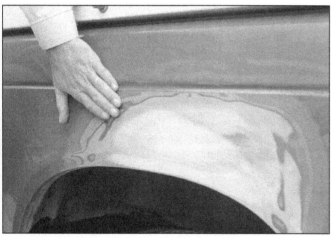

9 You shouldn't be able to feel any ridge at the transition from the filler to the bare metal or from the bare metal to the old paint. As soon as the repair is flat and uniform, remove the dust and mask off the adjacent panels or trim pieces

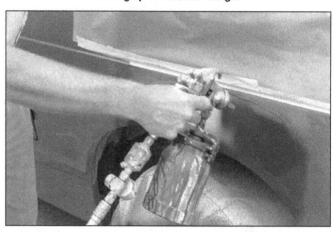

10 Apply several layers of primer to the area. Don't spray the primer on too heavy, so it sags or runs, and make sure each coat is dry before you spray on the next one. A professional-type spray gun is being used here, but aerosol spray primer is available inexpensively from auto parts stores

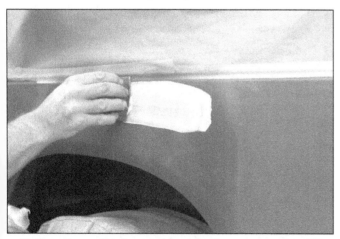

11 The primer will help reveal imperfections or scratches. Fill these with glazing compound. Follow the label instructions and sand it with 360 or 400-grit sandpaper until it's smooth. Repeat the glazing, sanding and respraying until the primer reveals a perfectly smooth surface

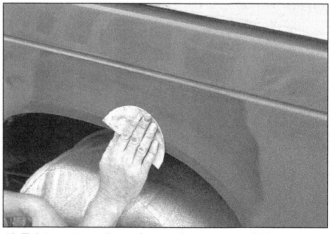

12 Finish sand the primer with very fine sandpaper (400 or 600-grit) to remove the primer overspray. Clean the area with water and allow it to dry. Use a tack rag to remove any dust, then apply the finish coat. Don't attempt to rub out or wax the repair area until the paint has dried completely (at least two weeks)

the scratch is level with the surrounding paint. Allow the new paint at least two weeks to harden, then blend it into the surrounding paint by rubbing with a very fine rubbing compound. Finally, apply a coat of wax to the scratch area.

3   If the scratch has penetrated the paint and exposed the metal of the body, causing the metal to rust, a different repair technique is required. Remove all loose rust from the bottom of the scratch with a pocket knife, then apply rust inhibiting paint to prevent the formation of rust in the future. Using a rubber or nylon applicator, coat the scratched area with glaze-type filler. If required, the filler can be mixed with thinner to provide a very thin paste, which is ideal for filling narrow scratches. Before the glaze filler in the scratch hardens, wrap a piece of smooth cotton cloth around the tip of a finger. Dip the cloth in thinner and then quickly wipe it along the surface of the scratch. This will ensure that the surface of the filler is slightly hollow. The scratch can now be painted over as described earlier in this section.

Repair of dents

4   When repairing dents, the first job is to pull the dent out until the affected area is as close as possible to its original shape. There is no point in trying to restore the original shape completely as the metal in the damaged area will have stretched on impact and cannot be restored to its original contours. It is better to bring the level of the dent up to a point which is about 1/8-inch below the level of the surrounding metal. In cases where the dent is very shallow, it is not worth trying to pull it out at all.

5   If the back side of the dent is accessible, it can be hammered out gently from behind using a soft-face hammer. While doing this, hold a block of wood firmly against the opposite side of the metal to absorb the hammer blows and prevent the metal from being stretched.

6   If the dent is in a section of the body which has double layers, or some other factor makes it inaccessible from behind, a different technique is required. Drill several small holes through the metal inside the damaged area, particularly in the deeper sections. Screw long, self tapping screws into the holes just enough for them to get a good grip in the metal. Now the dent can be pulled out by pulling on the protruding heads of the screws with locking pliers.

7   The next stage of repair is the removal of paint from the damaged area and from an inch or so of the surrounding metal. This is easily done with a wire brush or sanding disk in a drill motor, although it can be done just as effectively by hand with sandpaper. To complete the preparation for filling, score the surface of the bare metal with a screwdriver or the tang of a file or drill small holes in the affected area. This will provide a good grip for the filler material. To complete the repair, see the Section on filling and painting.

Repair of rust holes or gashes

8   Remove all paint from the affected area and from an inch or so of the surrounding metal using a sanding disk or wire brush mounted in a drill motor. If these are not available, a few sheets of sandpaper will do the job just as effectively.

9   With the paint removed, you will be able to determine the severity of the corrosion and decide whether to replace the whole panel, if possible, or repair the affected area. New body panels are not as expensive as most people think and it is often quicker to install a new panel than to repair large areas of rust.

10   Remove all trim pieces from the affected area except those which will act as a guide to the original shape of the damaged body, such as headlight shells, etc. Using metal snips or a hacksaw blade, remove all loose metal and any other metal that is badly affected by rust. Hammer the edges of the hole inward to create a slight depression for the filler material.

11   Wire brush the affected area to remove the powdery rust from the surface of the metal. If the back of the rusted area is accessible, treat it with rust inhibiting paint.

12   Before filling is done, block the hole in some way. This can be done with sheet metal riveted or screwed into place, or by stuffing the hole with wire mesh.

13   Once the hole is blocked off, the affected area can be filled and painted. See the following subsection on filling and painting.

Filling and painting

14   Many types of body fillers are available, but generally speaking, body repair kits which contain filler paste and a tube of resin hardener are best for this type of repair work. A wide, flexible plastic or nylon applicator will be necessary for imparting a smooth and contoured finish to the surface of the filler material. Mix up a small amount of filler on a clean piece of wood or cardboard (use the hardener sparingly). Follow the manufacturer's instructions on the package, otherwise the filler will set incorrectly.

15   Using the applicator, apply the filler paste to the prepared area. Draw the applicator across the surface of the filler to achieve the desired contour and to level the filler surface. As soon as a contour that approximates the original one is achieved, stop working the paste. If you continue, the paste will begin to stick to the applicator. Continue to add thin layers of paste at 20-minute intervals until the level of the filler is just above the surrounding metal.

16   Once the filler has hardened, the excess can be removed with a body file. From then on, progressively finer grades of sandpaper should be used, starting with a 180-grit paper and finishing with 600-grit wet or-dry paper. Always wrap the sandpaper around a flat rubber or wooden block, otherwise the surface of the filler will not be completely flat. During the sanding of the filler surface, the wet-or-dry paper should be periodically rinsed in

water. This will ensure that a very smooth finish is produced in the final stage.

17   At this point, the repair area should be surrounded by a ring of bare metal, which in turn should be encircled by the finely feathered edge of good paint. Rinse the repair area with clean water until all of the dust produced by the sanding operation is gone.

18   Spray the entire area with a light coat of primer. This will reveal any imperfections in the surface of the filler. Repair the imperfections with fresh filler paste or glaze filler and once more smooth the surface with sandpaper. Repeat this spray-and-repair procedure until you are satisfied that the surface of the filler and the feathered edge of the paint are perfect. Rinse the area with clean water and allow it to dry completely.

19   The repair area is now ready for painting. Spray painting must be carried out in a warm, dry, windless and dust free atmosphere. These conditions can be created if you have access to a large indoor work area, but if you are forced to work in the open, you will have to pick the day very carefully. If you are working indoors, dousing the floor in the work area with water will help settle the dust which would otherwise be in the air. If the repair area is confined to one body panel, mask off the surrounding panels. This will help minimize the effects of a slight mismatch in paint color. Trim pieces such as chrome strips, door handles, etc., will also need to be masked off or removed. Use masking tape and several thicknesses of newspaper for the masking operations.

20   Before spraying, shake the paint can thoroughly, then spray a test area until the spray painting technique is mastered. Cover the repair area with a thick coat of primer. The thickness should be built up using several thin layers of primer rather than one thick one. Using 600-grit wet-or-dry sandpaper, rub down the surface of the primer until it is very smooth. While doing this, the work area should be thoroughly rinsed with water and the wet-or-dry sandpaper periodically rinsed as well. Allow the primer to dry before spraying additional coats.

21   Spray on the top coat, again building up the thickness by using several thin layers of paint. Begin spraying in the center of the repair area and then, using a circular motion, work out until the whole repair area and about two inches of the surrounding original paint is covered. Remove all masking material 10 to 15 minutes after spraying on the final coat of paint. Allow the new paint at least two weeks to harden, then use a very fine rubbing compound to blend the edges of the new paint into the existing paint. Finally, apply a coat of wax.

6   Body repair - major damage

1   Major damage must be repaired by an auto body shop specifically equipped to perform unibody repairs. These shops have the

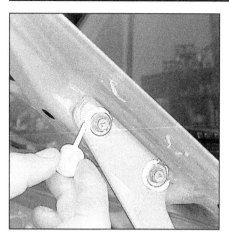

9.2 Use paint or a marker to mark the bolt locations

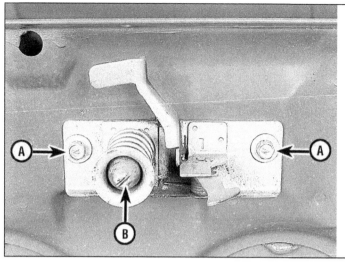

9.10 Loosen the latch bolts (A) when adjusting the hood side-to-side - to adjust the hood so it will be level with the fenders, screw the height adjuster screw (B) in or out after loosening the locknut at its base

specialized equipment required to do the job properly.

2   If the damage is extensive, the body must be checked for proper alignment or the vehicle's handling characteristics may be adversely affected and other components may wear at an accelerated rate.

3   Due to the fact that all of the major body components (hood, fenders, etc.) are separate and replaceable units, any seriously damaged components should be replaced rather than repaired. Sometimes the components can be found in a wrecking yard that specializes in used vehicle components, often at considerable savings over the cost of new parts.

7   Hinges and locks - maintenance

Once every 3000 miles, or every three months, the hinges and latch assemblies on the doors, hood and trunk should be given a few drops of light oil or lock lubricant. The door latch strikers should also be lubricated with a thin coat of grease to reduce wear and ensure free movement. Lubricate the door and trunk locks with spray-on graphite lubricant.

8   Windshield and fixed glass - replacement

Replacement of the windshield and fixed glass requires the use of special fast-setting adhesive/caulk materials and some specialized tools and techniques. These operations should be left to a dealer service department or a shop specializing in glass work.

9   Hood - removal, installation and adjustment

Refer to illustrations 9.2 and 9.10
Note: *The hood is heavy and somewhat awk-*

ward to remove and install - at least two people should perform this procedure.

Removal and installation

1   Use blankets or pads to cover the cowl area of the body and the fenders. This will protect the body and paint as the hood is lifted off.

2   Scribe or paint alignment marks around the bolt heads to insure proper alignment during installation **(see illustration)**.

3   Disconnect any cables or wire harnesses which will interfere with removal.

4   Have an assistant support the weight of the hood. Remove the hinge-to-hood nuts or bolts.

5   Lift off the hood.

6   Installation is the reverse of removal.

Adjustment

7   Fore-and-aft and side-to-side adjustment of the hood is done by moving the hood in relation to the hinge plate after loosening the bolts.

8   Scribe a line around the entire hinge plate so you can judge the amount of movement.

9   Loosen the bolts and move the hood into correct alignment. Move it only a little at a time. Tighten the hinge bolts and carefully lower the hood to check the alignment.

10   If necessary after installation, the entire hood latch assembly can be adjusted in-and-out as well as from side-to-side on the hood so the hood closes securely and is flush with the fenders. To do this, scribe a line around the hood latch mounting bolts to provide a reference point for the side-to-side movement. Then loosen the bolts and reposition the latch assembly as necessary to adjust the side-to-side movement and use a screwdriver to turn the height adjustment screw to adjust the hood up-and-down **(see illustration)**. Following adjustment, retighten the mounting bolts and adjustment screw locknut.

11   Finally, adjust the hood bumpers on the hood so the hood, when closed, is flush with the fenders.

12   The hood latch assembly, as well as the

hinges, should be periodically lubricated with white lithium-base grease to prevent sticking and wear.

10   Radiator grille - removal and installation

Refer to illustration 10.1

1   The radiator grill is held in place by clips and, on some models, screws. Remove any screws and disengage the grill retaining clips with a small screwdriver **(see illustration)**.

2   Once all the retaining clips are disengaged, pull the grille out and remove it.

3   To install the grille, press it in place until the clips lock it in position.

11   Bumpers - removal and installation

Refer to illustration 11.3

1   Disconnect any wiring or other components that would interfere with bumper removal.

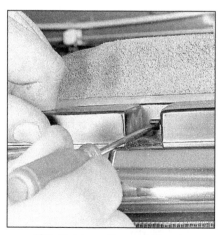

10.1 Use a small screwdriver to disengage the radiator grille retaining clips

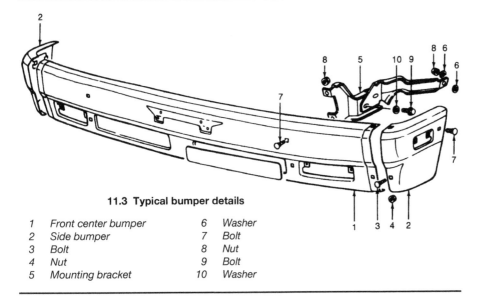

11.3 Typical bumper details

1	Front center bumper	6	Washer
2	Side bumper	7	Bolt
3	Bolt	8	Nut
4	Nut	9	Bolt
5	Mounting bracket	10	Washer

2   Support the bumper with a jack or jack-stand. Alternatively, have an assistant support the bumper as the bolts are removed.

3   Remove the retaining bolts and detach the bumper **(see illustration)**.

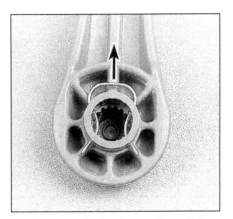

12.3 Use a wire hook or tool to pull the clip out of the window crank in the direction shown

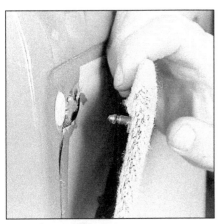

12.5 After detaching the clips at the corner with a tool, grasp the door trim panel and pull it out sharply to remove it from the door

4   Installation is the reverse of removal.

5   Tighten the retaining bolts securely.

12 Door trim panel - removal and installation

Refer to illustrations 12.2, 12.3, 12.5 and 12.6

1   Disconnect the negative cable from the battery.

2   Remove all door trim panel retaining screws and door pull/armrest assemblies **(see illustration)**.

3   Remove the window crank, using a wire hook or crank removal tool to pull out the retaining clip **(see illustration)**.

4   Insert a putty knife between the trim panel and the door and disengage the retaining clips. Work around the outer edge until the panel is free.

5   Once all of the clips are disengaged, detach the trim panel, unplug any wire harness connectors and remove the trim panel from the vehicle **(see illustration)**.

6   For access to the inner door, carefully

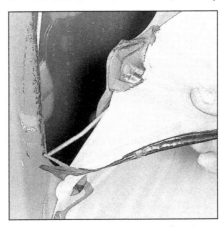

12.6 Be very careful when pulling the plastic watershield off – don't tear or distort it

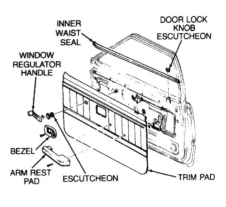

12.2 Door trim panel details

peel back the plastic watershield **(see illustration)**.

7   Prior to installation of the door panel, be sure to reinstall any clips in the panel which may have come out during the removal procedure and remain in the door itself.

8   Plug in the wire harness connectors and place the panel in position in the door. Press the door panel into place until the clips are seated and install the armrest/door pulls. Reinstall the clip and press the manual regulator window crank onto the shaft until it locks.

13 Door - removal, installation and adjustment

Refer to illustration 13.6

1   Remove the door trim panel. Disconnect any wire harness connectors and push them through the door opening so they won't interfere with door removal.

2   Place a jack or jackstand under the door or have an assistant on hand to support it when the hinge bolts are removed. **Note:** *If a jack or jackstand is used, place a rag between it and the door to protect the door's painted surfaces.*

3   Scribe around the door hinges.

4   Remove the hinge-to-door bolts and carefully lift off the door.

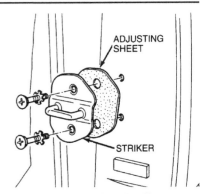

13.6 Door striker details

14.3 Mark the cargo door bolts with paint or a marker before loosening them

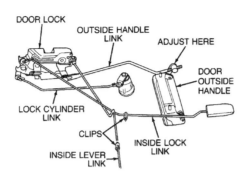

15.3 Door lock and handle assembly details

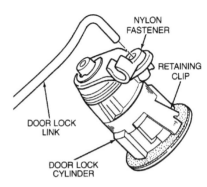

15.9 Depress the retaining clip and push the lock cylinder out of the door

5   Installation is the reverse of removal.
6   Following installation of the door, check the alignment and adjust it if necessary as follows:

a) *Up-and-down and forward-and-backward adjustments are made by loosening the hinge-to-body bolts and moving the door as necessary.*

b) *The door lock striker can also be adjusted both up-and-down and sideways to provide positive engagement with the lock mechanism. This is done by loosening the mounting screws and moving the striker as necessary (see illustration).*

14   Rear cargo door (Trooper models) - removal, installation and adjustment

Refer to illustration 14.3
1   Disconnect any wire harness connectors and push them through the door opening so they won't interfere with door removal.
2   Place a jack or jackstand under the door or have an assistant on hand to support it when the hinge bolts are removed. **Note:** *If a jack or jackstand is used, place a rag between it and the door to protect the door's painted surfaces.*
3   Scribe around the door bolts **(see illustration)**.
4   Remove the hinge-to-door bolts and carefully lift off the door.
5   Installation is the reverse of removal.
6   Following installation of the door, check the alignment and adjust it if necessary as follows:

a) *Up-and-down and side-to-side adjustments are made by loosening the hinge-to-body bolts and moving the door as necessary.*

b) *The door lock striker can also be adjusted both up-and-down and sideways to provide positive engagement with the lock mechanism. This is done by loosening the mounting screws and moving the striker as necessary.*

15   Door lock, lock cylinder and handles - removal and installation

Refer to illustrations 15.3 and 15.9
1   Remove the door trim panel as described in Section 12.
2   Remove the plastic sealing screen, taking care not to tear it.
3   Remove the inside lever link clip and detach the link from the door lock assembly **(see illustration)**.
4   Remove the screws that retain the interior handle assembly and lift it out.
5   Remove the glass run channel.
6   Disengage the door lock rod from the door lock assembly.
7   Remove the door lock assembly mounting screws and lift out the door lock assembly.

8
If necessary, remove the two nuts retaining the exterior handle and lift it out.
9   Squeeze the retaining clip with pliers and push the door lock cylinder through to the outside of the door and remove it **(see illustration)**.
10   Installation is the reverse of removal.
Note: *During installation, apply grease to the sliding surface of all levers and springs.*

16   Door window and regulator - removal and installation

Refer to illustrations 16.2, 16.3, 16.4, 16.5 and 16.6
1   Remove the door trim panel and plastic watershield (see Section 12).
2   Remove the screws from the window bottom channel assembly and lower the win-

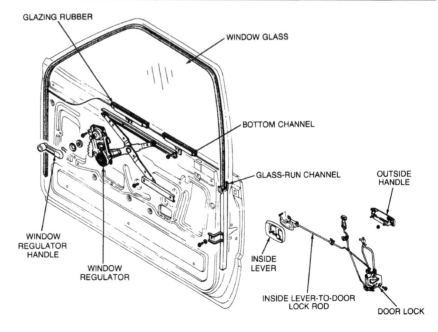

16.2 Door window glass and mechanism details

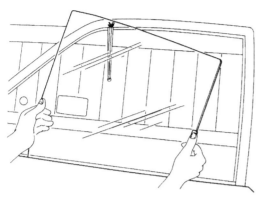

16.3 Tilt the glass as you lift it out of the door opening

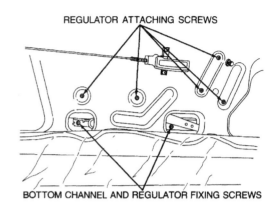

16.4 Typical window regulator screw locations

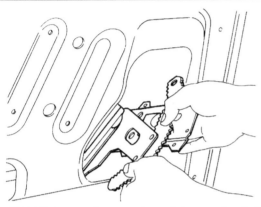

16.5 Lift the regulator out through the opening in the inner door panel

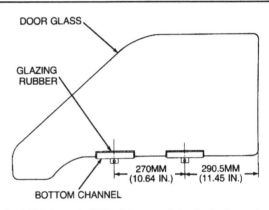

16.6 On 1981 through 1987 pickup models, the bottom channel must be installed on the glass as shown

dow glass **(see illustration)**.

3   Pry the two glass seals from the window opening. Remove the window glass by tilting it to detach the regulator arm from the glass channel and then sliding the glass up and out of the door **(see illustration)**.

4   Remove the regulator retaining screws **(see illustration)**.

5   Detach the regulator and guide it out of the opening in the door **(see illustration)**.

6   Installation is the reverse of removal. On 1981 through 1987 pickup models, make sure the bottom channel is installed the correct distance from the rear edge of the glass **(see illustration)**.

17   Outside mirror - removal and installation

Refer to illustrations 17.1, 17.2a and 17.2b

1   On later models, use a small screwdriver to pry the screw cover off the mirror base **(see illustration)**.

2   Remove the screws and lift the mirror off the door **(see illustrations)**.

3   Installation is the reverse of removal.

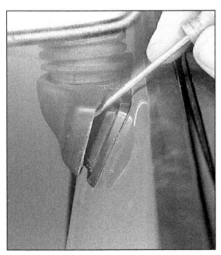

17.1 Pry the cover off the mirror base, using a small screwdriver

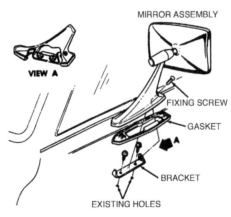

17.2a Early model outside mirror details

17.2b Later model mirror screw locations

Chapter 12
Chassis electrical system

Contents

1   General information

The electrical system is a 12-volt, negative ground type. Power for the lights and all electrical accessories is supplied by a lead/acid-type battery which is charged by the alternator.

This chapter covers repair and service procedures for the various electrical components not associated with the engine. Information on the battery, alternator, distributor and starter motor can be found in Chapter 5.

It should be noted that when portions of the electrical system are serviced, the negative battery cable should be disconnected from the battery to prevent electrical shorts and/or fires.

2   Electrical troubleshooting - general information

A typical electrical circuit consists of an electrical component, any switches, relays, motors, fuses, fusible links or circuit breakers related to that component and the wiring and connectors that link the component to both the battery and the chassis. To help you pinpoint an electrical circuit problem, wiring diagrams are included at the end of this book.

Before tackling any troublesome electrical circuit, first study the appropriate wiring diagrams to get a complete understanding of what makes up that individual circuit. Trouble spots, for instance, can often be narrowed down by noting if other components related to the circuit are operating properly. If several components or circuits fail at one time, chances are the problem is in a fuse or ground connection, because several circuits are often routed through the same fuse and ground connections.

Electrical problems usually stem from simple causes, such as loose or corroded connections, a blown fuse, a melted fusible link or a bad relay. Visually inspect the condition of all fuses, wires and connections in a problem circuit before troubleshooting it.

If testing instruments are going to be utilized, use the diagrams to plan ahead of time where you will make the necessary connections in order to accurately pinpoint the trouble spot.

The basic tools needed for electrical troubleshooting include a circuit tester or voltmeter (a 12-volt bulb with a set of test leads can also be used), a continuity tester, which includes a bulb, battery and set of test leads, and a jumper wire, preferably with a circuit breaker incorporated, which can be used to bypass electrical components. Before attempting to locate a problem with test instruments, use the wiring diagram(s) to decide where to make the connections.

Voltage checks

Voltage checks should be performed if a circuit is not functioning properly. Connect one lead of a circuit tester to either the negative battery terminal or a known good ground. Connect the other lead to a connector in the circuit being tested, preferably nearest to the battery or fuse. If the bulb of the tester lights, voltage is present, which means that the part of the circuit between the connector and the battery is problem free. Continue checking the rest of the circuit in the same fashion. When you reach a point at which no voltage is present, the problem lies between that point and the last test point with voltage. Most of the time the problem can be traced to a loose connection. **Note:** *Keep in mind that some circuits receive voltage only when the ignition key is in the Accessory or Run position.*

Finding a short

One method of finding shorts in a circuit is to remove the fuse and connect a test light or voltmeter in its place to the fuse terminals. There should be no voltage present in the circuit. Move the wiring harness from side-to-side while watching the test light. If the bulb goes on, there is a short to ground somewhere in that area, probably where the insulation has rubbed through. The same test can be performed on each component in the circuit, even a switch.

Ground check

Perform a ground test to check whether a component is properly grounded. Disconnect the battery and connect one lead of a self-powered test light, known as a continuity tester, to a known good ground. Connect the other lead to the wire or ground connection being tested. If the bulb goes on, the ground is good. If the bulb does not go on, the ground is not good.

Continuity check

A continuity check is done to determine if there are any breaks in a circuit - if it is passing electricity properly. With the circuit off (no power in the circuit), a self-powered continuity tester can be used to check the circuit. Connect the test leads to both ends of the circuit (or to the "power" end and a good ground), and if the test light comes on the circuit is passing current properly. If the light doesn't come on, there is a break somewhere in the circuit. The same procedure can be used to test a switch, by connecting the continuity tester to the switch terminals. With the switch turned On, the test light should come on.

Finding an open circuit

When diagnosing for possible open circuits, it is often difficult to locate them by sight because oxidation or terminal misalignment are hidden by the connectors. Merely wiggling a connector on a sensor or in the wiring harness may correct the open circuit condition. Remember this when an open circuit is indicated when troubleshooting a circuit. Intermittent problems may also be caused by oxidized or loose connections.

Electrical troubleshooting is simple if you keep in mind that all electrical circuits are basically electricity running from the battery, through the wires, switches, relays, fuses and fusible links to each electrical component (light bulb, motor, etc.) and to ground, from which it is passed back to the battery. Any electrical problem is an interruption in the flow of electricity to and from the battery.

3   Fuses - general information

Refer to illustrations 3.1 and 3.3

The electrical circuits of the vehicle are protected by a combination of fuses, circuit breakers and fusible links. The two fuse blocks are located under the instrument panel on the left side of the dashboard and in the engine compartment **(see illustrations)**.

Each of the fuses is designed to protect a specific circuit, and the various circuits are identified on the fuse panel itself.

Miniaturized fuses are employed in the fuse block. These compact fuses, with blade terminal design, allow fingertip removal and replacement. If an electrical component fails, always check the fuse first. A blown fuse is easily identified through the clear plastic body. Visually inspect the element for evidence of damage **(see illustration)**. If a continuity check is called for, the blade terminal tips are exposed in the fuse body.

Be sure to replace blown fuses with the correct type. Fuses of different ratings are physically interchangeable, but only fuses of the proper rating should be used. Replacing a fuse with one of a higher or lower value than specified is not recommended. Each electrical circuit needs a specific amount of protection. The amperage value of each fuse is molded into the fuse body.

If the replacement fuse immediately fails, don't replace it again until the cause of the problem is isolated and corrected. In most cases, the cause will be a short circuit in the wiring caused by a broken or deteriorated wire.

4   Fusible links - general information

Some circuits are protected by fusible links. The links are used in circuits which are not ordinarily fused, such as the ignition circuit.

Although the fusible links appear to be a heavier gauge than the wire they are protecting, the appearance is due to the thick insulation. All fusible links are several wire gauges smaller than the wire they are designed to protect.

Fusible links cannot be repaired, but a new link of the same size wire can be put in its place. The procedure is as follows:

a) *Disconnect the negative cable from the battery.*
b) *Disconnect the fusible link from the wiring harness.*
c) *Cut the damaged fusible link out of the wiring just behind the connector.*
d) *Strip the insulation back approximately 1/2 inch.*
e) *Position the connector on the new fusible link and crimp it into place.*
f) *Use rosin core solder at each end of the new link to obtain a good solder joint.*

3.1 The fuse box is located on the side of the engine compartment under a cover

g) *Use plenty of electrical tape around the soldered joint. No wires should be exposed.*
h) *Connect the battery ground cable. Test the circuit for proper operation.*

5   Circuit breakers - general information

Circuit breakers, which are located in the main fuse block on some models, protect accessories such as the heater and air conditioner.

An electrical overload in the system will cause the blower or air compressor to go off and come on or, in some cases, to remain off. If this happens, check the malfunctioning circuit. Refer to the wiring diagrams at the end of this book for the application of circuit breakers in your vehicle.

6   Relays - general information

Several electrical accessories in the vehicle use relays to transmit the electrical signal to the component. If the relay is defective, that component will not operate properly.

The various relays are grouped together in several locations. Some can be found attached to or near the dashboard fuse box. Others are in the engine compartment, near the battery.

If a faulty relay is suspected, it can be removed and tested by a dealer service department or a repair shop. Defective relays must be replaced as a unit.

7   Turn signal and hazard flashers - check and replacement

1   The turn signal flasher and hazard flashers are contained in a single small canister-shaped unit located in the wiring harness

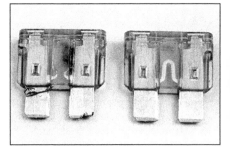

3.3 When a fuse blows, the metal between the element melts; the fuse on the right is good and the one on the left is blown

under the dash near the steering column. The turn signal flashes the turn signals and the hazard flasher flashes all four turn signals simultaneously when activated.

2   When the flasher unit is functioning properly, an audible click can be heard during its operation. If the turn signals fail on one side or the other and the flasher unit does not make its characteristic clicking sound, a faulty turn signal bulb is indicated.

3   If both turn signals fail to blink, the problem may be due to a blown fuse, a faulty flasher unit, a broken switch or a loose or open connection. If a quick check of the fuse box indicates that the turn signal fuse has blown, check the wiring for a short before, installing a new fuse.

4   To replace the flasher, simply pull it out of the wiring harness.

5   Make sure that the replacement unit is identical to the original. Compare the old one to the new one before installing it.

6   Installation is the reverse of removal.

8   Ignition switch and lock cylinder - check and installation

Refer to illustration 8.4

1   Disconnect the cable from the negative terminal of the battery.

2   Remove the steering column cover.

3   Unplug the ignition switch wiring harness.

4   Remove the two ignition switch/steering lock mounting bolts. On models with shear-head bolts it will be necessary to cut slots in the tops of the bolts with a chisel, then unscrew them with a screwdriver **(see illustration)**.

5   Remove the switch assembly from the steering column.

6   Installation is the reverse of the removal procedure.

9   Steering column switches - removal and installation

Refer to illustration 9.5

1   Disconnect the cable from the negative terminal of the battery.

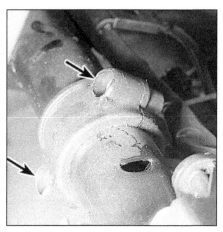

8.4 Use a hammer and chisel to make slots in the heads of the break-off bolts (arrows) which retain the ignition switch, and then unscrew them with a short screwdriver

2   Remove the steering wheel (see Chapter 10).
3   Remove the steering column covers.
4   Unplug the electrical connector from the combination switch.
5   Remove the retaining screws and pull the switch from the shaft **(see illustration)**.
6   To install the switch, reverse the removal procedure.

10   Headlights - removal and installation

1   Disconnect the negative cable from the battery.

Sealed beam type

Refer to illustration 10.5

2   Remove the radiator grille (Chapter 11)
3   Remove the headlight retainer screws, taking care not to disturb the adjustment screws.
4   Remove the retainer and pull the headlight out far enough to unplug the connector.
5   Remove the headlight **(see installation)**.
6   To install the headlight, plug the connector in, place the headlight in position and install the retainer and screws. Tighten the screws securely.
7   Install the radiator grille.

Halogen bulb-type

Refer to illustrations 10.8a, 108b and 10.9
Warning: *Halogen gas filled bulbs are under pressure and may shatter if the surface is scratched or the bulb is dropped. Wear eye protection and handle the bulbs carefully, grasping only the base whenever possible. Do not touch the surface of the bulb with your fingers because the oil from your skin could cause it to overheat and fail prematurely. If you do touch the bulb surface, clean it with rubbing alcohol.*
8   Remove the screws and pull the parking

9.5 After removing the steering wheel, the switch screws (arrows) can be removed

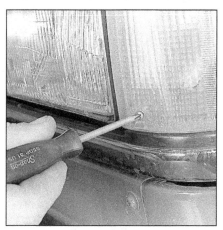

10.8a Remove the parking light screws . . .

light assembly out for access **(see illustrations)**.
9   Reach behind the headlight assembly, unplug the electrical connector and unscrew the plastic retaining collar **(see illustration)**.
10   Grasp the bulb holder securely and pull it out of the housing.
11   Insert the new bulb into the holder and secure it with the collar.
12   Plug in the electrical connector and install the parking light assembly.

11   Headlights - adjustment

Refer to illustrations 11.3, 11.6a and 11.6b
Note: *The headlights must be aimed correctly. If adjusted incorrectly they could blind the driver of an oncoming vehicle and cause a serious accident or seriously reduce your ability to see the road. The headlights should be checked for proper aim every 12 months and any time a new headlight is installed or front end body work is performed. It should be emphasized that the following procedure is only an interim step which will provide temporary adjustment until the headlights can be adjusted by a properly equipped shop.*

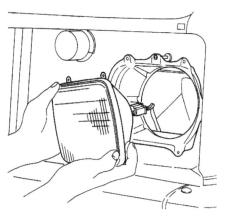

10.5 Pull the headlight out, support it and unplug the connector

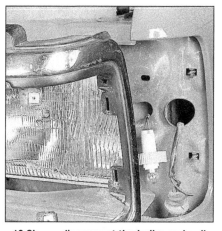

10.8b . . . disconnect the bulbs and pull the housing forward for access to the headlight bulb

10.9 Unscrew the retaining collar and slide it back so the bulb can be pulled out of the housing

1   Headlights have two spring-loaded adjusting screws, one on the bottom controlling up-and-down movement and one on the side controlling left-and-right movement **(see the accompanying illustration and illustration 8.4a)**.
2   There are several methods of adjusting

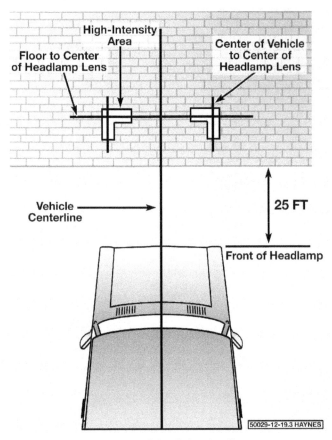

11.3 Headlight aiming details

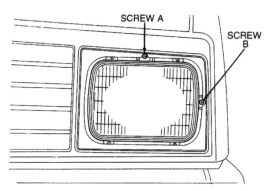

11.6a Turn screw A for vertical adjustment and screw B for horizontal adjustment (sealed beam headlights)

11.6b Halogen bulb-type headlight housing adjustment screw locations (arrows)

the headlights. The simplest method requires a blank wall 25 feet in front of the vehicle and a level floor.

3   Position masking tape vertically on the wall in reference to the vehicle centerline and the centerlines of both headlights.**(see illustration)**.

4   Position a horizontal tape line in reference to the centerline of all the headlights. **Note:** *It may be easier to position the tape on the wall with the vehicle parked only a few inches away.*

5   Adjustment should be made with the vehicle sitting level, the gas tank half-full and no unusually heavy load in the vehicle.

6   Starting with the low beam adjustment, position the high intensity zone so it is two inches below the horizontal line and two inches to the right of the headlight vertical line. Adjustment is made by turning the adjusting screw at the top or bottom of the headlight to raise or lower the beam. The adjusting screw on the side should be used in the same manner to move the beam left or right **(see illustrations)**.

7   With the high beams on, the high intensity zone should be vertically centered with the exact center just below the horizontal line. **Note:** *It may not be possible to position the headlight aim exactly for both high and low beams. If a compromise must be made, keep in mind that the low beams are the most used and have the greatest effect on safety.*

12.3a Push the bulb in, rotate it counterclockwise and withdraw it from the housing

8   Have the headlights adjusted by a dealer service department or service station at the earliest opportunity.

12   Bulb replacement

Refer to illustrations 12.3a, 12.3b, 12.3c, and 12.4

1   The lenses of many lights are held in place by screws, which makes it a simple

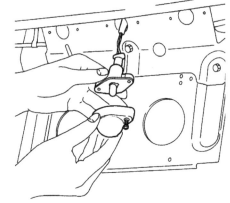

12.3b Remove the housing screws for access to the license plate light bulb

procedure to gain access to the bulbs.

2   On some lights the lenses are held in place by clips. The lenses can be removed either by unsnapping them or by using a small screwdriver to pry them off.

3   Several types of bulbs are used. Some are removed by pushing in and turning them counterclockwise **(see illustrations)**. Others can simply be unclipped from the terminals or pulled straight out of the socket.

4   To gain access to the instrument panel lights, the instrument cluster will have to be removed first **(see illustration)**.

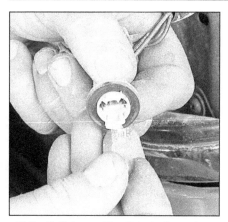

12.3c Pull the parking light bulb straight out to remove it (Trooper models)

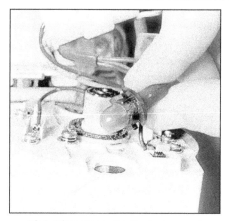

12.4 After removing the instrument cluster, turn the bulb holders, lift them out and withdraw the bulbs

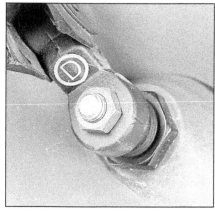

13.2 Flip the cover up, unscrew the nuts and detach the wiper arms

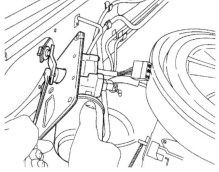

13.3 Pull the motor out of the firewall and remove the shaft nut

13   Wiper motor - removal and installation

Refer to illustrations 13.2 and 13.3

1   Disconnect the negative cable from the battery.
2   Remove the wiper arms **(see illustration)**.
3   Unplug the electrical connector, remove the four retaining bolts, then pull the motor out and remove the motor shaft nut.**(see illustration)**.
4   Detach the wiper motor from the linkage and remove it from the vehicle.
5   Installation is the reverse of removal.

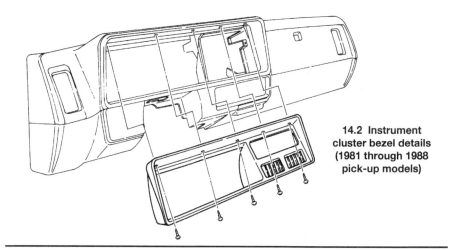

14.2 Instrument cluster bezel details (1981 through 1988 pick-up models)

14   Instrument cluster - removal and installation

Refer to illustrations 14.2, 14.3a, 14.3b and 14.4

1   Disconnect the cable from the negative terminal of the battery.
2   Remove the cluster bezel **(see illustration)**.
3   Remove the cluster assembly-to-dash

panel screws **(see illustrations)**.
4   Pull the cluster out, reach behind the cluster and disconnect the speedometer cable **(see illustration)**.
5   Pull the cluster assembly out as far as possible and unplug the multi-wire connector(s) from the back of the unit. Some models have additional wires that must be disconnected, so be sure to mark them before removing.

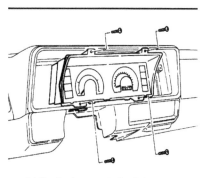

14.3a Instrument cluster screw locations (1989 and later pick-up models shown, others similar)

14.3b Trooper model instrument cluster screw locations

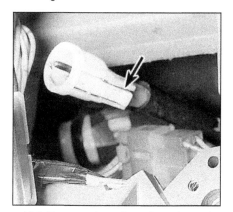

14.4 Press on the release lever (arrow) and detach the speedometer cable from the cluster

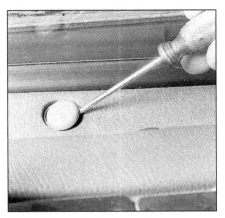

15.4a Pry out the instrument panel screw and bolt covers with a small screwdriver

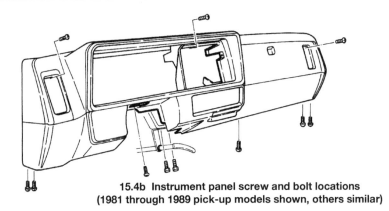

15.4b Instrument panel screw and bolt locations (1981 through 1989 pick-up models shown, others similar)

6   Remove the cluster from the dash panel.

7   The instrument cluster assembly can be disassembled to allow access to the enclosed gauges and indicators by removing the lens and plate retaining screws. On Trooper models, the gauge units can be removed separately from the cluster.

8   Installation is the reverse of the removal procedure.

15   Instrument panel - removal and installation

Refer to illustrations 15.4a and 15.4b

1   *Disconnect the cable from the negative terminal of the battery.*

2   Remove the steering wheel (see Chapter 10).

3   Remove steering column covers, instrument cluster, radio, heater controls, ventilation ducts, glove box, trim panels and any other components which would interfere with removal.

4   Remove the retaining bolts, nuts and screws, unplug any electrical connectors and pull the instrument panel to the rear to detach it **(see illustrations)**.

5   Installation is the reverse of removal.

16   Cruise control system - description and check

The cruise control system maintains vehicle speed with a vacuum actuated servo motor located in the engine compartment, which is connected to the throttle linkage by a cable. The system consists of the servo motor, clutch switch, brake switch, control switches, a relay and associated vacuum hoses.

Because of the complexity of the cruise control system and the special tools and techniques required for diagnosis, repair should be left to a dealer service department or a repair shop. However, it is possible for the home mechanic to make simple checks of the wiring and vacuum connections for minor faults which can be easily repaired. These include:

a) *Inspect the cruise control actuating switches for broken wires and loose connections.*

b) *Check the cruise control fuse.*

c) *The cruise control system is operated by vacuum so it's critical that all vacuum switches, hoses and connections are secure. Check the hoses in the engine compartment for tight connections, cracks and obvious vacuum leaks.*

17   Wiring diagrams - general information

Since it isn't possible to include all wiring diagrams for every year covered by this manual, the following diagrams are those that are typical and most commonly needed.

Prior to troubleshooting any circuits, check the fuse and circuit breakers (if equipped) to make sure they're in good condition. Make sure the battery is properly charged and check the cable connections (Chapter 1).

When checking a circuit, make sure that all connectors are clean, with no broken or loose terminals. When unplugging a connector, do not pull on the wires. Pull only on the connector housings themselves.

Refer to the accompanying table for the wire color codes applicable to your vehicle.

Circuit	Wire color and abbreviation	Wire color and abbreviation	Wire color and abbreviation	Wire color and abbreviation	Wire color and abbreviation	Wire color and abbreviation
Starter motor circuit	Black (B)	Black-white (BW)	Black-yellow (BY)	Black-red (BR)	Black-green (BG)	
Charging circuit	White (W)	White-red (WR)	White-black (WB)	White-blue (WL)	White-yellow (WY)	White-green (WG)
Lighting circuit	Red (R)	Red-white (RW)	Red-black (RB)	Red-yellow (RY)	Red-green (RG)	Red-blue (RL)
Turn signal circuit	Green (G)	Green-white (GW)	Green-red (GR)	Green-yellow (GY)	Green-black (GB)	Green-blue (GL)
Instrument circuit	Yellow (Y)	Yellow-red (YR)	Yellow-black (YB)	Yellow-green (YG)	Yellow-blue (YL)	Yellow-white (YW)
Wiper motor circuit	Blue (L)	Blue-white (LW)	Blue-red (LR)	Blue-black (LB)		
Other circuits	Light green (LG)	Brown-white (BrW)				

Wire color codes. All connections to ground are through a black wire unless noted otherwise. All wires connected to a wire are of same color unless identified otherwise.

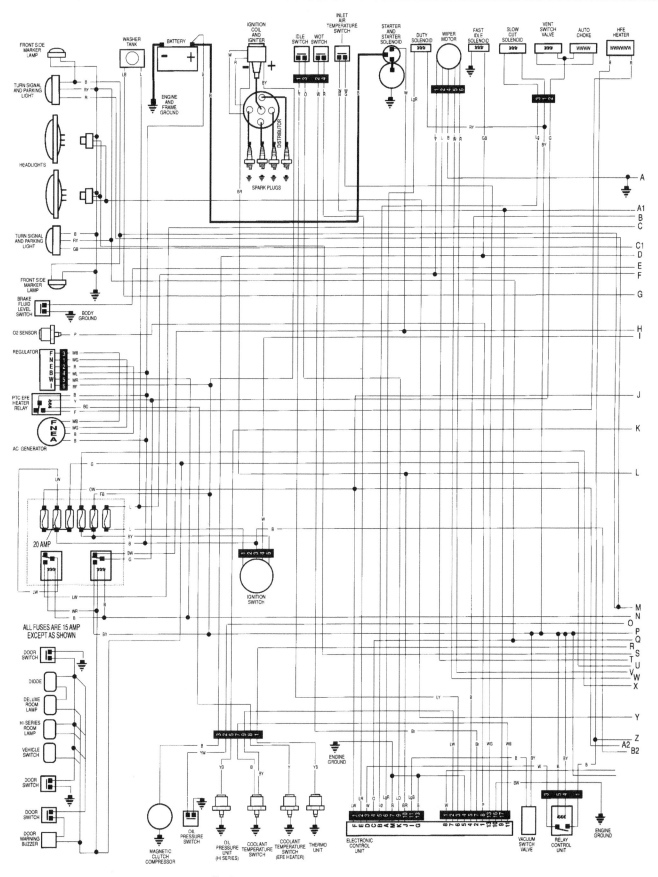

Typical wiring diagram (Pick-up models) (1 of 2)

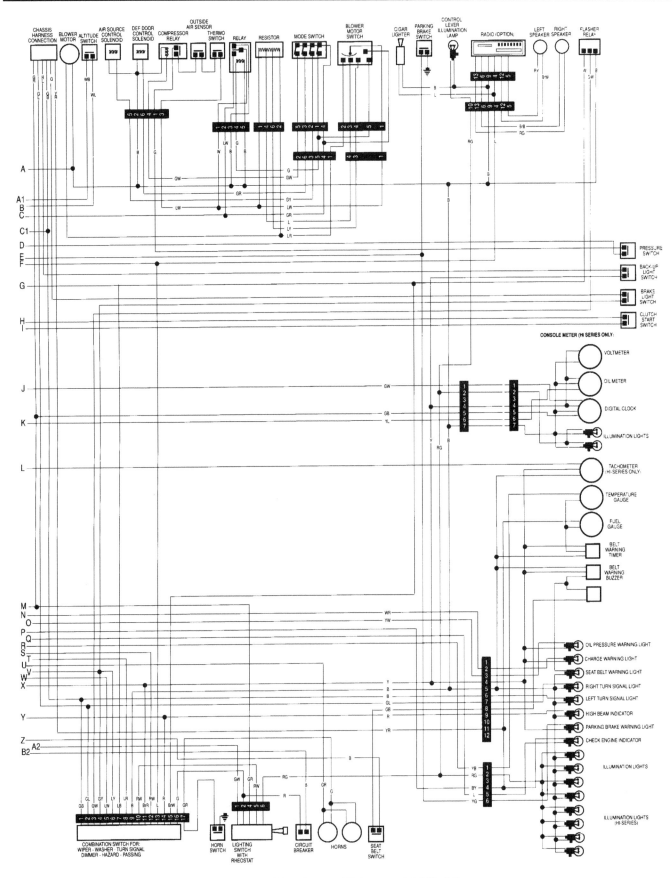

Typical wiring diagram (Pick-up models) (2 of 2)

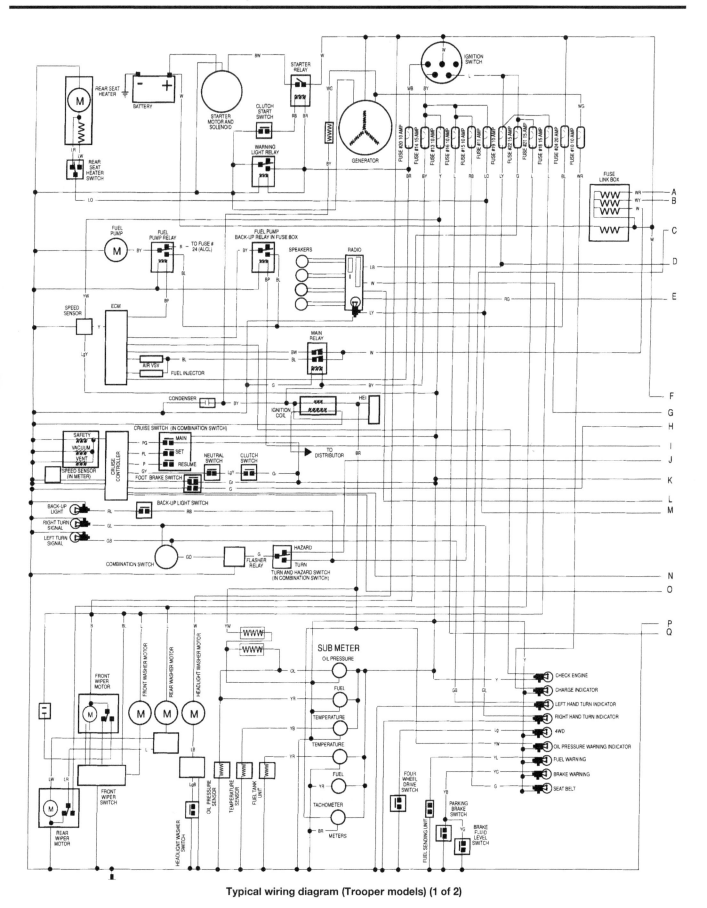

Typical wiring diagram (Trooper models) (1 of 2)

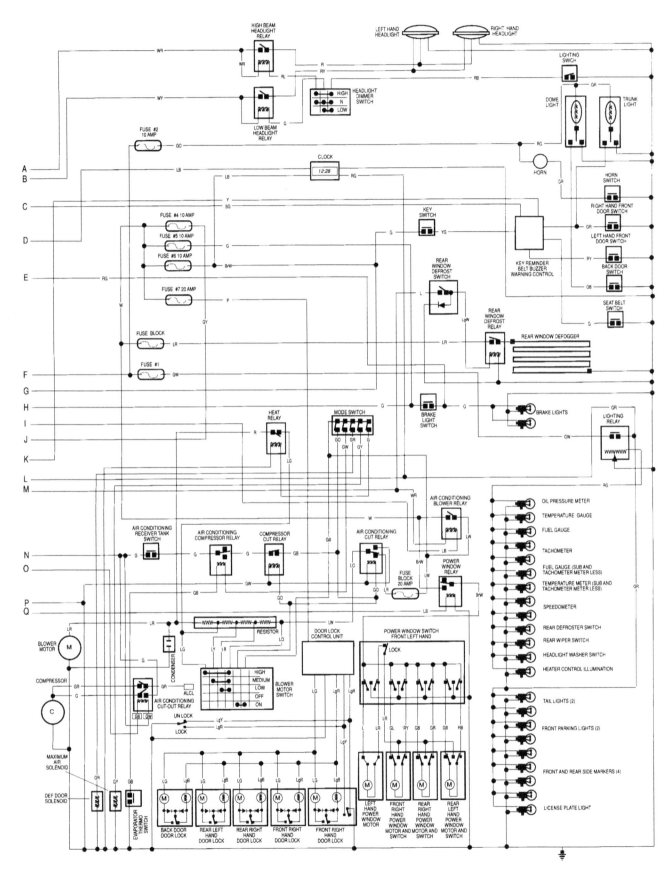

Typical wiring diagram (Trooper models) (2 of 2)

Index

A

B

C

Haynes Automotive Manuals

ACURA
- **12020** Integra '86 thru '89 & **Legend** '86 thru '90
- **12021** Integra '90 thru '93 & **Legend** '91 thru '95
  - Integra '94 thru '00 - see HONDA Civic (42025)
  - MDX '01 thru '07 - see HONDA Pilot (42037)
- **12050** Acura TL all models '99 thru '08

AMC
- **14020** Mid-size models '70 thru '83
- **14025** (Renault) Alliance & Encore '83 thru '87

AUDI
- **15020** 4000 all models '80 thru '87
- **15025** 5000 all models '77 thru '83
- **15026** 5000 all models '84 thru '88
  - Audi A4 '96 thru '01 - see VW Passat (96023)
- **15030** Audi A4 '02 thru '08

AUSTIN-HEALEY
- Sprite - see MG Midget (66015)

BMW
- **18020** 3/5 Series '82 thru '92
- **18021** 3-Series incl. Z3 models '92 thru '98
- **18022** 3-Series incl. Z4 models '99 thru '05
- **18023** 3-Series '06 thru '14
- **18025** 320i all 4-cylinder models '75 thru '83
- **18050** 1500 thru 2002 except Turbo '59 thru '77

BUICK
- **19010** Buick Century '97 thru '05
  - Century (front-wheel drive) - see GM (38005)
- **19020** Buick, Oldsmobile & Pontiac Full-size (Front-wheel drive) '85 thru '05
  - Buick Electra, LeSabre and Park Avenue; Oldsmobile Delta 88 Royale, Ninety Eight and Regency; Pontiac Bonneville
- **19025** Buick, Oldsmobile & Pontiac Full-size (Rear wheel drive) '70 thru '90
  - Buick Estate, Electra, LeSabre, Limited, Oldsmobile Custom Cruiser, Delta 88, Ninety-eight, Pontiac Bonneville, Catalina, Grandville, Parisienne
- **19027** Buick LaCrosse '05 thru '13
  - Enclave - see GENERAL MOTORS (38001)
  - Rainier - see CHEVROLET (24072)
  - Regal - see GENERAL MOTORS (38010)
  - Riviera - see GENERAL MOTORS (38030, 38031)
  - Roadmaster - see CHEVROLET (24046)
  - Skyhawk - see GENERAL MOTORS (38015)
  - Skylark - see GENERAL MOTORS (38020, 38025)
  - Somerset - see GENERAL MOTORS (38025)

CADILLAC
- **21015** CTS & CTS-V '03 thru '14
- **21030** Cadillac Rear Wheel Drive '70 thru '93
  - Cimarron - see GENERAL MOTORS (38015)
  - DeVille - see GENERAL MOTORS (38031 & 38032)
  - Eldorado - see GENERAL MOTORS (38030)
  - Fleetwood - see GENERAL MOTORS (38031)
  - Seville - see GM (38030, 38031 & 38032)

CHEVROLET
- **10305** Chevrolet Engine Overhaul Manual
- **24010** Astro & GMC Safari Mini-vans '85 thru '05
- **24013** Aveo '04 thru '11
- **24015** Camaro V8 all models '70 thru '81
- **24016** Camaro all models '82 thru '92
- **24017** Camaro & Firebird '93 thru '02
  - Cavalier - see GENERAL MOTORS (38016)
  - Celebrity - see GENERAL MOTORS (38005)
- **24018** Camaro '10 thru '15
- **24020** Chevelle, Malibu & El Camino '69 thru '87
  - Cobalt - see GENERAL MOTORS (38017)
- **24024** Chevette & Pontiac T1000 '76 thru '87
  - Citation - see GENERAL MOTORS (38020)
- **24027** Colorado & GMC Canyon '04 thru '12
- **24032** Corsica & Beretta all models '87 thru '96
- **24040** Corvette all V8 models '68 thru '82
- **24041** Corvette all models '84 thru '96
- **24042** Corvette all models '97 thru '13
- **24044** Cruze '11 thru '19
- **24045** Full-size Sedans Caprice, Impala, Biscayne, Bel Air & Wagons '69 thru '90
- **24046** Impala SS & Caprice and Buick Roadmaster '91 thru '96
  - Impala '00 thru '05 - see LUMINA (24048)
- **24047** Impala & Monte Carlo all models '06 thru '11
  - Lumina '90 thru '94 - see GM (38010)
- **24048** Lumina & Monte Carlo '95 thru '05
  - Lumina APV - see GM (38035)
- **24050** Luv Pick-up all 2WD & 4WD '72 thru '82
- **24051** Malibu '13 thru '19
- **24055** Monte Carlo all models '70 thru '88
  - Monte Carlo '95 thru '01 - see LUMINA (24048)
- **24059** Nova all V8 models '69 thru '79

- **24060** Nova and Geo Prizm '85 thru '92
- **24064** Pick-ups '67 thru '87 - Chevrolet & GMC
- **24065** Pick-ups '88 thru '98 - Chevrolet & GMC
- **24066** Pick-ups '99 thru '06 - Chevrolet & GMC
- **24067** Chevrolet Silverado & GMC Sierra '07 thru '14
- **24068** Chevrolet Silverado & GMC Sierra '14 thru '19
- **24070** S-10 & S-15 Pick-ups '82 thru '93, Blazer & Jimmy '83 thru '94,
- **24071** S-10 & Sonoma Pick-ups '94 thru '04, including Blazer, Jimmy & Hombre
- **24072** Chevrolet TrailBlazer, GMC Envoy & Oldsmobile Bravada '02 thru '09
- **24075** Sprint '85 thru '88 & Geo Metro '89 thru '01
- **24080** Vans - Chevrolet & GMC '68 thru '96
- **24081** Chevrolet Express & GMC Savana Full-size Vans '96 thru '19

CHRYSLER
- **10310** Chrysler Engine Overhaul Manual
- **25015** Chrysler Cirrus, Dodge Stratus, Plymouth Breeze '95 thru '00
- **25020** Full-size Front-Wheel Drive '88 thru '93
  - K-Cars - see DODGE Aries (30008)
  - Laser - see DODGE Daytona (30030)
- **25025** Chrysler LHS, Concorde, New Yorker, Dodge Intrepid, Eagle Vision, '93 thru '97
- **25026** Chrysler LHS, Concorde, 300M, Dodge Intrepid, '98 thru '04
- **25027** Chrysler 300 '05 thru '18, Dodge Charger '06 thru '18, Magnum '05 thru '08 & Challenger '08 thru '18
- **25030** Chrysler & Plymouth Mid-size front wheel drive '82 thru '95
  - Rear-wheel Drive - see Dodge (30050)
- **25035** PT Cruiser all models '01 thru '10
- **25040** Chrysler Sebring '95 thru '06, Dodge Stratus '01 thru '06 & Dodge Avenger '95 thru '00
- **25041** Chrysler Sebring '07 thru '10, 200 '11 thru '17 Dodge Avenger '08 thru '14

DATSUN
- **28005** 200SX all models '80 thru '83
- **28012** 240Z, 260Z & 280Z Coupe '70 thru '78
- **28014** 280ZX Coupe & 2+2 '79 thru '83
  - 300ZX - see NISSAN (72010)
- **28018** 510 & PL521 Pick-up '68 thru '73
- **28020** 510 all models '78 thru '81
- **28022** 620 Series Pick-up all models '73 thru '79
  - 720 Series Pick-up - see NISSAN (72030)

DODGE
- **400 & 600** - see CHRYSLER (25030)
- **30008** Aries & Plymouth Reliant '81 thru '89
- **30010** Caravan & Plymouth Voyager '84 thru '95
- **30011** Caravan & Plymouth Voyager '96 thru '02
- **30012** Challenger & Plymouth Sapporo '78 thru '83
- **30013** Caravan, Chrysler Voyager & Town & Country '03 thru '07
- **30014** Grand Caravan & Chrysler Town & Country '08 thru '18
- **30016** Colt & Plymouth Champ '78 thru '87
- **30020** Dakota Pick-ups all models '87 thru '96
- **30021** Durango '98 & '99 & Dakota '97 thru '99
- **30022** Durango '00 thru '03 & Dakota '00 thru '04
- **30023** Durango '04 thru '09 & Dakota '05 thru '11
- **30025** Dart, Demon, Plymouth Barracuda, Duster & Valiant 6-cylinder models '67 thru '76
- **30030** Daytona & Chrysler Laser '84 thru '89
  - Intrepid - see CHRYSLER (25025, 25026)
- **30034** Neon all models '95 thru '99
- **30035** Omni & Plymouth Horizon '78 thru '90
- **30036** Dodge & Plymouth Neon '00 thru '05
- **30040** Pick-ups full-size models '74 thru '93
- **30042** Pick-ups full-size models '94 thru '08
- **30043** Pick-ups full-size models '09 thru '18
- **30045** Ram 50/D50 Pick-ups & Raider and Plymouth Arrow Pick-ups '79 thru '93
- **30050** Dodge/Plymouth/Chrysler RWD '71 thru '89
- **30055** Shadow & Plymouth Sundance '87 thru '94
- **30060** Spirit & Plymouth Acclaim '89 thru '95
- **30065** Vans - Dodge & Plymouth '71 thru '03

EAGLE
- Talon - see MITSUBISHI (68030, 68031)
- Vision - see CHRYSLER (25025)

FIAT
- **34010** 124 Sport Coupe & Spider '68 thru '78
- **34025** X1/9 all models '74 thru '80

FORD
- **10320** Ford Engine Overhaul Manual
- **10355** Ford Automatic Transmission Overhaul
- **11500** Mustang '64-1/2 thru '70 Restoration Guide
- **36004** Aerostar Mini-vans all models '86 thru '97
- **36006** Contour & Mercury Mystique '95 thru '00
- **36008** Courier Pick-up all models '72 thru '82

- **36012** Crown Victoria & Mercury Grand Marquis '88 thru '11
- **36014** Edge '07 thru '19 & Lincoln MKX '07 thru '18
- **36016** Escort & Mercury Lynx all models '81 thru '90
- **36020** Escort & Mercury Tracer '91 thru '02
- **36022** Escape '01 thru '17, Mazda Tribute '01 thru '11, & Mercury Mariner '05 thru '11
- **36024** Explorer & Mazda Navajo '91 thru '01
- **36025** Explorer & Mercury Mountaineer '02 thru '10
- **36026** Explorer '11 thru '17
- **36028** Fairmont & Mercury Zephyr '78 thru '83
- **36030** Festiva & Aspire '88 thru '97
- **36032** Fiesta all models '77 thru '80
- **36034** Focus all models '00 thru '11
- **36035** Focus '12 thru '14
- **36045** Fusion '06 thru '14 & Mercury Milan '06 thru '11
- **36048** Mustang V8 all models '64-1/2 thru '73
- **36049** Mustang II 4-cylinder, V6 & V8 models '74 thru '78
- **36050** Mustang & Mercury Capri '79 thru '93
- **36051** Mustang all models '94 thru '04
- **36052** Mustang '05 thru '14
- **36054** Pick-ups & Bronco '73 thru '79
- **36058** Pick-ups & Bronco '80 thru '96
- **36059** F-150 '97 thru '03, Expedition '97 thru '17, F-250 '97 thru '99, F-150 Heritage '04 & Lincoln Navigator '98 thru '17
- **36060** Super Duty Pick-ups & Excursion '99 thru '10
- **36061** F-150 full-size '04 thru '14
- **36062** Pinto & Mercury Bobcat '75 thru '80
- **36063** F-150 full-size '15 thru '17
- **36064** Super Duty Pick-ups '11 thru '16
- **36066** Probe all models '89 thru '92
  - Probe '93 thru '97 - see MAZDA 626 (61042)
- **36070** Ranger & Bronco II gas models '83 thru '92
- **36071** Ranger '93 thru '11 & Mazda Pick-ups '94 thru '09
- **36074** Taurus & Mercury Sable '86 thru '95
- **36075** Taurus & Mercury Sable '96 thru '07
- **36076** Taurus '08 thru '14, Five Hundred '05 thru '07, Mercury Montego '05 thru '07 & Sable '08 thru '09
- **36078** Tempo & Mercury Topaz '84 thru '94
- **36082** Thunderbird & Mercury Cougar '83 thru '88
- **36086** Thunderbird & Mercury Cougar '89 thru '97
- **36090** Vans all V8 Econoline models '69 thru '91
- **36094** Vans full size '92 thru '14
- **36097** Windstar '95 thru '03, Freestar & Mercury Monterey Mini-van '04 thru '07

GENERAL MOTORS
- **10360** GM Automatic Transmission Overhaul
- **38001** GMC Acadia '07 thru '16, Buick Enclave '08 thru '17, Saturn Outlook '07 thru '10 & Chevrolet Traverse '09 thru '17
- **38005** Buick Century, Chevrolet Celebrity, Oldsmobile Cutlass Ciera & Pontiac 6000 all models '82 thru '96
- **38010** Buick Regal '88 thru '04, Chevrolet Lumina '88 thru '04, Oldsmobile Cutlass Supreme '88 thru '97 & Pontiac Grand Prix '88 thru '07
- **38015** Buick Skyhawk, Cadillac Cimarron, Chevrolet Cavalier, Oldsmobile Firenza, Pontiac J-2000 & Sunbird '82 thru '94
- **38016** Chevrolet Cavalier & Pontiac Sunfire '95 thru '05
- **38017** Chevrolet Cobalt '05 thru '10, HHR '06 thru '11, Pontiac G5 '07 thru '09, Pursuit '05 thru '06 & Saturn ION '03 thru '07
- **38020** Buick Skylark, Chevrolet Citation, Oldsmobile Omega, Pontiac Phoenix '80 thru '85
- **38025** Buick Skylark '86 thru '98, Somerset '85 thru '87, Oldsmobile Achieva '92 thru '98, Calais '85 thru '91, & Pontiac Grand Am all models '85 thru '98
- **38026** Chevrolet Malibu '97 thru '03, Classic '04 thru '05, Oldsmobile Alero '99 thru '03, Cutlass '97 thru '00, & Pontiac Grand Am '99 thru '03
- **38027** Chevrolet Malibu '04 thru '12, Pontiac G6 '05 thru '10 & Saturn Aura '07 thru '10
- **38030** Cadillac Eldorado, Seville, Oldsmobile Toronado & Buick Riviera '71 thru '85
- **38031** Cadillac Eldorado, Seville, DeVille, Fleetwood, Oldsmobile Toronado & Buick Riviera '86 thru '93
- **38032** Cadillac DeVille '94 thru '05, Seville '92 thru '04 & Cadillac DTS '06 thru '10
- **38035** Chevrolet Lumina APV, Oldsmobile Silhouette & Pontiac Trans Sport all models '90 thru '96
- **38036** Chevrolet Venture '97 thru '05, Oldsmobile Silhouette '97 thru '04, Pontiac Trans Sport '97 thru '98 & Montana '99 thru '05
- **38040** Chevrolet Equinox '05 thru '17, GMC Terrain '10 thru '17 & Pontiac Torrent '06 thru '09

GEO
- Metro - see CHEVROLET Sprint (24075)
- Prizm - '85 thru '92 see CHEVY (24060), '93 thru '02 see TOYOTA Corolla (92036)
- **40030** Storm all models '90 thru '93
- Tracker - see SUZUKI Samurai (90010)

(Continued on other side)

Haynes North America, Inc. • (805) 498-6703 • www.haynes.com

Haynes Automotive Manuals (continued)

*NOTE: If you do not see a listing for your vehicle, please visit **haynes.com** for the latest product information and check out our **Online Manuals!***

GMC
Acadia - *see GENERAL MOTORS (38001)*
Pick-ups - *see CHEVROLET (24027, 24068)*
Vans - *see CHEVROLET (24081)*

HONDA
42010 Accord CVCC all models '76 thru '83
42011 Accord all models '84 thru '89
42012 Accord all models '90 thru '93
42013 Accord all models '94 thru '97
42014 Accord all models '98 thru '02
42015 Accord '03 thru '12 & Crosstour '10 thru '14
42016 Accord '13 thru '17
42020 Civic 1200 all models '73 thru '79
42021 Civic 1300 & 1500 CVCC '80 thru '83
42022 Civic 1500 CVCC all models '75 thru '79
42023 Civic all models '84 thru '91
42024 Civic & del Sol '92 thru '95
42025 Civic '96 thru '00, CR-V '97 thru '01
& Acura Integra '94 thru '00
42026 Civic '01 thru '11 & CR-V '02 thru '11
42027 Civic '12 thru '15 & CR-V '12 thru '16
42030 Fit '07 thru '13
42035 Odyssey all models '99 thru '10
Passport - *see ISUZU Rodeo (47017)*
42037 Honda Pilot '03 thru '08, Ridgeline '06 thru '14
& Acura MDX '01 thru '07
42040 Prelude CVCC all models '79 thru '89

HYUNDAI
43010 Elantra all models '96 thru '19
43015 Excel & Accent all models '86 thru '13
43050 Santa Fe all models '01 thru '12
43055 Sonata all models '99 thru '14

INFINITI
G35 '03 thru '08 - *see NISSAN 350Z (72011)*

ISUZU
Hombre - *see CHEVROLET S-10 (24071)*
47017 Rodeo '91 thru '02, Amigo '89 thru '94 & '98 thru '02
& Honda Passport '95 thru '02
47020 Trooper '84 thru '91 & Pick-up '81 thru '93

JAGUAR
49010 XJ6 all 6-cylinder models '68 thru '86
49011 XJ6 all models '88 thru '94
49015 XJ12 & XJS all 12-cylinder models '72 thru '85

JEEP
50010 Cherokee, Comanche & Wagoneer Limited
all models '84 thru '01
50011 Cherokee '14 thru '19
50020 CJ all models '49 thru '86
50025 Grand Cherokee all models '93 thru '04
50026 Grand Cherokee '05 thru '19
& Dodge Durango '11 thru '19
50029 Grand Wagoneer & Pick-up '72 thru '91
Grand Wagoneer '84 thru '91, Cherokee &
Wagoneer '72 thru '83, Pick-up '72 thru '88
50030 Wrangler all models '87 thru '17
50035 Liberty '02 thru '12 & Dodge Nitro '07 thru '11
50050 Patriot & Compass '07 thru '17

KIA
54050 Optima '01 thru '10
54060 Sedona '02 thru '14
54070 Sephia '94 thru '01, Spectra '00 thru '09,
Sportage '05 thru '20
54077 Sorento '03 thru '13

LEXUS
ES 300/330 - *see TOYOTA Camry (92007, 92008)*
ES 350 - *see TOYOTA Camry (92009)*
RX 300/330/350 - *see TOYOTA Highlander (92095)*

LINCOLN
MKX - *see FORD (36014)*
Navigator - *see FORD Pick-up (36059)*
59010 Rear-Wheel Drive Continental '70 thru '87,
Mark Series '70 thru '92 & Town Car '81 thru '10

MAZDA
61010 GLC (rear-wheel drive) '77 thru '83
61011 GLC (front-wheel drive) '81 thru '85
61012 Mazda3 '04 thru '11
61015 323 & Protegé '90 thru '03
61016 MX-5 Miata '90 thru '14
61020 MPV all models '89 thru '98
Navajo - *see Ford Explorer (36024)*
61030 Pick-ups '72 thru '93
Pick-ups '94 thru '09 - *see Ford Ranger (36071)*
61035 RX-7 all models '79 thru '85
61036 RX-7 all models '86 thru '91
61040 626 (rear-wheel drive) all models '79 thru '82
61041 626 & MX-6 (front-wheel drive) '83 thru '92
61042 626 '93 thru '01 & MX-6/Ford Probe '93 thru '02
61043 Mazda6 '03 thru '13

MERCEDES-BENZ
63012 123 Series Diesel '76 thru '85
63015 190 Series 4-cylinder gas models '84 thru '88
63020 230/250/280 6-cylinder SOHC models '68 thru '72
63025 280 123 Series gas models '77 thru '81
63030 350 & 450 all models '71 thru '80
63040 C-Class: C230/C240/C280/C320/C350 '01 thru '07

MERCURY
64200 Villager & Nissan Quest '93 thru '01
All other titles, see FORD Listing.

MG
66010 MGB Roadster & GT Coupe '62 thru '80
66015 MG Midget, Austin Healey Sprite '58 thru '80

MINI
67020 Mini '02 thru '13

MITSUBISHI
68020 Cordia, Tredia, Galant, Precis & Mirage '83 thru '93
68030 Eclipse, Eagle Talon & Plymouth Laser '90 thru '94
68031 Eclipse '95 thru '05 & Eagle Talon '95 thru '98
68035 Galant '94 thru '12
68040 Pick-up '83 thru '96 & Montero '83 thru '93

NISSAN
72010 300ZX all models including Turbo '84 thru '89
72011 350Z & Infiniti G35 all models '03 thru '08
72015 Altima all models '93 thru '06
72016 Altima '07 thru '12
72020 Maxima all models '85 thru '92
72021 Maxima all models '93 thru '08
72025 Murano '03 thru '14
72030 Pick-ups '80 thru '97 & Pathfinder '87 thru '95
72031 Frontier '98 thru '04, Xterra '00 thru '04,
& Pathfinder '96 thru '04
72032 Frontier & Xterra '05 thru '14
72037 Pathfinder '05 thru '14
72040 Pulsar all models '83 thru '86
72042 Roque all models '08 thru '20
72050 Sentra all models '82 thru '94
72051 Sentra & 200SX all models '95 thru '06
72060 Stanza all models '82 thru '90
72070 Titan pick-ups '04 thru '10, Armada '05 thru '10
& Pathfinder Armada '04
72080 Versa all models '07 thru '19

OLDSMOBILE
73015 Cutlass V6 & V8 gas models '74 thru '88
*For other OLDSMOBILE titles, see BUICK,
CHEVROLET or GENERAL MOTORS listings.*

PLYMOUTH
For PLYMOUTH titles, see DODGE listing.

PONTIAC
79008 Fiero all models '84 thru '88
79018 Firebird V8 models except Turbo '70 thru '81
79019 Firebird all models '82 thru '92
79025 G6 all models '05 thru '09
79040 Mid-size Rear-wheel Drive '70 thru '87
Vibe '03 thru '10 - *see TOYOTA Corolla (92037)*
*For other PONTIAC titles, see BUICK,
CHEVROLET or GENERAL MOTORS listings.*

PORSCHE
80020 911 Coupe & Targa models '65 thru '89
80025 914 all 4-cylinder models '69 thru '76
80030 924 all models including Turbo '76 thru '82
80035 944 all models including Turbo '83 thru '89

RENAULT
Alliance & Encore - *see AMC (14025)*

SAAB
84010 900 all models including Turbo '79 thru '88

SATURN
87010 Saturn all S-series models '91 thru '02
Saturn Ion '03 thru '07- *see GM (38017)*
Saturn Outlook - *see GM (38001)*
87020 Saturn L-series all models '00 thru '04
87040 Saturn VUE '02 thru '09

SUBARU
89002 1100, 1300, 1400 & 1600 '71 thru '79
89003 1600 & 1800 2WD & 4WD '80 thru '94
89080 Impreza '02 thru '11, WRX '02 thru '14,
& WRX STI '04 thru '14
89100 Legacy all models '90 thru '99
89101 Legacy & Forester '00 thru '09
89102 Legacy '10 thru '16 & Forester '12 thru '16

SUZUKI
90010 Samurai/Sidekick & Geo Tracker '86 thru '01

TOYOTA
92005 Camry all models '83 thru '91
92006 Camry '92 thru '96 & Avalon '95 thru '96
92007 Camry, Avalon, Solara, Lexus ES 300 '97 thru '01

92008 Camry, Avalon, Lexus ES 300/330 '02 thru '06
& Solara '02 thru '08
92009 Camry, Avalon & Lexus ES 350 '07 thru '17
92015 Celica Rear-wheel Drive '71 thru '85
92020 Celica Front-wheel Drive '86 thru '99
92025 Celica Supra all models '79 thru '92
92030 Corolla all models '75 thru '79
92032 Corolla all rear-wheel drive models '80 thru '87
92035 Corolla all front-wheel drive models '84 thru '92
92036 Corolla & Geo/Chevrolet Prizm '93 thru '02
92037 Corolla '03 thru '19, Matrix '03 thru '14,
& Pontiac Vibe '03 thru '10
92040 Corolla Tercel all models '80 thru '82
92045 Corona all models '74 thru '82
92050 Cressida all models '78 thru '82
92055 Land Cruiser FJ40, 43, 45, 55 '68 thru '82
92056 Land Cruiser FJ60, 62, 80, FZJ80 '80 thru '96
92060 Matrix '03 thru '11 & Pontiac Vibe '03 thru '10
92065 MR2 all models '85 thru '87
92070 Pick-up all models '69 thru '78
92075 Pick-up all models '79 thru '95
92076 Tacoma '95 thru '04, 4Runner '96 thru '02
& T100 '93 thru '08
92077 Tacoma all models '05 thru '18
92078 Tundra '00 thru '06 & Sequoia '01 thru '07
92079 4Runner all models '03 thru '09
92080 Previa all models '91 thru '95
92081 Prius all models '01 thru '12
92082 RAV4 all models '96 thru '12
92085 Tercel all models '87 thru '94
92090 Sienna all models '98 thru '10
92095 Highlander '01 thru '19
& Lexus RX330/330/350 '99 thru '19
92179 Tundra '07 thru '19 & Sequoia '08 thru '19

TRIUMPH
94007 Spitfire all models '62 thru '81
94010 TR7 all models '75 thru '81

VW
96008 Beetle & Karmann Ghia '54 thru '79
96009 New Beetle '98 thru '10
96016 Rabbit, Jetta, Scirocco & Pick-up
gas models '75 thru '92 & Convertible '80 thru '92
96017 Golf, GTI & Jetta '93 thru '98, Cabrio '95 thru '02
96018 Golf, GTI, Jetta '99 thru '05
96019 Jetta, Rabbit, GLI, GTI & Golf '05 thru '11
96020 Rabbit, Jetta & Pick-up diesel '77 thru '84
96021 Jetta '11 thru '18 & Golf '15 thru '19
96023 Passat '98 thru '05 & Audi A4 '96 thru '01
96030 Transporter 1600 all models '68 thru '79
96035 Transporter 1700, 1800 & 2000 '72 thru '79
96040 Type 3 1500 & 1600 all models '63 thru '73
96045 Vanagon Air-Cooled all models '80 thru '83

VOLVO
97010 120, 130 Series & 1800 Sports '61 thru '73
97015 140 Series all models '66 thru '74
97020 240 Series all models '76 thru '93
97040 740 & 760 Series all models '82 thru '88
97050 850 Series all models '93 thru '97

TECHBOOK MANUALS
10205 Automotive Computer Codes
10206 OBD-II & Electronic Engine Management
10210 Automotive Emissions Control Manual
10215 Fuel Injection Manual '78 thru '85
10225 Holley Carburetor Manual
10230 Rochester Carburetor Manual
10305 Chevrolet Engine Overhaul Manual
10320 Ford Engine Overhaul Manual
10330 GM and Ford Diesel Engine Repair Manual
10331 Duramax Diesel Engines '01 thru '19
10332 Cummins Diesel Engine Performance Manual
10333 GM, Ford & Chrysler Engine Performance Manual
10334 GM Engine Performance Manual
10340 Small Engine Repair Manual, 5 HP & Less
10341 Small Engine Repair Manual, 5.5 thru 20 HP
10345 Suspension, Steering & Driveline Manual
10355 Ford Automatic Transmission Overhaul
10360 GM Automatic Transmission Overhaul
10405 Automotive Body Repair & Painting
10410 Automotive Brake Manual
10411 Automotive Anti-lock Brake (ABS) Systems
10420 Automotive Electrical Manual
10425 Automotive Heating & Air Conditioning
10435 Automotive Tools Manual
10445 Welding Manual
10450 ATV Basics

Over a 100 Haynes
motorcycle manuals
also available

10/22